ALGAE AND MAN

ALGAE AND MAN

Based on lectures presented at the
NATO Advanced Study Institute
July 22 - August 11, 1962
Louisville, Kentucky

Edited by
Daniel F. Jackson
Department of Civil Engineering
Syracuse University
Syracuse, New York

PLENUM PRESS
NEW YORK
1964

Library of Congress Catalog Card Number 63-21218

A Division of Consultants Bureau Enterprises, Inc.
227 West 17th Street
New York, N.Y. 10011

Printed in the United States of America

This book is dedicated to the
Science Committee of the
North Atlantic Treaty Organization
in tribute to its activities in
fostering international scientific thinking

Preface

With the continuous increase in human population and its constant demands on the aquatic environment, there has been a compounding of the interrelationships between algae and man. These relatively simple green plants not too long ago were often considered as merely biological curiosities.

Within the past twenty-five years, with advances in technology and the increased eutrophication of lakes and streams, the interplay between algae and man has become more complex and more important. Problems of taste, odor, toxicity, or obnoxious growth caused by algae are unfortunately quite familiar to the water supplier and to the public health worker. Algae have met their role in the space age as a possible source for food or as a gas exchanger. In order to explore any of these practical problems, it is essential to have adequate, basic knowledge of algal taxonomy, physiology, cytogenetics and ecology.

This book is the outgrowth of a North Atlantic Treaty Organization Advanced Study Institute in which authorities in both the applied and basic fields of phycology, as well as in cognate disciplines, met and discussed various topics related to algae. It is of significance to note that this was the first NATO Advanced Study Institute to be held in the United States and that it had for its theme a subject which is of import for the welfare of all mankind.

Special acknowledgment is due the Scientific Affairs Division of NATO for providing the funds with which to conduct the Advanced Study Institute. Dr. Hans Jørgen Helms and Dr. B. Coleby, both of the Scientific Affairs Division, were instrumental in sponsorship of the program. Recognition and thanks are extended to President Philip G. Davidson for his graciousness in permitting the Advanced Institute to be held on the University of Louisville campus.

Among the many other individuals who have helped in either conducting the Advanced Institute or in preparing this book are the members of the editorial staff of Plenum Press, who have been most helpful in all respects, and, as always, in any of my endeavors, my wife, Bettina Jackson, who is my unfailing partner.

A salute is due each of the contributors to this book for his cooperation as well as to his scientific achievements. Through the conscientious efforts of each, a more complete understanding of the relations between algae and man has been achieved.

DANIEL F. JACKSON
Syracuse, New York

Contents

Contributions of Current Research to Algal Systematics

G. W. Prescott

Department of Botany
Michigan State University
East Lansing, Michigan

It is fitting, and a compliment to phycologists, that this symposium is not entitled "Man and Algae"—those organisms which today represent the most venerable of all chlorophyll-bearing life on earth. Algae are certainly here to stay and of course we are less sure about Man, so it is appropriate that emphasis be placed on algae and that we think of what these plants mean to Man, this marauding upstart on the biological scene.

It is also appropriate that attention be directed toward systematics to throw up a backdrop for the consideration of the many aspects of phycology related to Man. In thinking about the contributions of modern research to systematics it is impossible, conversely, to divorce the role of systematics in research. Therefore, in following the subject at hand I may consciously or unconsciously take some license and reverse the emphasis. At the same time I shall take the opportunity to single out some phycological problems much in need of investigation. In so doing it may be that some fences will be broken down and I shall encroach, apologetically, upon the fields assigned to other participants in this conference. It will be impossible to avoid doing this, so intimately enmeshed are the facets of phycology.

The current conservative concept of the groups of organisms which constitute the "algae" recognizes the following nine phyla: 1) Chlorophyta; 2) Cyanophyta; 3) Chrysophyta; 4) Euglenophyta; 5) Pyrrhophyta; 6) Rhodophyta; 7) Phaeophyta; 8) "Cryptophyta"; 9) "Chloromonadophyta". Such a division into nine phyla is based mostly upon the delineations proposed by Pascher (1931) who, however, relegated the Cyanophyta to a position exterior to the algae "proper." (See also van Oye, 1961.) Whereas this scheme is subject to modification by some students of the algae it seems, in general, to be the most natural classification yet conceived. At the same time it must be recognized and kept in mind that research and the accumulation of knowledge require that the classification system be fluid and that the composition of the phyla may be continually shifting. This is disconcerting of course to those who look for a definite and static system of classification such as is more nearly approached in the arrangement of land or "higher" plants.

The familiarly recognized foundations for the definitions of algal phyla are: the chemistry of pigments and their relative amounts; the chemistry of food reserve; and cytology, including flagellation, flagella morphology and wall structure. Therefore, it is only to be expected that as our knowledge becomes more refined, the criteria for assignment of organisms to this or to that phylum, or the definitions of an algal group, should be revised. Especially is this true in respect to the varying degrees of importance attached to flagellation and to chemistry of food reserve. Recent and current information on such features no doubt will bring about some realignments and the present Pascherian system eventually will give way to new, modern schemes that will see a reconstitution of the algal phyla. This is especially predictable for the Chrysophyta which, as we know, includes three subphyla that, at best, are somewhat incompatible.

The majority of biologists accept the premise that systematics has the same fundamentally important relation to biology that mathematics have to the physical sciences. This is not to say, of course, that mathematical science is not fundamental as well to certain facets of biology. But the describing, pigeonholing and name-giving biologist, independent of mathematics, is something of a philosophical would-be artist who creates a fundamental backdrop for all other aspects of biology. His goal is to create order and to provide arrangement, a goal for which the giving of names to living and fossil organisms is only part of the task. It is also the systematist who collects facts, interrelates them, and accordingly attempts to derive concepts that may answer questions relative to phylogeny and evolution—questions which continually pique and tantalize the inquisitive mind.

Because of the basically important tasks and duties of the

systematist I am led to the first and introductory point in the present consideration; namely—let us appreciate and take cognizance of the great necessity of keeping alive the spirit of systematics in the face of great waves of interest directed these days to other aspects of phycology (and for that matter of botany in general). The systematist must turn a deaf ear to the sometimes derogatory criticisms leveled at him by those of his colleagues who are enthusiastically devoted to genes, nucleic acids, and test-tube biology. The systematist must remember that he himself has an important obligation to his science as others have to theirs. But at the same time the taxonomist must be alert to the tools, techniques, and information contributed by colleagues in genetics, cytology, biostatistics, and other areas of biology. It is a truism that a biologist is not yet arrived as a matured scientist until he can visualize, appreciate, and make use of the interlocking dependencies of one field of study with another. Of late, however, it seems that there is a danger that descriptive biology will be washed overboard by some of the aforementioned waves of modern interest.

There are just two kinds of biologists—those who can and wish to be systematists, and those who do not. To those who choose taxonomic disciplines as an outlet for their desire to understand and organize our knowledge of Nature falls the duty of maintaining this aspect of biology and of inviting oncoming botanists to co-operate by illuminating the opportunities of, and contributions which taxonomy can make to science. Algal taxonomists should advance their science so that when systematics is referred to, consideration of terrestrial plant taxonomy only is not implied. It can be shown, as has been pointed out before, how urgently we need 1) more field work; 2) more manuals and compendiums, and 3) more philosophical analyses and syntheses of information already accumulated. We need to take inventory of and to evaluate what is known. The esteemed Fritsch one time suggested that systematists defer entirely descriptive activities and species-naming, and that they devote attention to evaluation and interpretation. This is good advice, but we are too ignorant in systematics to make syntheses satisfactorily complete as yet.

To be sure, we already possess a body of taxonomic literature that is considerable—even if discouragingly scattered. We find ourselves overtaken; we find that we are unable to digest, individually, information coming from the endless production belt. This is of course neither a unique problem nor necessarily a current one, but the logarithmic increase in research during the past two decades has produced an unmanageable problem. Because of this, any and all taxonomic contributions that can be made of a summarizing nature will be most gratefully received by all. Let us hope that we shall have more and more monographic works in phycology.

It is recognized that such admonitions and expressions of hopes need not be addressed to most systematists, nor is it necessary to exhort systematists to devote greater effort to their tasks. The only persons who attend a lecture on Temperance are those who do not drink anyway. But I am sure that systematists join in the conviction that there is a danger that increasing obligations and duties of the systematist to other aspects of botany may not be met. Also there is a danger that systematists will lower their guard in the struggle for the conservation of species in Nature.

In respect to the latter, it is clear that the systematist of terrestrial biota at present has far greater responsibilities—where whole groups of species are being eliminated or depredated at an alarming rate, and where gene "pools" are dropping out of existence. Plants on land are under the most terrific competition and are undergoing the greatest struggle for survival that they have ever experienced in their long history. It is doubtful whether any great cataclysmic environmental change in the past has been so severe for plants as competition with Man. In spite of his touted intelligence Man insists upon indulging himself in the pleasurable whim of having offspring, so he must continually look to and prepare for the future. With Man's demands on terrestrial plants increasing, aquatic plants may be called upon in the foreseeable future to furnish and replace or at least augment food and other substances valuable or essential for human life. We must keep in mind that many and varied uses of aquatic plants are still unknown and that a great area of research is involved. It is for this reason if for no other that phycological taxonomy must contribute its role in systematizing knowledge.

Certainly one of the obligations of the systematist is to provide names and systems of classification on which there is maximum agreement among botanists—albeit there will always be differences of opinion and of interpretation. The systematist often comes under fire because of differences in systems of classification, and is accused of adding confusion. The advent of both national and international phycological societies within the past few years will do much to speed the interchange of knowledge, the amalgamation of opinions, and the recognition and adoption of more stable and fundamentally workable systems of classification. But honest differences of opinion and constructive criticism among systematists are highly desirable because these lead to improvements and the cancellation of errors. In this growth and improvement modern research in biology is profiting by numerous mechanical gadgets and by the refinement of techniques for handling and observing both living and dead organisms, or slices of them. Present-day research activities are many and may be categorized in several different ways. Because they are modern does not mean necessarily that they are new, but that they are current. Arbitrarily, the following areas of research in relation to systematics seem to be the most rewarding, and some of these will be selected for brief consideration.

1. Explorations in life histories, phylogeny and evolution.
2. Microscopy, light and electron.
3. Physiology; toxins, antibiotics and extracellular metabolites.
4. Limnology and related facets.
5. Ecology and distribution.
6. Physiology of biological assay and culture studies.
7. Physiology of oxidation of wastes; aeration.
8. Economics; commercially valuable products and useful aspects of algae.
9. Taxonomy *per se.*

Clearly, many of these categories are interlocked and overlapping. Many of these fields of research contribute directly to and aid systematics. Others are indirectly helpful because they stimulate and invite taxonomic considerations. And for all of these, it is not necessary to be reminded, taxonomy serves as a fundamental background. Let us consider some examples of the contributions which research areas make to systematics—remembering that these are but some of the many examples which might be cited.

First, in respect to research in life histories. Both the sophisticated scientist and the child of the aborigine asks the same question—where did we—where did life come from and how? Because of the fundamental position that algae have in the evolutionary scheme which we have synthesized, considerable attention is consistently and continually directed toward them—searching for other and more convincing evidences of evolution within and among the several phyla—and the connection between algae and higher plants. These searches for the most part use two lines of attack: one, on evolutionary patterns in the development of the plant body; two, on reproductive organs, reproductive methods, and the life history. Such searches have been in progress for a long time, but especially since the turn of the century—and have been very rewarding, for at least some consistent notions have been derived. But just now the search can be said to be modern and invitational to further research because of the great gaps in our knowledge which recently have become evident—gaps great enough to provoke us to increased study.

The accumulation of knowledge to fill these gaps is slow and comes from tedious work and the exercise of considerable patience. Because concepts, of necessity, must be built up fact by fact is one reason why we require, as mentioned before, more philosophical syntheses of the fruits of particularized, fact-contributing studies. It is encouraging to note the relative frequency of appearance and the completeness of studies which today illuminate the problem of phylogeny and which assist in systematics. Perhaps the most rewarding researches are those which have to do with life histories, these giving us bit by bit evidence of interrelationships, one group, one genus, one species with another—researches which often also involve anatomy and physiology.

The long-known fact that *Aglaozonia* and *Cutleria* were two life-history expressions of the same plant is paralleled by other recently determined situations: *Conchocelis* as *Porphyra*, *Codiolum petrocelidis* as *Spongomorpha coalita* (Fan, 1959), *Sterrocola decipiens* as a nemathecial, parasitic stage of *Ahnfeldtia plicata,* tetrasporic *Trailiella* with *Asparagopsis* as its gametophyte, *Actinococcus* as a nemathecial stage of *Gymnogongrus Griffithsae.* These are a few examples of realignments in systematics resulting from life history studies. We now know that *Gloeochrysis maritima* produces a motile stage that is identical with an organism described as *Ochromonas oblonga.*

With scores of genera in need of similar life history exploration the breadth of the research field open to phycologists is indicated. We have every right to hope and expect that such investigations will continue to yield other instances of duplicity among both marine and freshwater algae which, as discovered, will help to clarify taxonomic problems and strengthen phylogenetic concepts. At the same time, of course, life history studies aid in further characterizing and differentiating genera which have been erected primarily on morphological features alone.

There is continuing uncertainty and no complete agreement on the taxonomic position, composition, and phylogenetic status of certain orders, families, and genera in the Chlorophyta (Fritsch, 1949); an example of such unresolved questions is that of how the Cladophorales are related to the Ulotrichales on the one hand and to the Siphonocladales and Siphonales on the other. Several recent studies on both life history and cellular morphology in these orders have given us a fairly clear (although possibly incorrect) picture—or perhaps pictures, because all students do not visualize the same detailed scheme of interrelationships. But at least now we can conceive a connection between a ulotrichaceous ancestor to *Urospora*—to *Rhizoclonium*—to *Cladophora*—and to the Valoniaceae.

Fritsch (1947, 1949), Chapman (1953), Hillis (1959), Egerod (1952), and others are responsible for clarifying such an evolutionary pattern. Schussnig (1935), however, some time ago outlined a system of classification which conceivably should be agreeable to a majority of systematists, especially those who philosophize on the disposition of the Siphonales and the Siphonocladales (such as Chapman who, of late, has made very positive suggestions). As far as can be learned from the literature, however, Schussnig's scheme, as such, has not been adopted or incorporated in evolutionary systematics. He recognized three classes of green algae: the Euchlorophyceae, the Siphonophyceae, and the Charophyceae. Under the first (Euchlorophyceae) he established, using some of the older group names, the Dicontae, the Stephanocontae, and the Acontae. Then under the second (Siphonophyceae), by using three subclasses, he presented the Dicontae Siphonophyceae, the Stephanocontae Siphonophyceae (Order Derbesiales), and the Holocontae Siphonophyceae (Order Vaucheriales). The latter order, of course, has now been removed completely from the Chlorophyta to a natural position in the Chrysophyta. Schussnig's plan has a very logical appeal because it both subdivides and groups the genera and families according to such important characters as life history, food reserve, and flagellation. In arranging this scheme Schussnig simply used orders and suborders to group tribes and families so that in the end the disputed and ill-treated Cladophorales (Cladophoraceae) are in one suborder, and the Siphonocladaceae are in another, all under

Siphonocladales, and there is ample opportunity to visualize phylogenetic relationships.

Chapman (1954, 1954a) has more recently elucidated his views on the systematics of the same orders. He incorporates the Cladophorales of earlier students within the Siphonocladales but in an ancestral position. Thus he has primitive *Cladophora* giving rise to three lines of development, one ending in the Boodleaceae, one in the Valoniaceae, while the third merely developed the "higher" *Cladophora* species.

Incidentally, the recent discovery of still another species of *Cladophora* showing an isomorphic alternation of generations is another contribution which the systematist can use in fortifying a scheme which will give the Cladophorales a less tenuous position. This is an invitation to the systematist to examine freshwater species of *Cladophora* cytologically and in terms of their life history.

Another section of the Chlorophyta in which there is considerable evolutionary interest is the Volvocalus, an order which has been recently elevated to a class by Ettl (1958). Evolution and phylogenetic trends within the Volvocales have long been recognized and interrelationships between genera and families have been clearly evident. But the almost conventional, usual interpretation of the relationship of the Volvocales to the Tetrasporales and the ancestral position of these in reference to the "lower" Ulotrichales is still worthy of critical and exhaustive studies to elaborate on those which first clarified the probable connection between volvocalean and tetrasporalean genera.

Culture studies by which life history stages are determined have been helping us to see more clearly the probable connections between *Palmella*-stages of motile unicells and *Tetraspora,* then to *Palmodictyon— Geminella— Hormidium* and on to *Ulothrix* and the ulotrichine line. But we need to locate, among the extant genera in these supposed ancestral classes, some yet undiscovered, interconnecting forms, or some significant life histories. It is of interest to establish the veracity of this evolutionary series (this volvocine to ulotrichine evolution) to confirm our notions concerning the flagellate ancestry of the Chlorophyta, a notion which most students hold for the ancestry of other algal phyla as well.

Further, systematics has been aided by recent critical and definitive studies in the Ulotrichales, Prasiolales (Schizogoniales of some authors). Such studies have pointed up similarities and differences in the members of these orders from which one may make reasonable inferences. This evolution of filamentous to membranous as interpreted by some students has led to the erection of more genera on the basis of morphological characteristics. Whereas this may be deplored by some, at least there is an advantage in that orders, families, and genera become more precisely defined (if multiplied). Hence a clearer view of a natural classification system and of evolutionary intergradations results. At the same time interpretations of these intergradations provide some lively speculation and differences of opinions which are slow to become resolved among phycologists. Thus Chapman, for example, considering cell anatomy as well as reproductive habits and life histories, recognizes but one order, Ulotrichales, to include the three mentioned above. He designates five families in this overall order: Ulotrichaceae, Monostromaceae, Capsosiphonaceae, Prasiolaceae, and Ulvaceae. *Monostroma* (Monostromaceae) has been separated from the previous position with the Ulvales because life history studies have demonstrated (convincingly to some students at least) that an incipient sporophyte generation exists during the development of the zygospore to form zoospores and that a *Ulothrix*-stage develops from the zoospores. This so-called sporophyte, however, amounts only to a conspicuous enlargement of the zygote. Thus the haplobiontic life history is used to segregate the Monostromaceae. Studies of all species of *Monostroma* and carefully worked out life histories are needed for an understanding of the systematics of the membranous and tubular ulotrichoids.

The enigmatic systematic and phylogenetic position of the Characeae has been helpfully commented on in a paper by Desikachary and Sundaralingam (1962) in which the primitive position of *Chara* is postulated. The ecorticate species of *Chara* and the other genera are more advanced, having evolved from the same ancestor which gave rise to corticate *Chara*.

Another family in much need of additional systematic classification is the Trentepohliaceae. The recent studies of Thompson on the life history of epiphyllous genera such as *Cephaleuros* are going to aid not only in respect to the taxonomy of the family but also in establishing phylogenetic connections.

As is well known, systematics within the Rhodophyta, at almost all levels—orders, families, genera, and species—is based principally on reproductive organs and reproductive habitats, morphology of the thallus being less useful in taxonomy.

It is within the researches on the red algae that we find perhaps the best examples of the contributions that life history studies can make to systematics. The literature is voluminous and many fruitful investigations have been made by Svedelius (1931), Kylin (1923-1935), Drew (1954, 1954a, 1956, 1958), Papenfuss (1951), and others. But as a matter of fact, there have been more taxonomic questions raised than answered in reference to a generally acceptable scheme of classification, and first one student and then another has come forward with a classification scheme based on reproductive habits and elements.

It is the life history which long ago formed the basis for Kylin's division of the red algae into two groups: 1) orders without auxiliary cells (Nemalionales-Gelidiales), the Protoflorideae or Bangioideae; and 2) orders in which auxiliary cells are used in the production of carpospores (Cryptonemiales, Gigartinales, Rhodymeniales, and Ceramiales). But since his time of writing additional information has shown that Kylin's system was only partly consistent. Members of the Gelidiales, for example, have been shown to possess auxiliary cells. Therefore, as Papenfuss (1951) has pointed out, it is now possible to regroup the Rhodophyta according to their life history types. It is well recognized, however, that much more information is needed before a wholly acceptable classification system can be developed. The most recent classification recognizes: 1) forms which are haplobiontic and 2) orders which are diplobiontic. The first is composed of the Nemalionales. In the second class are first, the Gelidiales, in which gonimoblasts develop from ooblast filaments after fusing with auxiliary cells; and second, a group of four orders in which gonimoblasts develop from a generative auxiliary cell (Cryptonemiales, Gigartinales, Rhodymeniales, and Ceramiales). As Papenfuss points out, however, the grouping agrees with practically all situations, but there are exceptions. This of course throws the door wide open and invites further study for reconsidering the system of classification proposed.

Anatomical and life history studies are certain to give an increasingly clearer picture of Rhodophycean systematics. Already the heretofore clear-cut separation of the Bangioideae from the Florideae has been shaken by Fan's (1960) demonstration that connecting pits exist in the former, thus throwing out this vegetative feature as one of the criteria for classification. This of course opens the question as to whether such connecting pits in other primitive genera may not be demonstrable by electron microscopy. In any event, stress is now placed more than ever upon other characters such as life history and carpogonium features to maintain and separate the two groups in the Rhodophyta.

Krishnamurthy's (1962) recent study of *Compsopogon* and *Compsopogonopsis* has helped to establish the genera in the Erythropeltidales of the Bangiophyceae on the basis of life history as well as morphology. He advocates that the genera not be placed in a separate order. His studies serve to strengthen the position of the Protoflorideae.

The description by Herndon of a curious new saccate freshwater red alga (*Boldia*) poses an interesting question involving the systematics of the primitive Rhodophyta. Just what the new genus implies cannot be evaluated fully at present, but such an intriguing plant invites speculation and suggests the possibility of another order within the Protoflorideae.

There are countless papers, of course, dealing with general morphology and anatomy of the algae which contribute to or which influence systematics. A few examples of recent work may be cited to illustrate the theme of this review. Hillis (1959) revised the genus *Halimeda,* using both gross vegetative features and cytological details such as inner cortical utricles as a basis for species delineation, thus permitting the recognition of species.

Another example of how critical and detailed observations of morphological features, howsoever minute, can aid in systematics is the work of Silva (1951) on *Codium.* His study showed that the recognized Adhaerentia and Bursae groups of species within this genus possess respective wall features in the utricles, thus providing bases for discrimination. [See also Christensen's (1952) morphological study of *Chlorosaccus* which leads to the conclusion that two species are identical.]

CYTOLOGY

The electron microscope is a relatively new tool which is both a boon and on the other hand a worry to the systematist. Although results of electron microscopy sometimes are open to query and sometimes produce inconsistent results because of artifacts and the refinements that are required in the interpretation of images, this instrument has already yielded many and promises to illuminate many more, precise morphological features useful to the systematist. In fact, electron microscopy may lead to a complete realignment of not only minor taxa but phyla as well. But the instrument raises at least two specters for the taxonomist: 1) The average worker cannot have an electron microscope at hand for ready use in identifying organisms. 2) Even when the systematist is fortunate enough to be able to employ electron microscopy he is not sure whether the minute, ultramicroscopic features which conceivably might be used taxonomically really constitute species or generic differences, or both; or are only variables, local, ecologic, and incidental. Such uncertainties will probably be eliminated in time. The electron microscope finds its greatest and most successful contribution to systematics in studies of organisms with siliceous walls, scales, or appendages. This means that mostly members of the Chrysophyta lend themselves to electron microscopy (*Synura, Mallomonas,* the Diatomaceae). This is providential because this phylum as it is now constituted presents such an association of dissimilar organisms that a reconstitution is required. Chrysophycean taxonomy is based strongly on flagellar number and structure. Hence electron-microscopic studies of minute but fundamentally important characters doubtless are going to make it possible to rearrange the Chrysophyta, especially in the Chrysophyceae (Parke, 1961), and possibly to see a redefinition of other classes and phyla.

The Chrysophyceae are a group illustrating advanced evolution in some respects while in other features there is primitiveness. In all, it is a highly intriguing group for systematic work but one which, until recently, has been relatively neglected.

The recent and current work of Parke using the electron microscope has confirmed the status of two taxonomic series in the Chrysophyceae based on flagellation and upon life history stages. One series suggests a pathway toward the Diatoms, whereas the other appears to have led to the development of the advanced family, the Coccolithophoridaceae. She has pointed out that life histories and developmental stages, rather than morphology, will probably have to serve as the basis for classification within the family; and that the life history type may or may not parallel morphological features. Let us hope that at least in some instances this parallelism will be developed, because the task of the taxonomist is difficult enough without having to resort to culture of coccolithophorids. But it may be that here, as in the Chlorococcales in the Chlorophyta, the taxonomist will be confronted with a high hurdle, namely, culture techniques and ultramicroscopy.

Synura species have been re-examined by Asmud (1956, 1961), Harris and Bradley (1957), Petersen and Hansen (1956, 1958), and others. In this genus individuals of the same species may be shown to have similar features as observed by the light microscope, but quite different when demonstrated by the electron instrument. It is then necessary to consider two questions or problems: 1) Are the variations indicators of different species or are they normal variations of the same species (ecological or otherwise)? 2) Are there morphological features or other characteristics which can be demonstrated by the light microscope which accompany the differentiating characters which are shown by the electron microscope? In any event, the description of a number of new species of *Mallomonas* has been possible by using the electron microscope and, as already demonstrated by Fott, Petersen, and others, we can expect this tool to be a material aid in systematics.

Dawes, Scott, and Bowler (1961) have reported on the electron-microscopic structure of some brown and red algal walls and cell products. The plants studied showed uniformity of detailed features characterizing the Rhodophyta on the one hand and the Phaeophyta on the other. Such studies when extended may be expected to aid in taxa differentiation among members of the various algal phyla.

Turned towards the Diatoms, the electron microscope has revealed many refined details of frustules, etchings, and pittings. Accordingly, Hustedt has revised the terminology for wall markings and other anatomical features for making taxonomic distinctions. Also, the studies of Desikachary (1952, 1957), Helmcke, Krieger, Kolbe, Manton, and others have made it possible and necessary to change the nomenclature of some species. Hustedt, however, while appreciating the value of the electron microscope in delineating species, seems not to believe that taxonomy of the Diatoms will be much changed by this technique (at least not by the electron microscope alone). The newer and more finely drawn diagnoses of species, however, will permit precise verification of new taxa—one of the chief advantages of electron microscope technique.

The Coccolithophoridaceae were mentioned previously. The taxonomy within this group is based on coccoliths, which are minute but extremely complex and variable in shape. Hence the light microscope is often scarcely adequate to elucidate the features clearly. The electron microscope, on the other hand, defines taxonomic features very clearly--so much so that one species described as *Pontosphaera huxleyi* was found in reality to involve three species in two other genera, the named species actually not belonging to *Pontosphaera* at all. In view of such adjustments and other similar species transfers it is clear that the Coccolithophoridaceae are scheduled for a complete taxonomic overhaul by the electron microscope.

Cytological studies of the nucleus, pyrenoid, and chloroplast by electron and light microscopy have furnished us in the past, and continue to furnish us, with taxonomic information—or at least to contribute to the solution of taxonomic problems. Numerous examples of the taxonomic role of cytology might be cited, but space does not permit reference to more than a very few.

Dodge in 1960 made a study of the nucleus in certain Pyrrhophyta, making several contributions which not only have taxonomic potentialities but also help toward a more complete definition of the group. He found, for example, that chromosomes of Dinophyceae are chromophilic in the interphase as well as during mitosis. Further he found that whereas *Prorocentron* and *Exuviella* have similar nuclei in respect to chromosome composition, *Oxyrrhis* and *Marina,* placed in separate families, have similar features (permanent endosome) thus supporting a closer relationship that has been given them, and indicating that a transfer would be appropriate. Likewise, *Amphidinium* and *Massartia* have similar spherical chromosomes and probably should be separated from *Gymnodinium* and *Gyrodinium,* which two genera have similar chromosomes but different from the former. *Ceratium, Gonyaulax,* and *Peridinium* (each in a separate family) have similar nuclei during interphase. But the possible closer relationship of *Ceratium* and *Gonyaulax* is demonstrated by nuclear similarities during mitosis. To what extent differentiation based on nuclear characteristics can be used in the taxonomy of the group will be determined by the recognition of more evident and readily detected features ordinarily used in Pyrrhophyta systematics.

Another cytological aid to taxonomy is found in the Cave and Pocock (1956) study of the nuclei of the unusual volvocoid genus *Astrephomena.* Here it was found that *A. gubernaculifera* had two expressions, one with 4 chromosomes and one with a variable number (6-8). Hence there is suggested two distinct species passing under one name. Confirming studies will induce changes in the systematics of the genus.

It is not likely that many systematists can have the opportunity to expose flagellated organisms to electron microscopy, but studies on *Chlamydomonas, Euglena, Phacus,* and a few other genera show distinct if ultramicroscopic differences in the morphology and attachment of the flagella. Chen (1949) should be mentioned in this connection, as should the work of Lewin and Meinhart (1953), who found eight strains of one species of *Chlamydomonas* to have different flagella characters, thus raising the question as to whether such strains or at least some of them might not have other correlated differences which could constitute definable species.

Judging from the attention that is being directed toward the cytology of the Cyanophyta (e.g., Rabinovich, 1956), we have hopes that systematics will be further aided, that notions concerning the internal phylogeny of the group will be derived, and that present notions will be strengthened. Rabinovich found three types of nuclear organization in the central body of blue-green cells: a diffuse primitive type, as in the Homocystineae; a reticular caryosome type, as in *Dermocarpa* (Chamaesiphonales); and a primitive mitotic type, as in *Synechocystis* (Chroococcales). (See also the study of Leak and Wilson, 1960.) The fact that the Cyanophyta possess characters unique among chlorophyll-bearing plants, and although evidence accumulates to support their inclusion with the Schizomycetes, they are in no way relegated to a position outside the interest and the realm of the phycologist. Since Pascher (1931) suggested his classification system long ago, the Cyanophyta have been under the wing of the bacteriologist and they are treated in current manuals as a class under the Schizophyta.

TOXINS, ANTIBIOTICS, GROWTH INHIBITORS

Perhaps the more usual concept of the "social status" of the algae is that they are lowly, primitive plants, retarded and unrefined—relegated to an environment from which the more erudite and more progressive forms of life have evolved to higher planes in the ladder of complexity, in respect to both form and refinement of ecological adjustments. The phycologist does not regard the algae with this point of view. On the contrary we look upon the algae of today as the ultimate divarications of evolutionary processes within the oldest, the most ancient of extant plants. Evolution for them has not led to the development of high degrees of specialization of tissues to be sure. There are no elaborate conducting and mechanical systems and this, of course, is correlated with the fact that survival and natural selection have not operated in these respects. An alga with collenchyma, sclerenchyma, xylem, etc., would be a misfit. Absorbtive root system and guard cells are superfluous and the seed habit would involve a prolonged and unneccessary rigmarole uncalled for in an environment where dormancy is not essential. We see the algae as highly evolved and well-adapted organisms in respect to their morphology and internal anatomy, and if they appear to be conservative in these respects they can well afford to be so. Their conservatism is paralleled by a more stable and less variable environment than is found in terrestrial ecology.

But it is in their physiology and in the finesse of their adjustments (as in the Bacteria) that evolution has been actively operating. The refinements in this connection have only begun to be appreciated. It is only comparatively recently for example, that their enzyme systems have been critically studied and diagnosed, and such factors and substances as toxins, antibiotics, growth inhibitors and promoters and other extracellular metabolites recognized.

We know that photosynthesis is as complex a process and perhaps involves more varied factors in some groups of the algae than in "higher" plants. The products of photosynthesis are more varied chemically, their growth hormones are probably similar to those of higher plants, their response to indoleacetic acid is the same as in higher plants. In studies of physiological attributes, systematics plays an important role and taxonomists have an increasingly important obligation to provide accurate and clarified identifications. Here systematics is not so much aided by modern physiological researches as modern research is aided by the taxonomist. But if physiological researches do not enhance systematics directly they surely do so indirectly. For the specificity of metabolites and of endo- and exotoxins and the selectivity of vitamins demand of the systematist the most exacting and discriminative taxonomy. As a result, taxonomic positions become clarified and new specific and generic taxa are recognized.

For example, the toxins produced by the infamous *Microcystis* are detectably different in respective species and even in types or strains. It behooves the systematist to search for morphological or growth habits associated with toxin variables, such characters being useful in making proper identifications. Shelubsky (1951, 1951a), incidentally, has found that *Microcystis aeruginosa* strains are able to develop precipitins in rabbits but that toxins have no antigenic properties.

Work continues in many laboratories on *Microcystis, Anabaena,* and other genera, especially in Canada under Gorham (1960). More and more toxin properties are becoming known and are being identified with variety of alga, age, and presence or absence of bacteria in the habitat (or culture). We may hope to have important toxins isolated in the near future as a few have been already. As the work goes forward the taxonomist of course is called upon to co-operate. For *Microcystis*, so far only three species (and their "strains" or varieties) seem to have been identified with toxins: *M. aeruginosa, M. flos-aquae*, and *M. toxica*, the latter being known only from Africa (Stephens, 1949; Steyn, 1945), where it is responsible for extensive and costly cattle deaths. But there well may be other species in the Chroococcaceae guilty of toxin production. Now that suitable, more or less successful culture media have been developed, thanks to the work of Gerloff, Skoog, and their associates in Wisconsin, laboratory investigations can proceed more rapidly in an examination of toxic blue-greens, and of their taxonomy. We need more of the type of study conducted in Canadian laboratories by Gorham and associates, and by Shelubsky in Israel. In reference to toxic species it may be in order here to mention the relation of toxicity to species which contain pseudovacuoles. Rarely if ever in my observations have toxic reactions been observed from algae which did not possess these vacuoles (*Microcystis aeruginosa, Anabaena flos-aquae, Aphanizomenon flos-aquae*). The toxicity of the latter species is still open to question. But the taxonomist and physiologist might bear this observation in mind as relationships are developed between blue-green species and toxins. Our knowledge of the properties of toxins and their related algal species have been summarized recently by D. F. Jackson and not so long ago by Ingram and myself (1954).

Similarly, the devastating toxins induced by certain Dinoflagellate species merit taxonomic studies with reciprocal advantages to the physiologist and the taxonomist. Species of *Gonyaulax* and

Gymnodinium (Ballantine, 1956; Ray and Wilson, 1957; Davis, 1948) should be investigated, as well as the chrysomonad genus *Prymnesium* (McLaughlin, 1958), in which three varietal strains of *P. parvum* have been detected. This flagellated species produces lethal effect among fish and other gilled organisms, and the genus invites more taxonomic and morphological study, especially in respect to the use of the haptonema in nutrition.

In respect to the Pyrrhophyta, studies of Howell on the Red Tide in the Gulf of Mexico point up the need of revision of the genus *Gonyaulax* in respect to specific diagnostic characters and toxin production. In any event it is clear that inasmuch as the occurrence of or level of toxicity is highly specific, the taxonomist has his work cut out for him.

Another study of interest to the systematist is that of Allen and Dawson (1959) on antibiotics produced by bottom-living marine algae, as well as the reports of Fogg and Boalch (1958) and Burkholder and Almodovar (1960).

Related to toxins, and possibly the same as toxins for some organisms, are antibiotics and growth inhibitors. Extracellular metabolites of *Polysiphonia* operate against several species of Bacteria. What other species of this genus likewise produce antibiotics? The antibiotic action of *Azolla* toward Zygnematales may be produced by the endophytic *Anabaena* (Sun, 1943).

The antibiosis of *Phaeocystis* (Sieburth, 1959) in the Antarctic is an interesting example which invites the attention of the systematist. It was discovered that penguins feeding on the crustacean *Euphasia,* which in turn forages on the phytoplankter *Phaeocystis,* have bacteria-free intestinal tracts. In reference to this it is interesting to recall that sea water is notoriously scant in bacterial flora, especially in areas heavily populated with phytoplankton species—another incentive for the taxonomist.

The relationship of antigenic protein production to strains and species of *Chlorella* is another example of specificity, a critical study of which is an aid to the systematist who must rely on culture techniques for final identification of species in this genus.

Only a few algal species have been tested for mutual antibiosis and a great deal of work needs to be done by the taxonomist and the physiologist. Related to this is the phenomenon of algal species succession in the plankton, both fresh-water and marine, and the sequence of forms in sewage oxidation ponds.

LIMNOLOGY, ECOLOGY

The limnologist is consistently calling upon the systematist for identification of organisms which are important in the numerous interrelationships. Limnologists show encouraging signs of giving more attention to specific names of algae in their reports, rather than generic alone. Numerous limnological studies of ecological factors vs. phytoplankton composition and quantity emphasize the need of taxonomic analyses. This is true because biological analyses and correlations must be made species by species. Search is still in progress for index organisms: indices of chemical nature, pH, of electrolytes, of contamination (pollution). The problem of the indirect or direct role, or both, of phytoplankton in the food chain of microfauna is still open to investigation. This subject (with taxonomic involvements) has been repeatedly investigated, but results are conflicting and the taxonomist often finds himself in the midst of contentions. Suffice to say, it is reasonable that both plankton organisms as well as detritus and dissolved organic matter enter the food chain at one point or another. Scagel (1959, 1961) has recently treated the matter succinctly. Obviously such plankton-food-chain studies aid the systematist and vice versa. Also taxonomy is stimulated by the examination of aquatic organisms (especially plants of the sea) for concentration of radioactive elements such as strontium, titanium and zirconium. In reference to this area, the taxonomist must needs go to work, for the specificity of uptake may provide some significant clues to relationships and differences among algae. Certainly the taxonomist has an obligation to other aquatic biologists, for obviously the problems of the limnologist are not aided by confusion or discrepancies, or both, in the interpretation of species. Some limnological studies related to plankton may deal specifically with two different names for the same organism because taxonomists had not clarified the nomenclature.

One of the contributions of limnology and oceanography to systematics stems from the fact that many new and interesting, sometimes evolutionarily significant, organisms are brought to light. Limnological studies are sweeping and far-ranging, both vertically and horizontally. Unfortunately many of the algae collected by limnologists are not recognized for what they are, but are squeezed conveniently into existing categories.

That we are slow in recognizing index organisms among the

algae is probably related directly to the fact that not enough attention has been given to analyzing the plethora of literature and the mountains of reports on plankton. We are driven to collect more and more data. Also the fact that plankton studies so frequently are not accompanied by full limnological data and vice versa is a deterrent to the recognition of index organisms. When scrutiny is applied, index organisms or meaningful associations of species are often detectable. Thus it seems possible that *Melosira granulata* is definitely an index of a eutrophic type of lake and of most of the features that accompany such a lake, whereas *M. islandica* is an index of an oligotrophic or alpine lake. The late Rawson compiled lists of organisms which he regarded as indices of eutrophic, of mesotrophic, and of oligotrophic lakes. Here of course, we are dealing more with aid that systematics give to limnological research rather than research aids to taxonomy.

The presently much-used C-14 test for productivity may prove to be an aid to the taxonomist—when and if this technique and its full significance are clarified or standardized. Productivity is of course a specific matter and the relationship to systematics is mutual or reversible. Limnologists, however, are usually concerned only with total productivity and not with physiology of species.

The involved relationships between pulses of phytoplankton and pulses of microfauna and the relationships of these to light, temperature, and nutrients call for species identification, much more so than has been true in the past. The voluminous literature includes results which are of questionable value because specific determinations of algae have not been made, although animals related to such studies have been specifically identified. Correlations of limnological and ecological features to algal groups or phyla may be justified in drawing very broad generalizations, but can be used only for generalizations when ecological factors are also considered broadly and philosophically. Thus it is possible to state that hardwater, eutrophic lakes are populated by Cyanophyta-Diatom floras.

Ecological studies naturally aid systematics, since taxonomy is fundamental in ecology. In fact, of course, ecology *per se* is not a succinct science, for an ecologist is in turn, or all at one time, a taxonomist, a morphologist, a physiologist, a chemist, a mathematician and statistician, a physicist, a geologist, and a meteorologist. An approach to some of the aspects of ecology cannot be undertaken except from one or more points of view provided by these several disciplines. Likewise the taxonomist cannot follow his art without drawing upon other related aspects of biology. It naturally follows therefore that systematic phycology is greatly aided when the several aspects of ecology are applied to either species or groups or both, of algae.

Distribution studies, vertical, horizontal and geographical, serve as active stimuli to systematics, although distributional ecology may not in itself contribute directly to clarification of taxonomic problems. Womersley has expressed the opinion that ecological data and distribution data cannot be used as bases for species determination, although ecology is an essential aid to a species concept.

An example of an approach to the ecological basis for taxonomy is a recent study on the distribution of the genus *Halimeda.* The known range of species helps in identification, for *H. cuneata* is a subtropical species only, and *H. discoidea,* with a wider range, may be mistaken for the former species. This means that a *Halimeda* not occurring in the tropics should be referred to *H. discoidea.* Further, the occurrence of species in vertical zonation provides clues to if not positive identification of species, while at the same time recognition of species helps in the establishment of zones. Having related zones to species and vice versa, it is then more nearly possible to single out and to examine or to detect (or both) critical factors in the environment leading to meaningful physiological explanations of distribution.

As an aside it might be mentioned that the science and art of scuba diving is proving to be a great ecological aid to systematics. In all latitudes, but especially in the little-explored Arctic and Antarctic waters, such underwater operations have brought to light many unsuspected or ecologically unsuspected species of brown and red algae; much more can be expected from this type of field operation.

Ecological studies in unusual or little-known habitats are aiding systematics. More observations are being made on glaciers and ice fields, partly because of military interests in alpine and polar situations. Cryoplankton and cryoconite offer ample opportunities to the systematists.

Studies on floras of hot springs are apparently stymied at present, but in the past these have been an aid to systematics since new, critical, and phylogenetically interesting species have been brought to light, especially among Cyanophyta.

Soil (although long studied) and nitrogen-fixation investigations are leading to specific and generic delineations, especially helpful in the differentiation of confusingly similar unicellular forms (Deason and Bold, 1960). Aerial distribution likewise is an area of modern research which aids systematics. The recent work of Schlichting (1961) has provided us (in one study) with a list of 23 species of algae from a series of "air catches."

Endophytic and endozoic and parasitic algal studies, although not modern, certainly are providing us with aids to systematics and vice versa. Algae (*Zoochlorella*) in the tissues of marine invertebrates, for example, have been investigated recently (McLaughlin and Zahl, 1959). Particular species or strains of species or both, are used in these symbiotic relationships. Incidentally, it has been noted and is of interest to the taxonomist that corals possessed of *Zoochlorella* extract calcium from sea water only during daylight or photosynthetic hours.

It has been found that marine animals (sea anemones) position themselves in reference to light intensity which is adventitiously suitable to *Zoochlorella*. The precise systematics of these endozoic species are poorly known; hence another invitation is issued to the taxonomist, especially since animals containing *Zoochlorella* have been obtained from as deep as 116 fathoms in Florida waters. We would like to know whether shore-inhabiting species and deep water forms are the same. Also the taxonomist must identify and differentiate species which relate themselves specifically to animal hosts apparently as adaptations to an otherwise nutrient-poor medium. The confusion that has existed and which still exists in the systematics of endozoic algae has been lessened by axenic culture methods.

Other ecological niches which offer opportunities for the systematist are the utricles of *Utricularia*, the intestinal tract and other tissues and organs of animals, including Man, the cutaneous and bones of fish, tissues of higher plants. The recent finding of *Stigeoclonium* and *Cladophora* (by Vinyard) growing attached on and penetrating the mouth bones of fish is of interest. Incidentally, algal taxonomy is occasionally an aid in fish systematics. Nigrelli reported that a species of fish had been passing under two names, *Mugil poecilius* and *M. troscheli*, separated by the occurrence of dark spots on one. The differentiation of these two was broken down when it was discovered that the spots were caused by *Stigeoclonium* and chlorococcalean cells.

CULTURE, PHYSIOLOGY

Research in and improvements to the point of refinement in culture media have been of inestimable value to systematists, particularly in respect to the taxonomy of unicellular forms in the various phyla, including flagellated genera.

Bold (1960) and his students and Starr (1955) have clarified the taxonomic position of some genera and at the same time have described several new ones, especially soil algae. The task of giving names to little green spheres has long baffled taxonomists and has led to a confusion of names when students have had the courage to attach names to green cells without cultural observations.

At the same time, some (not all) culturing of algae has introduced taxonomic confusion when characterization of species has been attempted only from material under artificial conditions. We all know, of course, that algae in culture do not maintain their natural or original morphology and that they do not pass through normal life history stages. Also they vary in their particular vegetative or asexual reproductive techniques in cultivation, according to the type of medium used and according to variables in other ecological factors. Hence it is not easy to characterize algae under artificial conditions and in many instances it is not feasible, for these expressions on one medium in one laboratory seldom are reproducible in others.

What we need is a key to algal species on particular media under specific culture conditions of temperature and light. In any event the taxonomist had better turn his eyes away from unicellular green algae and pretend that he does not see them unless he has the time and facilities for making a long culture study—and even then he will be stymied because there is no manual or key to species. The widely scattered, piecemeal literature is of little help.

The splitting or lumping of species based on culture studies is, in my opinion, very risky. Some students are disposed to throw two species together when it is shown that variable, and what might be called "abnormal," expressions of both are similar, if not identical. We should expect two species descended from the same parent to have at least some features in common. It is not surprising that variations from the normal would appear similar when placed in artificial culture. Likewise, one species is "split" or defined, when in a culture medium, if it demonstrates a modified reproductive detail different from that previously observed in nature or on some other medium.

Reproductive processes and sexual behavior are induced by high concentrations of biological substances or trace elements in culture media which are not present or are very dilute in nature. Probably other and as yet unknown substances or elements, or both, play a role in determining physiological behavior which is of value in classification and identification.

As demonstrated by McLaughlin and Zahl (1959) and Hutner and co-workers (1949) axenic culture provides techniques for isolating extracellular metabolites, leading to taxonomic differentiation of organisms. Such extracellular metabolites very well might have a bearing on or an explanation of algal succession in nature, as mentioned previously.

Another type of culture research is that of testing for mutual growth-stimulation or -inhibition of two species; such studies involve precise taxonomic distinctions, and in some instances serve as aids to taxonomy. An example is the work of Parker and Bold (1961), who demonstrated mutual growth stimulation between autotrophic and heterotrophic organisms of the soil. It appears that stages in one organism's life cycle are evoked by physiological activities of adjacent cells of the other. Reference should be made to the work of Proctor (1957), whose culture studies on antibiosis are stimulating to the systematist. He found antibiotic reaction using *Haematococcus* and *Chlamydomonas*.

Light has been thrown on the probable origin of colorless flagellates from chlorophyll-bearing ancestors by the culture studies in the Haskins laboratories (Provasoli, Hutner, and Schatz, 1948). Streptomycin induces a chlorotic condition in *Euglena* which becomes permanent and genetic. Another aspect of culture work in the Haskins laboratory which invites systematic studies is the discovery that *Euglena* (Hutner *et al.*, 1949) responds in its growth to crystalline antipernicious anemia factor. This opens the possibility of using various species of *Euglena* for assay of biologicals.

Culture studies involving abnormal forms of *Cosmarium* suggest modes by which triradiate species of Desmids may have evolved. It is thought that the high percentage of triradiate or quadriradiate forms in a clone culture is explainable by diploidy or tetraploidy. Taxonomic studies and cytological studies of chromosome number are invited, with related systematics involved.

OXIDATION, AERATION

The lowly, one-celled algae are more than ever growing in stature and respect. Military research laboratories have been conducting intensive research on the physiology of algae with a view to using them as producers of oxygen and as aids in the oxidation of wastes. Similarly, algae are used successfully in small cities for the oxidation of sewage.

Airplane manufacturers are conducting their own programs of research on algae for use in aeration of possible space flights. Whereas such modern research has not as yet aided systematics, the taxonomist certainly has a role in it, and systematic phycology is stimulated thereby.

The behavior of algal genera and species in sewage treatment ponds points up a number of problems other than taxonomic. The recurring sequence of *Euglena, Chlamydomonas, Scenedesmus*—and then back to *Euglena*—suggests the presence of, among other substances, growth promoters.

In Alaska this series, followed finally by the appearance of *Closterium,* is thought to represent a full cycle of oxidation and digestion of sewage, with the development of an acid condition in the pond.

ECONOMICS

In economic research systematics aids and is aided. Analyses of algae for various carbohydrates, proteins, and fatty substances help to further characterize genera and orders, if they actually do not form bases for taxonomic distinctions. During the past decade or so a great variety of investigations have been conducted by the Scottish Seaweed Research Institute. Substances of commercial importance, such as alginates for paint, foodstuffs, and fertilizers, have been extracted. Furthermore, in that country pilot plants are producing *Chlorella pyrenoidosa* and its various strains by the ton for animal feeds and possibly for human food. As much as 16 tons of *Chlorella* per acre per year are harvested at a cost of about 30 cents per pound. The specificity of quantities of protein and the differences in kinds of chemical substances produced by *Chlorella* may be an aid in differentiating species and varieties. Certainly there are few if any morphological features useful in identifying these species. The same substantiation for the separation and characterization of genera in the marine algae is derived from the analyses of gels produced, for example, by *Gracilaria* Spp. and *Hypnea musciformis;* these show different specific qualities of the same gels.

The long-time study of N-fixation with its economic implications continues and of course calls for the co-operation of the taxonomist. Blue-green algae, some species in particular, are concerned (Williams and Burris, 1952). *Nostoc muscorum* and *Calothrix parietana* unexplainably fix N, whereas other species placed in culture fail to do so.

SYMBIOTISM

Systematic studies in respect to symbiotism seem to be somewhat lethargic judging from the paucity of recent literature on the problem. We are looking for contributions to systematic phycology from the analyses of lichens and in those studies which separate the fungal and algal elements.

Another interesting problem which demands more attention is the true nature of *Gloeochaete* and *Glaucocystis,* recently studied by Pringsheim (1958) and by Skuja (1954). Also the endozoic relationships between *Zoochlorella* and anemones, sponges, protozoa, and between *Casiopeia* and Dinoflagellata (previously mentioned), when fully understood, will aid systematics at the genus and species level, if at no other.

OTHER FIELDS OF RESEARCH

In the space available it is impossible to refer to all areas of research which contribute to algal systematics. The roles of palynology and paleontology, for example, could properly be discussed. The major fields have been touched upon, but it would be inappropriate if one other important research area were not considered, if only briefly—namely, systematic research itself. This is a facet of investigation which is older than phycology as a science, for systematics is an outgrowth of the favorite and necessary habit of Man to name objects, and Man has been naming algae for a long, long time.

Recently and currently, since we are considering modern research influences, taxonomy is aided by studies and publications from scores of workers throughout the world—so many that if one were to mention a few, others seemingly would be neglected. But it is proper to recall the intensive work of Taylor and Papenfuss on the marine algae of North and Tropical America; Chapman and Womersley in New Zealand and Australia; Skuja, Christensen, Thomasson, and Teiling in Scandinavia; Gautier and Pocock in Africa; Bourrelly in France and Africa; Parke, Brook, Lund, and many others in Great Britain; Fott in Czechoslovakia; Drouet, Patrick, Reimer, Thompson, and Whitford on fresh-water algae in the United States. Recent interest in the Arctic and Antarctic has been accompanied by several reports, including exhaustive papers by Croasdale. Scott (lately) of New Orleans has written critically and extensively on Desmids from many parts of the world. There is increased productivity of taxonomic research on both fresh-water and marine algae in the Orient: Okamura, Hirano, Segawa, and Okada in Japan; Jao in China; Velasquez in the Philippines; Randhawa and Desikachary in India; Islam in Pakistan, to mention just a few. Gonzales has written on Iberian algae and Hustedt continues his work on the Diatomaceae of the world. Geitler is one of the most prolific contributors to algal systematics in Germany, as is Schussnig in Austria. Kol in Hungary has given us more information on cryoplankton than perhaps any other student.

No intent is involved in the apparent slighting of the numerous other workers in algal taxonomy. Although there seems to be an army of taxonomists, and while algal classification seems to be at full tide, yet an inventory of the problems we are facing reminds us that we need so very much more work. I return to my primary point, that if we are to keep abreast of the demands that other facets of phycology are making, and of all of biology for that matter, there must be increasing attention given to algal systematics.

LITERATURE CITED

Allen, M. B., and Dawson, E. Y. 1959. Production of antibacterial substances by benthic tropical marine algae. *Jour. Bact.* 79:459.

Asmud, B. 1956. Electron microscope observations on *Mallomonas* species and remarks on their occurrence in some Danish ponds. *Bot. Tidss.* 53:75-85.

Asmud, B. 1961. Studies on Chrysophyceae from some ponds and lakes in Alaska. I. *Mallomonas* species examined with the electron microscope. *Hydrobiol.* 17(3): 237-258.

Ballantine, Dorothy. 1956. Two new marine species of *Gymnodinium* isolated from the Plymouth area. *Jour. Mar. Biol. Assoc. U.K.* 35:467-474.

Beckwith, T. D. 1933. Metabolic studies upon certain Chlorellas and allied forms. *Univ. Calif. Los Angeles Biol. Sci.* 1(1):1-34.
Bold, H.C., and Parker, B.C. 1960. Some new attributes for clarifying species of Chlorococcales. *Phyc. News Bull.* 40:63-64.
Burkholder, P. R., and Almodovar, L. 1960. Antibiotic activity of some marine algae of Puerto Rico. *Bot. Mar.* 2:149-156.
Cave, Marion S., and Pocock, Mary A. 1956. The variable chromosome number in *Astrephomena gubernaculifera. Am. J. Bot.* 43(2):122-134.
Chapman, V. J. 1953. Phylogenetic problems in the Chlorophyceae. Rep. 7th Sci. Congr., N. Z. (Roy. Soc. N. Z.), 1953:55-68.
Chapman, V. J. 1954. The classification of the Ulvales and Siphonocladiales. *Rapp. Comm. 8-me Congr. Inter. Bot.* 17:93-95.
Chapman, V. J. 1954a. The Siphonocladiales. *Bull. Torr. Bot. Club* 81(1):76-82.
Chen, Y. T. 1949. The flagella structure of some Protista. *Proc. Conf. Electron Micr.*, pp. 156-158.
Christensen, T. 1952. On *Chlorosaccus* Luther. *Bot. Tidss.* 49(1):33-38.
Davis, C. C. 1948. *Gymnodinium brevis,* Sp. Nov., a cause of discolored water and animal mortality in the Gulf of Mexico. *Bot. Gaz.* 109(3):358-360.
Dawes, C. J., Scott, F. M., and Bowler, E. 1961. A light- and electron-microscopic survey of algal cell walls. I. Phaeophyta and Rhodophyta. *Am. J. Bot.* 48(10): 925-934.
Deason, T. R., and Bold, H. C. 1960. Exploratory studies of Texas soil algae. *Univ. Texas Publ.* 6022. 70 pp.
Desikachary, T. V. 1952. Electron microscope study of the Diatom-wall structure. *J. Sci. and Indust. Res.* 11B(11):491-500.
Desikachary, T. V. 1957. Electron microscope studies on Diatoms. *J. Roy. Micr. Soc.* 76(1/2):9-36(1956).
Desikachary, T. V., and Sundaralingam, V. S. 1962. Affinities and interrelationships of the Characeae. *Phycologia* 2(1):9-16.
Dodge, J. D. 1960. Nuclei, nuclear division and taxonomy in the Dinophyceae (abst.). *British Phycol. Bull.* 2(1):14-15.
Drew, K. M. 1954. The organization and interrelationships of the carposporophytes of living Florideae. *Phytomorphol.* 4(1/2):55-69.
Drew, K. M. 1954a. Studies in the Bangioideae. III. The life history of *Porphyra umbilicalis* (L.) Kütz. var. *laciniata* (Lightf.). *J. Agr. Annal. Bot.* 18(70):183-211.
Drew, K. M. 1956. Reproduction in the Bangiophycidae. *Bot. Rev.* 22(8):553-611.
Drew, K. M. 1958. Studies in the Bangiophycidae. *Publ. Staz. Zool. Nap.* 30(3): 358-372.
Egerod, L. E. 1952. An analysis of the Siphonous Chlorophyta with special reference to the Siphonocladales, Siphonales, and Dasycladales of Hawaii. *Univ. Calif. Publ. Bot.* 28(5):325-454.
Ettl, H. 1958. Zur Kenntnis der Klasse Volvophyceae. *Algol. Stud. Prague,* 1958: 206-289.
Fan, K.-C. 1959. Studies on the life histories of marine algae. I. *Codiolum* and *Spongomorpha. Bull. Torr. Bot. Club* 86(1):1-12.
Fan, K.-C. 1960. On pit connections in Bangiophycidae. *Nova Hedwigia* 1(3/4): 305-307.
Flint, L. H., and Moreland, C. F. 1946. Antibiosis in blue-green algae. *Am. J. Bot.* 33:218.
Fogg, G. E., and Boalch, G. T. 1958. Extracellular production in pure culture of a brown alga. *Nature* 181:789.
Fritsch, F. E. 1947. The status of the Siphonocladales. *J. Ind. Bot. Soc.* (Iyengar Comm. Vol., 1946):29-50.

Fritsch, F. E. 1949. The lines of algal advance. *Biol. Rev.* 24:94-124.
Gorham, P. R. 1960. Toxic waterblooms of blue-green algae. *Canad. Vet. J.* 1(16): 235-245.
Harris, K., and Bradley, D. E. 1957. An examination of the scales and bristles of *Mallomonas* in the electron microscope using carbon replicas. *J. Roy. Micr. Soc.* 76(1/2):37-46 (1956).
Hillis, L. W. 1959. A revision of the genus *Halimeda* (Order Siphonales). *Inst. Mar. Sci.* 6:321-403.
Hutner, S. H., et al. 1949. Assay of anti-pernicious anemia factor with *Euglena. Proc. Soc. Exptl. Biol. Med.* 70:118-120.
Ingram, W. M., and Prescott, G. W. 1954. Toxic fresh-water algae. *Am. Mid. Nat.* 52(1):75-87.
Krishnamurthy, V. 1962. The morphology and taxonomy of the genus *Compsopogon* Montagne. *J. Linn. Soc. Bot.* 58:207-222.
Kylin, H. 1923. Studien über die Entwicklungsgeschichte der Florideen. *Kongl. Sv. Vet.-Akad. Handl.* 63(11):1-139.
Kylin, H. 1928. Entwicklungsgeschichte Florideenstudien. *Lunds Univ. Arsskr., n.f.,* 24(Avd. 2, No. 4):1-127.
Kylin, H. 1930. Some physiological remarks on the relationship of the Bangiales. *Bot. Not.* 1930:417-420.
Kylin, H. 1930a. Ueber die Entwicklungsgeschichte der Florideen. *Lunds Univ. Arsskr., n. f.,* 2,26(6):1-104.
Kylin, H. 1935. Remarks on the life history of the Rhodophyceae. *Bot. Rev.* 1(4): 138-148.
Leak, L. V., and Wilson, G. B. 1960. The distribution of chromatin in blue-green alga *Anabaena variabilis* Kütz. *Canad. J. Gen. Cytol.* 2(4).
Lewin, R. A., and Meinhart, J. O. 1953. Studies on the flagella of algae. III. Electron micrographs of *Chlamydomonas moewusii. Canad. J. Bot.* 31:711-717.
McLaughlin, J. J. A., 1958. Euryhaline Chrysomonads: Nutrition and toxigenesis in *Prymnesium parvum,* with notes on *Isochrysis galbana* and *Monochrysis lutheri. J. Protozool.* 5(1):75-81.
McLaughlin, J. J. A., and Zahl, P. A. 1959. Axenic Zooxanthellae from various invertebrate hosts. *Ann. N. Y. Acad. Sci.* 77(2):55-72.
Papenfuss, G. F. 1951. Problems in the classification of the marine algae. *Svensk Bot. Tids.* 45(1):4-11.
Parke, Mary. 1961. Some remarks concerning the Class Chrysophyceae. *British Phycol. Bull.* 2(2):46-55.
Parke, Mary, Manton, I., and Clark, B. 1955. Studies on marine flagellates. II. Three new species of *Chrysochromulina. J. Mar. Biol. Assoc. U. K.* 34:579-609.
Parker, B. C., and Bold, H. C. 1961. Biotic relationships between soil algae and other microorganisms. *Am. J. Bot.* 48(2):185-197.
Pascher, A. 1931. Systematische Übersicht über die mit Flagellaten in Zusanimengang stehenden Algenreihen und Versuch einer Einreihung deiser Algenstämme in die Stämme des Pflanzenreiches. *Beih. Bot. Centralbl.,* 2Abt., 48:317-332.
Petersen, J. B., and Hansen, J. B. 1956. On the scales of some *Synura* species. *Biol. Medd. Kongl. Danske Vidensk. Selskab* 23(2):1-27 (reprint).
Petersen, J. B., and Hansen, J. B. 1958. On the scales of some *Synura* species. *Ibid.* 23(7):1-13 (reprint).
Pringsheim, E. G. 1958. Organismen mit blaugrünen Assimilatoren. *Stud. in Plant Physiol.,* pp. 165-184.
Proctor, V. W. 1957. Studies of algal antibiosis using *Haematococcus* and *Chlamydomonas. Limnol. Oceanog.* 2:125-139.

Provasoli, L., Hutner, S. H., and Schatz, A. 1948. Streptomycin-induced chlorophyll-less races of *Euglena. Proc. Soc. Exptl. Biol. Med.* 69:279-282.
Rabinovich, Delia. 1956. Estudios citológicos sobre la presencia de sustancia nuclear en algunas Schizophyta. *Contrib. Ceintif.* 1(3):91-156.
Ray, S. M., and Wilson, W. B. 1957. Effects of unialgal and bacteria-free cultures of *Gymnodidium brevis* on fish and notes on related studies with bacteria. *Fish. and Wildlife Serv., Fish. Bull.* 123:469-496.
Scagel, R. F. 1959. The role of plants in relation to animals in the marine environment. *Mar. Biol., Biol. Colloq., Oregon State College,* No. 20:9-29.
Scagel, R. F. 1961. Marine plant resources of British Columbia. *Fish. Res. Board Canad., Bull.* 127. 39 pp.
Schlitchting, H. E. 1961. Viable species of algae and protozoa in the atmosphere. *Lloydia* 24(2):81-88.
Shelubsky, M. 1951. Observations on the properties of a toxin produced by *Microcystis. Proc. Inter. Assoc. Theor. and Appl. Limnol.* 11:362-366.
Shelubsky, M. 1951a. Toxic blue-green algae in fish-ponds in Israel. *Bamidgeh* 3(49/50):146-154.
Schussnig, B. 1935. Neure Vorstellungen über die Phylogenie der Grünalgen. *Biol. Gen.* 11(2):192-210.
Sieburth, J. M. 1959. Antibacterial activity of Antarctic marine phytoplankton. *Limnol. Oceanog.* 4(4):419-424.
Silva, P. C. 1951. The genus *Codium* in California with observations on the structure of the walls of the utricles. *Univ. Calif. Publ. Bot.* 25(2):79-114.
Skuja, H. 1954. Phylogenetische Stellung der Glaucophyceen. *Proc. 7th Congr. Intern. Bot.,* 1954:823-825.
Starr, R. C. 1955. A comparative study of *Chlorococcum* Meneghini and other spherical, zoospore-producing genera of the Chlorococcales. *Indiana Univ. Publ.* 20. 111 pp.
Stephens, Edith L. 1949. *Microcystis toxica* Sp. Nov.: A poisonous alga from the Transvaal and Orange Free State. *Trans. Roy. Soc. So. Africa* 32(1):105-112.
Steyn, D. G. 1945. Poisoning of animals by algae (scum or waterbloom) in dams and pans. *Dept. Agr. and Forestry, Union So. Africa,* 1945:1-9 (reprint).
Sun, C.-N. 1943. A preliminary study of a substance in *Azolla* affecting the growth of algae. *Sci. Record* 1:539-601.
Svedelius, N. 1931. Nuclear phases and alternation in the Rhodophyceae. *Beih. Bot. Centralbl.* 48:38-89.
Van Oye, P. 1961. Les Myxophyceae sont-elles des algues? *Hydrobiol.* 18(4):293-299.
Williams, A. E., and Burris, R. H. 1952. Nitrogen fixation by blue-green algae and their nitrogenous compounds. *Am. J. Bot.* 39(5):340-842.

Criteria and Procedures in Present-Day Algal Taxonomy

C. van den Hoek

Rijksherbarium,
Leiden, Netherlands

Although algal taxonomists may consult a considerable number of determination works, some of them quite recent, the present state of our knowledge in this field of science remains in some respects unsatisfactory, as few if any critical investigations on the validity of current taxonomic criteria used in many groups have been conducted.

Rather curiously, the most promising progress is being made in the taxonomy of certain difficult unicellular and colonial groups, for which it is now a more generally accepted point of view that only comparative studies on unialgal cultures and on copious living material from nature can yield reasonably satisfactory results. In these groups taxonomic study is becoming more and more a comparative study of life histories.

Recent contributions to the taxonomy of the Volvocales by Butcher (1959), Ettl (1958, 1959, 1960, 1961; and Ettl, 1959a,b), Gerloff (1962), Pocock (1959a) and Stein (1958a,b, 1959), and to that of the Chlorococcales by Starr (1955), Bold (1958), Deason (1959), Herndon (1958a,b), and Pocock (1960) are of a very high standard and should in some measure serve as examples to taxonomists working on other algal groups.

Critical investigations of dried and conserved material, however, can yield valuable results in groups of algae with more complicated, differentiated structures, e.g., the Charophyceae and many marine groups of Chlorophyta, Rhodophyta, and Phaeophyta. Silva's monographic series on the taxonomy of Codium (Silva, 1951, 1955, 1957, 1959, 1960; Silva and Womersley, 1956), Egerod's (1952)

Analysis of the siphonous Chlorophycophyta, Hillis's (1959) *Revision of the genus Halimeda,* and Dixon's (1960a,b) studies on the genus Ceramium are recent examples of fundamentally important papers entirely or mainly based on the study of preserved material.

Several algal groups have the honor to have been monographed more than once, e.g., the Cyanophyta by Geitler (1925, 1930-32), Elenkin (1936, 1949), and Desikachary (1959); the Zygnemales by Czurda (1932), Kolkwitz and Krieger (1941-44), Transeau (1951), and Randhawa (1959); the Oedogoniales by Hirn (1900), Tiffany (1930), and Gemeinhardt (1939). These monographs, which are indispensable frameworks for the taxonomy of the groups concerned, are at least partly compilations, to which the author's personal views and observations have been added. The disadvantage of such monographs is that further taxonomic studies in the same groups often rely too much on the infallibility of the criteria and taxa found in these volumes. We must not forget that works of such a very wide scope can hardly be—for practical reasons—the result of methodical and critical comparative taxonomic study of abundant living and dead material. For this reason the later monographs resemble the older ones to quite a high degree.

It is a rather current procedure to describe a specimen as a new species, when, in the author's eye, it does not tally sufficiently with any of the descriptions present in the monograph at hand. As the range of variability and the validity of many criteria in most cases have not been critically investigated, many such new taxa seem to be insufficiently defined. Such a procedure is often inspired by the understandable need, in case of a local flora or a biocoenotic study, to label all collected material with some name, be it even a new one, when the determination works refuse to give a satisfactory answer.

In the Cyanophyta, perhaps the most difficult of all phyla, there is a continuous inflow of such new species (see, for example, Claus, 1962; Dedusenko-Ščegoleva, 1959; Gayral and Seizilles de Mazancourt 1958; Kukk, 1959; Moruzi, 1960; Muzafarov, 1959; Pignatti, 1958; Welsh, 1961; Petrov, 1961; Novičkova, 1960). In a few cases material has been cultured (Schwabe, 1960a,b; Friedmann, 1961), but the results remain questionable, as these authors did not carry out comparative cultures of allied species grown under similar conditions in order to establish with certainty the differences between their new species and the already known related ones.

Drouet and Daily's (1956) *Revision of the coccoid Myxophyceae,* an audacious approach to the taxonomy of this most difficult group, is one of the few recent, methodical taxonomic studies in the Cyanophyta. It has been much criticized by Skuja (1956), Geitler (1960), and Bourrelly (1957), whose criticisms, however, are partly rather vague and sometimes more vindictive than factual. More measured criticism has been put forward by Powell and Powell (1957), Feldmann (1958), Komárek (1957), and Koster (1961), the latter author, in an instructive article about the difficulties in cyanophycean taxonomy, taking a positive attitude toward Drouet and Daily's work.

Since it is of course impossible to give a complete review of the criteria and methods in the present-day taxonomy of all algal groups, I have chosen to discuss in particular the recent developments in the taxonomy of two groups—the Ulvaceae and the genus Cladophora. The problems exhibited in these two examples, however, seem to have a more general applicability.

ULVACEAE

The taxonomy of the genera Enteromorpha and Ulva developed almost explosively, at least in Europe, during the last few years. This development started in 1938 with Bliding's work on the systematics of *Enteromorpha minima* (Bliding, 1938). In this paper Bliding drew attention to the impossibility of defining the interspecific delimitations by purely morphological and anatomical characteristics of dried or otherwise conserved material. This author studied essentially the following points: 1) the nature of the reproductive bodies; 2) the interfertility (or intersterility) of sexually reproducing species; 3) the development of zygotes or asexual swarmers into adult plants in culture; 4) the number of pyrenoids in the vegetative cells.

Working along these lines, Bliding succeeded in defining the delimitations and the variability of several Enteromorpha species (Bliding, 1939, 1944, 1948a,b, 1955).

A number of French phycologists, mainly guided or inspired by Dangeard, adopted the methods indicated by Bliding, and a few years ago started extensive investigations on the taxonomy of Enteromorpha and Ulva.

The combined efforts of these authors resulted in reinterpretations of nine already existing Enteromorpha and three Ulva species, and in the description of twenty-five new species, varieties and forms in the genus Enteromorpha, whereas only seventeen new species and infraspecific taxa were added to the genus Ulva. Two of these new species were described by the Norwegian investigator Föyn (1955, 1958).

The following criteria appeared to be taxonomically valid with respect to Ulva and Enteromorpha: 1) the external morphology of the thallus; 2) the form and arrangement of the cells; 3) the number of pyrenoids in a vegetative cell; 4) the size of the cells; 5) the mode of reproduction; 6) the interbreeding behavior of sexually reproducing species; 7) the morphology of young germlings; 8) the diameter of the thallus in cross section and the height-to-width ratio of the cells.

Of all these criteria, the mode of reproduction and the interbreeding behavior are the most constant for each species, although some variation may be present. The other criteria, however, can be extremely variable, as has been demonstrated by Bliding for several Enteromorpha species.

1. External Morphology of the Thallus. The variation can be very wide, particularly in the profusely branching species. In *Enteromorpha prolifera* (Müll.) J. Ag., *E. compressa* (L.) Grev., *E. ahlneriana* Bliding, and *E. intermedia* Bliding (Bliding, 1939, 1944, 1948a, 1955), the thalli may be unbranched or sparsely branched to profusely branched, and the branches may be filiform to broadly flattened. *Enteromorpha clathrata* (Roth) Grev. is profusely branched, but the branches can be delicate, filiform, to broadly ribbon-like.

Enteromorpha intestinalis (L.) Link and *E. linza* (L.) J. Ag., on the other hand, are unbranched or very rarely provided with some proliferations from the base. In *E. intestinalis* the plants may be delicate and tubular to broad, and more or less intestine-like or flattened. In *E. linza* the external form may be narrowly lanceolate to broadly suborbicular, Ulva-like.

Unbranched plants of *E. prolifera* and *E. compressa* may be deceivingly similar to *E. intestinalis,* the more so as their cytological characteristics resemble each other to a high degree. Only in *E. prolifera* are the cells mostly, but not always, arranged in longitudinal rows.

In these cases only culture experiments and interbreeding experiments could give the solution. Bliding in this way succeeded in separating morphologically identical *E. compressa* and *E. intestinalis* plants growing beside each other in a sheltered, eutrophic locality (Bliding, 1948a). In the same way he was also able to distinguish between extremely similar *E. prolifera* and *E. intestinalis* plants growing also in such a locality (Bliding, 1939).

Several authors (Gayral, 1960a; Beaudrimont, 1961) obtained in cultures plants with a morphology which was mostly abnormal or at least very different from that of the original material, and they doubted the validity of the morphology of these culture plants for taxonomic purposes. (The cytological characteristics agreed fairly well with those of the original material.)

Föyn (1955, 1958, 1959, 1960, 1961, 1962), on the other hand, obtained perfectly normal Ulva plants in his cultures. Morphological differences, characteristic for spontaneous mutants which arose in his cultures, could be very well recognized.

Bliding's work has demonstrated the existence of species with a pronounced tendency for ramification; plants of these species may be, incidentally or under the influence of certain environmental conditions, unbranched. Other species have a much reduced tendency for ramification, but they may proliferate, again incidentally or under the influence of certain environmental conditions (cf. Dangeard, 1959e, who describes dense proliferation of *E. intestinalis*).

Although the mode of ramification is genotypically defined, and therefore valid as a taxonomic criterion, it is highly variable in relation to age and environment, so that its evaluation as such a criterion should be a matter of the utmost caution (cf. also Burrows, 1959).

2. Form and Arrangement of the Cells. The cells may be either more or less rectangular and arranged in longitudinal series, or polygonal and without any definite arrangement.

Although some variation may be present, this criterion seems to be rather constant and taxonomically useful. Quite often, however, the arrangement may be distinct in one part of the thallus and indistinct in another part. To distinguish between orderly or disorderly arrangements is not always easy. Dangeard (1959a), for example, states that his species *E. rivularis* has its cells generally arranged without order; in his figure, however, the cells are orderly arranged.

Relatively thick cell walls and rather rounded cells are sometimes thought to be characteristic for certain species, e.g., for *Ulva dangeardii* Gayral and Seizilles de Mazancourt (1959), but on the other hand Bliding (1944, 1948a, 1955) records highly variable cell-wall diameters for *E. intestinalis, E. ahlneriana, E. intermedia,* and *E. prolifera.*

3. Number of Pyrenoids in a Vegetative Cell. Although the number of pyrenoids is used as a fundamental character, it may vary considerably in those species with more than one pyrenoid in each cell. In *E. clathrata* plants, notably those from sunny sheltered localities, the number per cell generally is 2-3, whereas it is 3-5 in plants from shady eutrophic localities (Bliding, 1944). In *E. intermedia* Bliding var. *biflagellata* Bliding (cf. Bliding, 1944) a similar phenomenon can be observed (1-2 and 3-> pyrenoids, respectively). It is, therefore, possible to distinguish between species with one pyrenoid per cell and species with more than one pyrenoid per cell. It seems to be a rather hazardous affair, however, to make distinctions based on the number of pyrenoids per cell in the latter group of species.

4. Size of the Cells. Although the cell size may vary considerably in a single thallus and in relation to environment, it is an important characteristic. The overlap of the cell sizes of the various species is large, as can be seen in Table I, in which the measurements of several thoroughly investigated species are given.

Bliding (1944) observed that the cells in slender, pale green *E. clathrata* plants from shallow localities were much larger (viz., about 30-50 × 20 μ) than in more robust, dark green plants from shady, eutrophic localities (viz., about 13 × 8 μ). The number of pyrenoids, however, was greater in the smaller-celled plants.

TABLE I
(Data from Bliding, 1944, 1948)

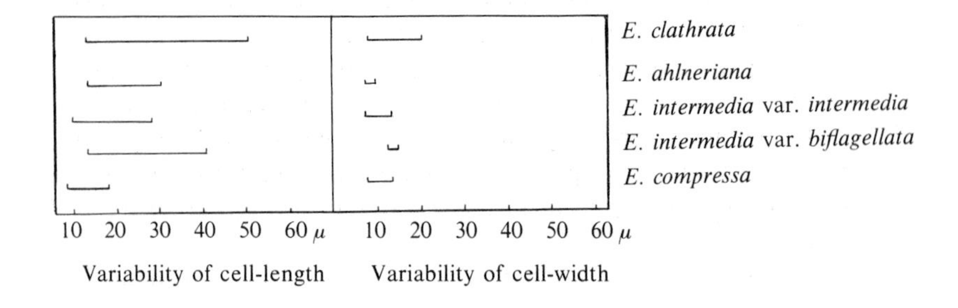

Enteromorpha ahlneriana (Bliding, 1944) comprises comparable dark green, robust forms which are extremely similar to *E. prolifera* plants, the more so since the cells each contain only one pyrenoid and have a size of about 8-13 $\times$ 6-8 μ (in pale green, slender plants 23-30 $\times$ 10 μ). The cells of *E. ahlneriana* are arranged in longitudinal rows; in *E. compressa* the cells are disorderly arranged. As to the regular cell arrangement, however, *E. ahlneriana* resembles *E. prolifera*, from which species it is, as a small-celled form, almost indistinguishable. Only extensive culture and interbreeding experiments could unravel this tangle.

According to Föyn (1934) the cells of diploid *Ulva "lactuca"* plants are larger than the cells of haploid plants; Moewus (1938) records the same difference in size for diploid and haploid *Enteromorpha intestinalis* plants.

5. Mode of Reproduction. Some species reproduce only by biflagellate asexual zoospores (neutral zoospores), some by quadriflagellate neutral zoospores, and other species by biflagellate gametes and quadriflagellate zoospores in a regular alternation of generations.

Although the mode of reproduction is generally constant for one species, differences exist; these differences, however, do not justify further splitting.

Enteromorpha compressa, for example, was observed by Bliding to comprise races with a regular alternation of generations and races with predominantly parthenogenetically developing gametes. *Enteromorpha intestinalis* was observed to include, apart from races with a regular alternation of generations, also a race with parthenogenetically developing gametes and a race with reproduction by only neutral quadriflagellate zoospores (Bliding, 1944). Races with parthenogenetically developing gametes have been recorded by Moewus (1938), but as no descriptions were given by this author, it is rather hazardous to compare his results with those of Bliding and other authors.

In Moewus's experiments, parthenogametes of *E. "compressa"* and *E. "intestinalis"* developed into plants, which became diploid by so-called "Aufregulierung" of the number of chromosomes from haploid to diploid. (Moewus, however, had no cytological evidence!) Such regulated diploid plants reproduced by neutral quadriflagellate zoospores, which arose without meiosis. These zoospores, however, were about twice as large as normal quadriflagellate zoospores. Such abnormally large zoospores were never observed by Bliding in material from nature (Bliding, 1948a).

Föyn (1955), mainly on account of interbreeding experiments, divided the common *Ulva lactuca* into two species, of which the one from northern Europe—for which the name *U. lactuca* was reserved—appeared to be slightly anisogamous, and that from southern Europe—named *U. thuretii*—was observed to be isogamous. Dangeard (1960b) and Feldmann (1960), however, have recorded anisogamously reproducing *Ulva "lactuca"* from the coasts of southwestern France and Brittany, where it should be isogamous, according to Föyn; Moewus (1938) recorded an isogamously reproducing *U. lactuca* race from Helgoland and an anisogamously reproducing one from Naples.

Ulva olivacea (*olivascens*) Dangeard (1951, 1960b) was observed to reproduce by neutral quadriflagellate zoospores only, but Gayral (1961a) described, apart from quadriflagellate zoospores, biflagellate zooids interpreted as gametes.

Dangeard (1960a) collected from the Mediterranean, near Banyuls, several plants in all respects very similar to *E. linza*. This species is known to reproduce only by means of neutral quadriflagellate zoospores, but Dangeard's material produced, apart from quadriflagellate zoospores, biflagellate zooids (interpreted as gametes). On this account he considered his material to be different from *E. linza* and conspecific with *Solenia bertoloni* Ag., for which the new combination *Enteromorpha bertoloni* (Ag.) P. Dangeard was created. No data about the reproduction of Agardh's Ulva are available, however, and as far as I know Dangeard did not investigate the type material.

We know also from groups with a more constant and a more easily recognizable morphology that differences in the mode of reproduction may exist within specific limits. *Codium vermilara* (Olivi) Delle Chiaje and *Codium fragile* (Sur.) Hariot, for example, comprise races or varieties reproducing by anisogametes arising by meiosis and races reproducing by biflagellate neutral zoospores (Dangeard, 1959b; Feldmann, 1956; Dangeard and Parriaud, 1956; Delépine, 1959), which are often interpreted as parthenogametes, although they have never been observed to copulate.

6. Interbreeding Behavior of Sexually Reproducing Species. This criterion plays an important role in the distinction of very similar modifications of various Enteromorpha species. Bliding tested all his strains extensively and on this ground he was able to separate certain, almost identical growth-forms of *E. compressa*, *E. intestinalis*, and *E. prolifera*.

This criterion, however, should not be considered an absolute one, since we know that intraspecific interfertility barriers exist. Coleman (1959), for *Pandorina morum*, and Stein, for *Gonium pectorale* (1958b) and *Astrephomene gubernaculifera* (1958a), demonstrated the existence of sexually incompatible strains. Köhler (1956) collected sexually incompatible strains of *Chaetomorpha linum* (Müll.) Kütz. from the Mediterranean and from the German North Sea coast. Exchange of gene-material is also impossible between sexually and purely asexually reproducing strains or varieties of one species.

On the other hand, hybridization between well-established algal species is known; especially in the genus Fucus are such hybrids known (cf. Parriaud, 1954).

Gayral (1960b) claims to have observed copulations between gametes of *U. fasciata* and *U. elegans* (a newly described species), but a further development of the zygotes was not observed. It is a curious coincidence that the biflagellate zooids of *U. fasciata* have never been observed to copulate among each other, but are only known to develop directly into new plants. The new species *Ulva elegans* resembles *U. fasciata*, differing from it by its thinner thallus and by lower cells in cross section. As far as I know, the range of morphological and anatomical variation of *U. fasciata* has not been established, so that the evidence for the validity of the new species *U. elegans* is not very convincing.

Fusion of zooids cannot always be interpreted as copulations; zoospores can incidentally also aggregate and fuse in a way rather different from the active and intensive typical copulation reaction. Pocock (1960) observed such noncopulation fusions among zoospores of Hydrodictyon; I quite often observed such fusions among zoospores of several Cladophora species.

Ulva crispa P. Dangeard (cf. Dangeard, 1959a) has been described as a new species mainly on account of its peculiar morphology. It forms small, intricately folded cushions or pompons on exposed coasts in Morocco. Cytologically it is very similar to *U. "lactuca."* Its gametes appeared to copulate with *Ulva lactuca* gametes. Are we here dealing with hybridization or with evidence that *U. crispa* and *U. lactuca* are conspecific? At any rate, the evidence for the establishment of a new species is not unequivocal, the more so because we can expect *Ulva "lactuca"* to have an extremely plastic morphology in relation to age and environment. Föyn (1955, 1958) recently discovered that the very common alga which is generally named *Ulva lactuca*—and not on indisputable grounds (cf. Papenfuss, 1960)—is not homogeneous, but consists of three different units, of which *Ulva thuretii nov. sp.* and *Ulva mutabilis nov. sp.* occur along the coasts of southern Europe as far north as Brittany, and *Ulva lactuca* s. s. only on the northern European coasts. These three species are sexually incompatible and show further differences in the morphology of young plants (cf. p. 35, p. 38). Adult plants of the three species are similar. *Ulva mutabilis* is characterized by the fact that spontaneous mutants arise regularly in cultures. The gametes of *U. lactuca* and *U. thuretii* copulate, but the zygotes grow very slowly and die at a few-celled stage.

7. Morphology of Young Germlings. Differences in the youngest developmental phases may be of taxonomic value. In *Enteromorpha minima* Näg. ex Kütz. and *E. marginata* J. Ag. [now *Blidingia minima* (Näg. ex Kütz.) Kylin and *Blidingia marginata* (J. Ag.) P. Dangeard] the developing spore grows into a basal, cushionlike stratum, from which upright tubular thalli later arise (Bliding, 1938; Dangeard, 1958, 1961b).

Other species may have a primary development somewhat similar to that of *E. minima* and *E. marginata*: the spore or the zygote grows into a small primary disc, from which one or more upright thalli arise at an early stage. This primary disc is overgrown by coalescent rhizoids growing downward from the basal part of the upright thallus forming a secondary disc. This type of primary development is considered a taxonomically valid characteristic for *E. sancti-joannis* P. Dangeard (1960c), *E. hendayensis* P. Dangeard et Parriaud (1960), *E. "tubulosa"* Kütz. (Dangeard, 1961c) and *Ulva gayralii* Cauro (1959).

The relation between the primary development of the attachment organ and of the upright shoot, however, may depend on culture conditions (Cauro, 1959) and on the nature of the substratum (Baudrimont, 1961). *Ulva linearis* P. Dangeard, for example, may form, depending on the "richness" of the culture medium, either primary filaments or primary discs.

Enteromorpha minima differs from the very similar *E. marginata* by the formation of stolons and secondary prostrate discs; in *E.*

marginata no stolons arise from the primary basal disc (Dangeard, 1961b).

8. Diameter of the Thallus in Cross Section and the Height-to-Width Ratio of the Cells. Thick thalli and a relatively large height-to-width ratio of the cells are considered characteristic for *Ulva rigida.* It appears from Table II that the measurements available hardly suggest any consistent division between the species represented on the ground of the thallus diameter. Especially in *U. olivacea* is the variation wide.

It is necessary to establish the variation range of this character in relation to age and environment before it can be used as a reliable taxonomic criterion.

* * *

It may be clear from all these considerations that the taxonomy of the Ulvaceae is far from easy.

For the characterization of an Ulvacean species one must have data as complete as possible. As we may expect a very wide morphological range, we cannot base the establishment of a new species on some striking morphological feature only.

For this reason taxa like *E. fasciculata* P. Dangeard (cf. Dangeard, 1959a), *E. compressa* var. *blidingioides* P. Dangeard (1959a), *Ulva denticulata* P. Dangeard (1959d), and *U. popenguinense* P. Dangeard (1959d) seem to be insufficiently defined.

It is interesting to record the large morphological range of the Ulvacean *Letterstedtia insignis* Aresch., as depicted by Pocock (1959b).

TABLE II

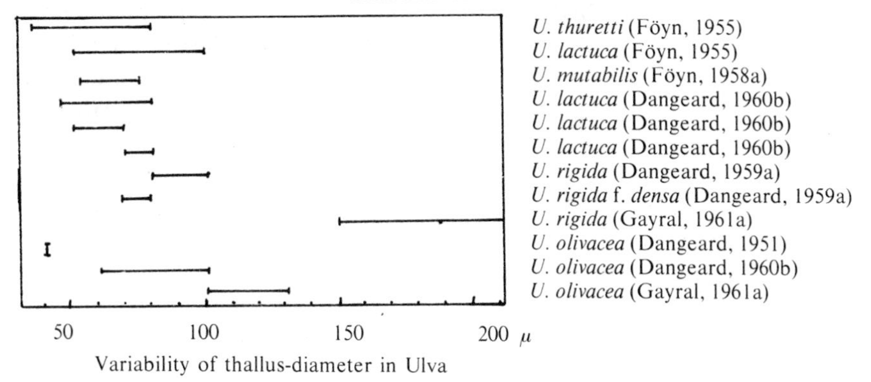

Variability of thallus-diameter in Ulva

Enteromorpha ahlneriana Blid. var. *roscoffensis* Beaudrimont (1960) has been recognized as a variety separate from *E. ahlneriana* Blid. on account of somewhat broader maximum diameters of the thallus. It is not clear whether Beaudrimont considers his new taxon a genotypic or only an environmental variety. In the latter case I would have preferred not to describe such a modification as a separate taxon, for a subdivision of a highly variable species such as *E. ahlneriana* into a number of modificatory infraspecific taxa is rather arbitrary. It may be possible that the broader thalli in his material constitute only an extension of the wide morphological range of *E. ahlneriana*, as this has been described by Bliding (1944).

For taxa like *E. flabellata* var. *mytilorum* P. Dangeard, *E. flabellata* var. *scopulorum* P. Dangeard (1959a), and *E. intestinalis* var. *musciformis* P. Dangeard (1959e), it is not clear whether or not we are dealing with genotypically defined units or modifications, although I have the impression that modifications are meant.

Most species on which these authors have been working were originally described by themselves. Nevertheless, these investigators have identified a small number of species with those of previous authors. Now, as we do not possess adequate taxonomic literature for the determination of these "old" species, such identifications should of course be a matter of caution; and it is clear that only by at least investigating the original material, including the type-material, can reliable results be achieved. I have the impression that the historical side of the problems—perhaps a rather tedious but inevitable aspect of all taxonomic work—has been rather neglected by most of the investigators of the Ulvaceae. For instance, two recently described Ulva species, viz., *U. dangeardii* Gayral and Seizilles de Mazancourt (1958a) and *U. incurvata* Parriaud (1958), appear to be conspecific with *Ulva curvata* (Kütz.) De-Toni, a nomenclatural synonym of *Phycoseris curvata* Kützing (v. d. Hoek, 1963).

I agree with Dangeard (1961c) that identification with dried type specimens is difficult, but it is nevertheless preferable to study type-material rather than to rely only on the old literature.

The type-material of *Enteromorpha tubulosa* (Kütz.) Kütz. (1855, p. 11, t. 32) [*E. intestinalis* (L.) Link γ *tubulosa* Kütz., 1849, p. 478] is present in the collection of the Rijksherbarium, Leiden (n. 938/7/132).

This is a delicate, tubular (up to 2 mm in diameter) Enteromorpha, some tubes not or hardly branching, others profusely branched from base to apex. The cells are 5×8 to 11×13.5 μ in diameter in the most delicate branches and 8×11 to 16×27 μ in the more robust branches and axes. The number of pyrenoids is one per cell (sometimes two). It has been collected from a salty source near Nauheim (Germany), far from the seacoast. These plants are rather different from those described by Dangeard under this name.

No pure taxonomic work on the genus Monostroma has been tackled in recent years, as far as I know. However, more and more data about the reproduction in this genus are becoming available. Some species are haplontic, the diploid phase being represented by a resting zygote, the contents of which divide meiotically so as to produce zoospores (Schreiber, 1941; Moewus, 1938; Arasaki and Tokuda, 1961; Segi, 1956; Segi and Gotô, 1956); in one species a regular alternation of haplontic and diplontic isomorphic generations seems to exist (Gayral, 1961c); other species reproduce by asexual neutral zoospores only (Bliding, 1935; Arasaki and Tokuda, 1961).

Recently, Arasaki and Tokuda (1961) and Chihara (1962) reported the existence of a haplontic Monostroma, the zygotic phase of which penetrates into mollusc shells adopting the morphology of *Gomontia polyrrhiza.* Also recently, Kornmann (1959) has demonstrated that *Gomontia polyrrhiza* is the sporophyte of a Chlorophycea with a disciform multicellular gametophyte. The zoospores of Monostroma, however, also grow at first into a multicellular disc (Segi, 1956; Segi and Gotô, 1956).

According to Chihara (1962) the chlorophycean *Collinsiella cava* also has zygospores which adopt a Gomontia-like morphology when penetrating into shells.

It is suggested here that all these rather complicated data be included in any future taxonomic work on Monostroma, but I am afraid that this task will be far from easy.

CLADOPHORA

During the last two years I have been investigating the taxonomy of the European representatives of the genus Cladophora, working mainly along the following lines:

1. Investigation of living material in its natural habitat, if possible in different seasons; comparative morphological studies

on plants of different species growing in the same locality and exhibiting comparable modifications in relation to the environment.

2. Comparative morphological study of unialgal cultures grown under approximately constant conditions (viz., in a constant temperature room of 12°C, in a 16-hour photoperiod, and in approximately the same culture-medium); strains isolated from as varied habitats as possible and from as distant localities as possible.

3. Study of the mode of reproduction and of the morphology of the reproductive bodies.

4. Investigation of herbarium material.

5. Formulation of valid taxonomic criteria and delimitation of the species and varieties.

6. Investigation of as many type-specimens as possible in order to establish the correct names and the synonymy.

As to the last-mentioned point, it is my opinion that the investigation of type-material should be the last step in a taxonomic study. Even now such a study is sometimes started with the investigation of type-material, perhaps because the investigator consciously or unconsciously attributes a fundamental taxonomic importance to the type specimen. The remaining material is then often ranged around the type specimen. Examples of such an approach to algal taxonomy are contributions to the systematics of Japanese Polysiphonia by Segi (1951, 1959, 1960). Obviously the expression "type," though improved by addition of the adjective "nomenclatural," remains a source of confusion. There is considerable advantage in Simpson's (1961) proposal to use the expression "onomatophore" (= name-bearer) instead. Another disadvantage of starting a taxonomic study with investigation of the type-material is the easily conceivable inability to interpret the taxonomic position of plants belonging to a group for which valid taxonomic criteria have not yet been formulated.

The following criteria appeared to be taxonomically valid:

1. Color. A limited number of species are characterized by a more or less constant dark green color, e.g., the fresh-water species *Cl. aegagropila* (L.) Rabenhorst and *Cl. okamurai* (Ueda) v. d. Hoek, and the marine species *Cl. rupestris* (L.) Kütz. Most species, however, may vary in color from pale whitish-green to dark green, depending on the exposure to sunshine.

Plants of *Cl. glomerata* growing in sunny localities, for example, are pale green, and those growing in shady localities are dark green. Eutrophy also increases the density of the chromatophore reticulum and intensifies the color of the individual chloroplasts, and hence gives the plants a darker green appearance.

Very often the intensity of the color has been incorrectly used for the characterization of certain species, e.g., of *Cl. crystallina* (Roth) Kütz., the name of which suggests a whitish "crystalline" color. Now we may find pale green, slender specimens of different species as *Cl. vagabunda* (L.) v. d. Hoek, *Cl. dalmatica* Kütz., *Cl. glomerata* (L.) Kütz., *Cl. sericea* (Huds.) Kütz., and *Cl. albida* (Huds.) Kütz., under this name.

2. Diameter and the Length-to-Width Ratio of the Cells. These measurements are very important for the taxonomy of Cladophora, although they may show an extreme variation in relation to the environment.

The diameters of the apical cells in particular are important, as they are the most constant in one plant. The diameters of the main axes are of much less value, since these depend very much on the age of a plant.

The exposure to sunlight greatly affects the diameter and the length-to-width ratio of the cells in many species. Pale green plants growing in sunny localities are extremely slender and the cells exhibit a large length-to-width ratio. Dark green plants from shady or eutrophic localities, on the other hand, are much more robust.

Closely related species, such as those of the section *Glomeratae*, widely overlap each other in the variation of their apical cell diameters (Table III).

When growing in the same locality, however, a *Cl. glomerata* plant is always more robust than a *Cl. fracta* plant; or a *Cl. laetevirens* plant more robust than a *Cl. vagabunda* plant, which in turn is more robust than a *Cl. dalmatica* plant. Yet the two first mentioned fresh-water species may be much alike, and the latter three marine species may be deceivingly similar.

In cultures the overlaps are much less pronounced, and it is possible to distinguish clearly between *Cl. lehmanniana* (which never gives much difficulty)—*Cl. laetevirens*—*Cl. vadorum*—*Cl. vagabunda, Cl. glomerata* var. *glomerata, Cl. glomerata* var. *crassior*—*Cl. dalmatica, Cl. parriaudii, Cl. fracta* var. *intricata, Cl. fracta* var. *fracta.*

TABLE III

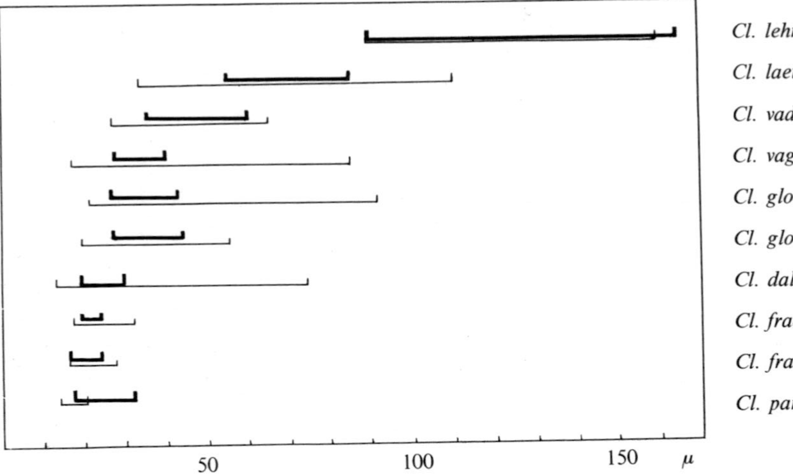

Variability of apical cell-diameter in the section *Glomeratae* (Cladophora)

The species can be subdivided into four groups according to the diameters of their apical cells. Determination of individual specimens is, of course, not always easy.

3. Ramification, Mode of Growth, and Thallus Organization. The mode of growth, ramification, and resulting thallus organization are important taxonomic criteria, though in most species also subject to considerable variation.

Species growing only by apical cell divisions and subsequent cell elongation exhibit a very regular thallus organization, since new branches arise in acropetal sequence, each on the apical pole of a newly cut-off segment directly under the apical cell or at a distance of several cells from the apex. Often the branches are unilaterally inserted on the axis. Such a thallus organization I have called an *acropetal organization.* An entirely undisturbed acropetal organization exists only in several marine species, e.g., *Cl. prolifera.* Here it is also a very constant feature in contradistinction to the sections *Glomeratae, Cladophora,* and *Rupestres.*

In most of the species of the section *Glomeratae* apical cell divisions are dominant in the apical region of the plant, while intercalary cell divisions start at some distance from the apex and increase in the basipetal direction. This mode of growth results in pseudodichotomously branching, intercalary growing main axes ending in dense clusters of acropetally organized terminal branch-systems: the characteristic thallus organization of *Cl. glomerata* (Fig. 1).

The relation between apical growth and intercalary growth depends to a high degree on environmental factors. For example, in *Cl. glomerata* var. *glomerata* plants growing in agitated water—the wash of canals, the surf of lakes, the stream of brooks and rivers—the apical growth dominates and an acropetal organization is generally beautifully constituted. In sheltered localities, however, the intercalary growth largely dominates the apical growth, so that the thallus organization becomes very indistinctly acropetal or rather irregular. The ultimate branches may disintegrate by sporulation, reducing the plants to the main axes, which may continue intercalary cell divisions for a long time, thus growing into long unbranched or scarcely branched filaments (Fig. 2).

This intricate pattern of a highly plastic morphology is still more complicated by the existence of a variety of *Cl. glomerata*—the variety *crassior* (Ag.) v. d. Hoek—in which the tendency for

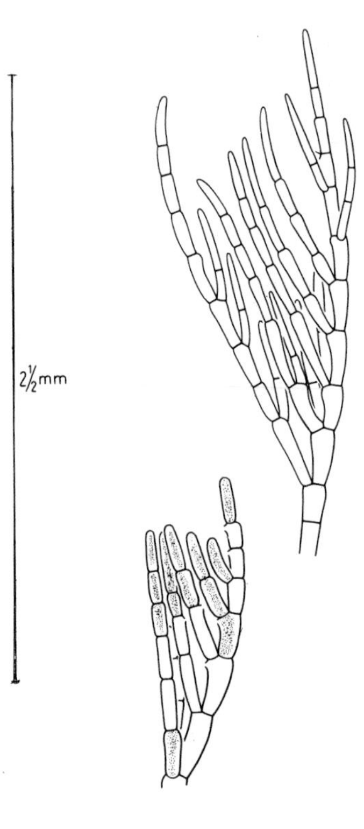

FIG. 1. *Cladophora glomerata.* Terminal, acropetally organized branch-systems; the lower plant sporulating.

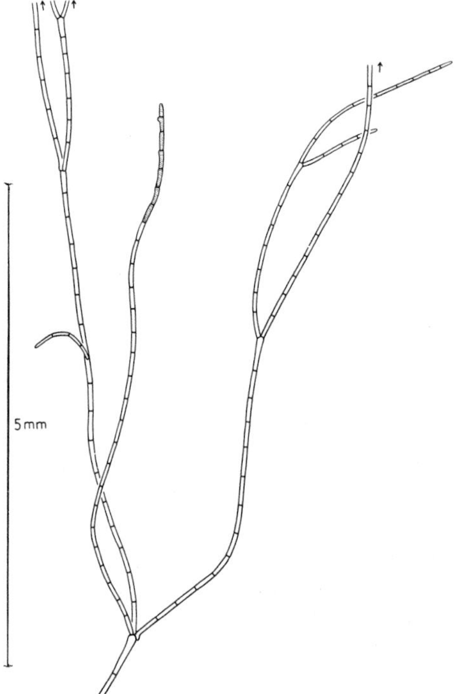

FIG. 2. *Cladophora glomerata.* Plant reduced to sparsely branched main axes, in which growth is predominantly intercalary, and in which the organization is irregular.

intercalary growth and for vegetative proliferation is more pronounced than in the var. *glomerata.* This variety seems to be characteristic for stagnant water-bodies. It is rather suggestive of what is called an "ecotype" in higher plants.

At first I had the impression that floating, irregularly or poorly acropetally organized plants from stagnant water constituted only a quiet water modification of *Cl. glomerata.* Extensive comparative culture experiments, however, gave evidence of a still more pronounced tendency for intercalary growth and a still more reduced tendency for ramification than seen in the quiet water form of *Cl. glomerata* var. *glomerata.*

The morphological range of *Cl. glomerata* var. *crassior* is entirely comprised within, and slightly more restricted than that of, *Cl. glomerata* var. *glomerata.*

The poorly branched forms of *Cl. glomerata* can be quite similar to *Cl. rivularis* (L.) v. d. Hoek, a fresh-water species which is generally unbranched or almost so, but which under certain conditions (e.g., by transfer into fresh medium, or when akinete filaments proliferate) may become lined with a large number of branches. In *Cl. rivularis* the tendency for ramification is still much more reduced than in *Cl. glomerata* var. *crassior. Cladophora rivularis* may vary extremely in diameter, viz., from about 30 to 150 μ, depending on environment and the age of a filament. When it is unbranched—which is often the case—the narrower plants may be easily confused with Rhizoclonium, the broader plants with Chaetomorpha.

Cladophora globulina (Kütz.) Kütz, when growing under optimum conditions, is also unbranched and deceptively similar to *Rhizoclonium hieroglyphicum* (Ag.) Kütz. The diameter of the filaments is smaller and much less variable than in *Cl. rivularis*, viz., mostly between 16 and 25 μ. Under certain conditions, e.g., when a fragment is transferred into a fresh culture medium, or when akinete filaments proliferate, the filaments may be quite densely branched, thus constituting a modification almost indistinguishable from *Cl. fracta* (Müll.) Kütz. The latter species, as a consolation, does not have, as far as I know, unbranched modifications in its turn!

4. Insertion of the Branches. The insertion of the branches also provides a valid taxonomic criterion.

In the section *Aegagropila* [e.g., in its fresh-water representative *Cl. aegagropila* (L.) Rabenhorst], the branches are laterally and mostly subterminally inserted under the apical poles of the cells, each of them cut off from the mother cell by a vertical cell wall. Older ramifications, however, may simulate apical insertion by considerable apical swelling of the mother cell.

In the section *Glomeratae,* for instance, each branch is apically inserted on the mother cell from which it is cut off by an oblique wall, which, with increasing age of the ramification, may grow into an almost horizontal position. At the same time the axis is slightly pushed aside and a pseudodichotomy arises. This process—often called evection (after Brand, 1899)—depends on the age of the branch as compared to that of the axis. In acropetally organized branch-systems, where a branch arises close to the apex, the branch is soon as vigorous as the axis and a pseudodichotomy is readily formed. A branch arising from an old, vigorous axis, however, forms a pseudodichotomy only after a longer period. The evection is still often used as a taxonomically important criterion. Although there are probably slight differences due to genetic factors, the variation within each species is so wide that this criterion had better be avoided as a differential character.

5. Features of the Attachment Organ. Three species are characterized by the absence of attachment organs, viz., the marine *Cl. retroflexa* (Bonnem. ex Crouan) Crouan and *Cl. battersii* v. d. Hoek, and the fresh-water species *Cl. cornuta* Brand, so far only known from one lake in southwestern Germany (Würmsee).

The species of the section *Affines* are attached by simple, hyaline basal discs, formed by the lower end of the basal cell, and in this respect they very much resemble Rhizoclonium.

The spores of the species of the section *Basicladia* [viz., the fresh-water *Cl. kosterae* v. d. Hoek and *Cl. okamurai* (Ueda) v. d. Hoek] grow first into a basal stratum, from which upright shoots later arise.

The spores of the species of the sections *Glomeratae* and *Rupestres* each grow into a primary filament with a primary basal rhizoid. Later, secondary rhizoids arise from the basal poles of cells in the basal region of the plant.

6. Mode of Reproduction. Some species and varieties most probably multiply only vegetatively, since no trace of sporulation has ever been observed in them. Examples are, with regard to fresh-water representatives, *Cl. fracta* var. *fracta, Cl. aegagropila,* and *Cl. globulina.*

Several species and varieties reproduce only by asexual (neutral) biflagellate zoospores, e.g., *Cl. glomerata, Cl. fracta* var. *intricata, Cl. sericea* var. *biflagellata,* and *Cl. albida* var. *biflagellata.*

One species, *Cl. parriaudii* v. d. Hoek, is known to reproduce only by quadriflagellata asexual zoospores.

Most species, however, exhibit a regular alternation between a zoospore and a gamete-producing generation. Of the fresh-water species, however, only *Cl. kosterae* and *Cl. okamurai* are known to have such an alternation of generations.

Schussnig, in a long series of publications, tried to demonstrate the existence of a diplontic life cycle in *Cl. glomerata.* According to this author, in the spring *Cl. glomerata* produces gametes, the formation of which is preceded by an abortive meiosis, as a result of which they seldom copulate. Now Schussnig has never been able to put forward much convincing evidence for the gamete nature of the zooids concerned. His pupil List (1930) succeeded only once in observing a rather peculiar sort of copulation; and a difference in size between zoospores and gametes—a very common phenomenon throughout many groups of Chlorophyta—was never indicated.

This theory was at first formulated at a time when little was yet known about life cycles in Chlorophyta, and it is understandable that Schussnig (1928) tried to find results corresponding to the facts then available, which suggested the general existence of a diplontic life cycle in Chlorophyta.

At the present time, however, we know that considerable diversity in the mode of reproduction may exist within the limits of one genus, some diversity even within specific limits.

The theory of diplontic *Cl. glomerata* should not be adopted until more convincing evidence has been gathered. (For a summary of Schussnig's ideas, see Schussnig, 1960.)

The fresh-water species *Cl. glomerata* is very similar to the marine *Cl. vagabunda,* which has often been designated under various names such as *Cl. fracta* var. *marina, Cl. expansa, Cl. sericea, Cl. crystallina,* and *Cl. glomerata* var. *marina.* The cell diameters and the morphological range of both species tally, although *Cl. vagabunda* never forms unbranched filaments as long as those of *Cl. glomerata.* The fresh-water species, however, reproduces by asexual biflagellate zooids, and the marine species by biflagellate gametes and quadriflagellate zoospores, in a regular alternation of generations.

Both species penetrate into brackish water. I have collected *Cl. vagabunda* from a locality in southern France, where salinity amounted to about 7‰. Both species occur in the vicinity of Kiel, on the German Baltic coast, where salinity amounts to about 15‰. In brackish water habitats, therefore, only observations on the reproduction can produce a correct determination.

* * *

As a general conclusion I should like to express my opinion that investigations of cultures grown under controlled conditions combined with observations on living material from nature and, if useful, on herbarium collections should be the base for taxonomic studies, not only of unicellular and colonial forms, but of many pluricellular forms as well. In many pluricellular groups such as the genera Stigeoclonium, Chaetophora, Draparnaldia, Draparnaldiopsis, and Cloniophora, we may expect a considerable morphological plasticity (cf. e.g. Uspenskaja, 1929) and possibly large overlaps of morphological characteristics between the species. It appears from the papers of Kuo Chi-Fang (1958) and Godward (1942) that possibly multiple modes of reproduction exist in the genus Stigeoclonium. For this reason such generic revisions as those of Draparnaldia by Forest (1956) and Cloniophora by Islam (1961), which are mainly based on preserved material, leave a feeling of uncertainty (evidently experienced by Forest), since the validity of the criteria employed in them remains problematic. For the same reason a new species like *Draparnaldiopsis intermedia* Obuchova (1959) seems to be insufficiently defined.

Taxonomy inevitably has its historical implications, as regards both nomenclature and the conceptions expressed by the older authors. These historical aspects, therefore, must be investigated thoroughly by each phycological taxonomist.

It may appear from papers such as those by Starr (1955) and Pocock (1959a) that this is possible, even for unicellular groups, without recourse to later starting points.

ACKNOWLEDGMENTS

I wish to record my indebtedness to Dr. Joséphine Th. Koster for her helpful criticism and for her permission to use her indispensable literature card-index; to Prof. H. J. Lam, Dr. F. Drouet, and Drs. J. and R. T. Wilce for critically reading through the manuscript; and to Mrs. D. J. Ondei for the correction of the English text.

LITERATURE CITED

Arasaki, S., and Tokuda, H. On the life history of Monostroma, especially on the Gomontia-like developments of certain species. Abstr. (Distributed during 10th Pac. Sci. Congr., Honolulu, August 1961.)

Beaudrimont, R. 1960. Sur quelques Ulvacées de la région de Roscoff. *Cahiers biol. marine* 1:251-258.

Beaudrimont, R. 1961. Influence de divers milieux de culture sur le développement de quelques Ulvacées. *Botaniste* 44:77-192.

Bliding, C. 1933. Über die Sexualität und Entwicklung bei der Gattung Enteromorpha. *Svensk Bot. Tidskr.* 27:233-256.

Bliding, C. 1935. Sexualität und Entwicklung bei einigen marinen Chlorophyceen. *Svensk Bot. Tidskr.* 29:57-64.

Bliding, C. 1938. Studien über Entwicklung und Systematik in der Gattung Enteromorpha I. *Bot. Notiser,* pp. 83-90.

Bliding, C. 1939. Studien über Entwicklung und Systematik in der Gattung Enteromorpha II. *Bot. Notiser,* pp. 134-144.

Bliding, C. 1944. Zur Systematik der schwedischen Enteromorphen. *Bot. Notiser,* pp. 331-356.

Bliding, C. 1948a. Über *Enteromorpha intestinalis* und *compressa. Bot. Notiser,* pp. 123-136.

Bliding, C. 1948b. *Enteromorpha kylini,* eine neue Art aus der schwedischen Westküste. *K. fysiogr. Sällsk, Lund Förh.* 18:1-6.

Bliding, C. 1955. *Enteromorpha intermedia,* a new species from the coasts of Sweden, England and Wales, *Bot. Notiser,* pp. 253-262.

Bliding, C. 1960. A preliminary report on some new Mediterranean green algae. *Bot. Notiser* 113:172-184.

Bold, H. 1958. Three new chlorophycean algae. *Amer. J. Bot.* 45:737-743.

Bourrelly, P. 1957. Review of: Drouet, F., and Daily, W. A. Revision of the coccoid Myxophyceae. *Rev. Algol. n.s.* 2:279-280.

Brand, F. 1899. Cladophoren Studien. *Bot. Zbl.* 79:145-152, 177-186, 209-221, 287.

Burrows, E. M. 1959. Growth form and environment in Enteromorpha. *J. Linn. Soc. (Bot.)* 56:204-206.

Butcher, R. W. 1959. An introductory account of the smaller algae of British coastal waters. Part I: Introduction and Chlorophyceae. *Gt. Brit., Min. Agr. Fish. and Food, Fish. Invest. ser IV,* 1(2). 74 pp.

Cauro, R. 1959. Sur la reproduction et le développement de quatre Ulvacées du Maroc. *Botaniste* 42(1958):89-130.

Chihara, M. 1958. Studies on the life history of the green algae in the warm seas around Japan 7; on the sexual reproduction of Collinsiella. *J. Jap. Bot.* 33:307-313.

Chihara, M. 1962. Occurrence of the Gomontia-like phase in the life history of certain species belonging to Collinsiella and Monostroma (a preliminary note). *J. Jap. Bot.* 37:44-45.
Claus, G. 1962. Beiträge zur Kenntnis der Algenflora der Abaligeter Höhle. *Hydrobiologia* 19:192-222.
Coleman, A. W. 1959, Sexual isolation in *Pandorina morum. J. Protozool.* 6:249-264.
Czurda, V. 1932. Zygnemales. *In* Pascher, A. Die Süsswasser-Flora Mitteleuropas, No. 9. Jena.
Dangeard, P. 1951. Sur une espèce nouvelle d'Ulva de nos côtes atlantiques (*Ulva olivacea* sp. n.). *Botaniste* 35:27-34.
Dangeard, P. 1957. Faculté de régénération et de multiplication végétative chez les Entéromorphes. *C. R. Acad. Sci., Paris* 244:2454-2457.
Dangeard, P. 1958. La reproduction et le développement de l'*Enteromorpha marginata* Ag. et le rattachement de cette espèce au genre Blidingia. *C. R. Acad. Sci., Paris* 246:347.
Dangeard, P. 1959a. Observations sur quelques Ulvacées du Maroc. *Botaniste* 42(1958):5-64.
Dangeard, P. 1959b. Recherches sur quelques "Codium." Leur reproduction et leur parthénogénèse. *Botaniste* 42(1958):66-88.
Dangeard, P. 1959c. L'*Ulva dangeardii* P. Gayral et J. S. de Mazancourt existe-t-elle en Gironde? *Botaniste* 42(1958):153-161.
Dangeard, P. 1959d. Sur quelques espèces d'Ulva de la région de Dakar. *Botaniste* 42(1958):163-171.
Dangeard, P. 1959e. Observations sur une Enteromorphe des prés salés de la Gironde et sur sa propriété de donner des proliférations. *In* Écologie des algues marines. *Colloques int. du C.N.R.S.* 81. Paris.
Dangeard, P. 1960a. L'"*Enteromorpha linza*" (Linné) J. Ag. *Botaniste* 43(1959-1960): 103-118.
Dangeard, P. 1960b. Recherches sur quelques "Ulva" des côtes françaises. *Botaniste* 43(1959-1960):119-147.
Dangeard, P. 1960c. Une Entéromorphe nouvelle de la région de Saint-Jean-de-Luz (*Enteromorpha sancti-joannis* nov. sp.). *C. R. Acad. Sci., Paris* 251:1603-1606.
Dangeard, P. 1961a. Le problème de l'espèce avec référence au groupe des Ulvacées. *Botaniste* 44:21-36.
Dangeard, P. 1961b. Quelques particularités du genre "Blidingia." *Botaniste* 44: 193-208.
Dangeard, P. 1961c. Observations sur l'*Enteromorpha tubulosa* Kützing. *Bull. Res. Counc. Israel* 10:29-34.
Dangeard, P. and Parriaud, H. 1956. Sur quelques cas de développement apogamique chez deux espèces de Codium de la région Sud-Ouest. *C. R. Acad. Sci., Paris* 243:1981-1983.
Dangeard, P., and Parriaud, H. 1960. Sur une Entéromorphe nouvelle (*E. hendayensis* nov. sp.) à développement du type tubulosa. *C. R. Acad. Sci., Paris* 250:2972-2975.
Dangeard, P., and Parriaud, H. 1961. Sur la présence en France de l'*Enteromorpha kylini* Bliding et sur les caractères de son développement. *C. R. Acad. Sci., Paris* 252:2975-2977.
Deason, T. R. 1959. Three Chlorophyceae from Alabama Soil. *Am. J. Bot.* 46:572-578.
Dedusenko-Ščegoleva, N. P. 1959. Novye vidy vodoroslej iz vodoemov char'kovskoj oblasti. *Bot. mat. otd. spor. rast.* 12:44-46.
Delépine, R. 1959. Observations sur quelques Codium (Chlorophycées) des côtes françaises. *Rev. gén. Bot.* 66:366-394.

Desikachary, T. V. 1959. Cyanophyta. New Delhi.
Dixon, P. S. 1960a. Studies on marine algae of the British Isles; the genus Ceramium. *J. Mar. Biol. Ass. U. K.* 39:331-374.
Dixon, P. S. 1960b. Studies on marine algae of the British Isles: *Ceramium shuttleworthianum* (Kütz.) Silva. *J. Mar. Biol. Ass. U. K.* 39:375-390.
Drouet, F., and Daily, W. A. 1956. Revision of the coccoid Myxophyceae. *Butler Univ. Bot. Stud.* 12:218 pp.
Elenkin, A. A. 1936. Sinezelenye vodorosli S.S.S.R., vol. I. Moscow—Leningrad.
Elenkin, A. A. 1949. Sinezelenye vodorosli S.S.S.R., vol. II. Moscow—Leningrad.
Ettl, H. 1958. Zur Kenntnis der Klasse Volvophyceae. *In* Komárek, J., and Ettl, H. Algologische Studien. Prague.
Ettl, H. 1959. Bemerkungen zur Artabgrenzung einiger Chlamydomonaden. *Nova Hedwigia* 1:167-193.
Ettl, H. 1960. Die Algenflora des Schönhengstes und seiner Umgebung I. *Nova Hedwigia* 2:509-544.
Ettl, H. 1961. Über pulsierende Vakuolen bei Chlorophyceen. *Flora. Jena* 151:88-98.
Ettl, H., and Ettl, O. 1959a. Zur Kenntnis der Klasse Volvophyceae II (Neue oder wenig bekannte Chlamydomonadalen). *Arch. Protistenk.* 104:51-112.
Ettl, H., and Ettl, O. 1959b. Einige Bemerkungen zur Gattung Gloeomonas Klebs (zur Kenntnis der Klasse Volvophyceae III). *Arch. Protistenk.* 104:113-132.
Feldmann, J. 1956. Sur la parthénogénèse du *Codium fragile* (Sur.) Hariot dans la Méditerranée. *C. R. Acad. Sci., Paris* 243:305-307.
Feldmann, J. 1958. Les Cyanophycées marines de la Guadeloupe. *Rev. Algol. n.s.* 4:25-40.
Feldmann, J. 1960. Les Ulves de Roscoff. *Soc. Phycol. France, Bull.* No. 6. 2 pp.
Forest, H. S. 1956. A study of the genera Draparnaldia Bory and Draparnaldiopsis Smith and Klyver. *Castagnea.* 21:1-29.
Forest, H. S. 1957. The remarkable Draparnaldia species of Lake Baikal, Siberia. *Castagnea* 22:126-134.
Föyn, B. 1934. Lebenszyklus und Sexualität der Chlorophycee *Ulva lactuca* L. *Arch. Protistenk.* 83:154-177.
Föyn, B. 1955. Specific differences between northern and southern populations of the green alga *Ulva lactuca. Pubbl. Staz. zool. Napoli* 27:261-270.
Föyn, B. 1958. Über die Sexualität und der Generationswechsel von *Ulva mutabilis* sp. n. *Arch. Protistenk.* 102:473-481.
Föyn, B. 1959. Geschlechtskontrollierte Vererbung bei den marinen Grünalgen *Ulva mutabilis. Arch. Protistenk.* 104:236-253.
Föyn, B. 1960. Sex-linked inheritance in Ulva. *Biol. Bull.* 118:407-411.
Föyn, B. 1961. Globose, a recessive mutant in *Ulva mutabilis. Botanica marina* 3:60-64.
Föyn, B. 1962. Two linked autosomal genes in *Ulva mutabilis. Botanica marina* 4:156-162.
Friedmann, I. 1961. *Chroococcidiopsis kashaii* sp. n. and the genus Chroococcidiopsis (studies on cave algae from Israel III). *Öst. bot. Z.* 108:354-367.
Friedmann, I. 1962. The ecology of the atmophytic nitrate alga *Chroococcidiopsis kashaii* Friedmann. *Arch. Mikrobiol.* 42:42-45.
Gayral, P. 1960a. Premières observations et réflexions sur les Ulvacées en culture. *Botaniste* 43(1959-60):85-100.
Gayral, P. 1960b. Une Ulve nouvelle, *Ulva elegans*: description et observations biologiques. *C. R. Acad. Sci., Paris* 251:768-770.
Gayral, P. 1960c. Sur la présence d'*Ulva dangeardii* en Bretagne. *Rev. Algol. n.s.* 5:211-212.

Gayral, P. 1961a. Sur deux Ulvacées de la côte marocaine du détroit de Gibraltar: *Ulva rigida* C. Agardh et *Ulva olivacea* P. Dangeard. *Botaniste* 44:223-228.
Gayral, P. 1961b. Description, reproduction et développement d'une variété d'*Enteromorpha hendayensis* P. Dangeard et H. Parriaud: *Enteromorpha hendayensis* var. *salensis* nov. var. *Botaniste* 44:229-239.
Gayral, P. 1961c. Sur la reproduction de *Monostroma obscurum* (Kütz.) J. Agardh. *C. R. Acad. Sci., Paris* 252:1642-1644.
Gayral, P., and Seizilles de Mazancourt, J. 1958. Algues microscopiques nouvelles provenant d'un sol d'estuaire (Oued Bou Regreg, Maroc). *Bull. Soc. Bot. Fr.* 105:344-350.
Gayral, P., and Seizilles de Mazancourt, J. 1959. Sur une espèce nouvelle d'Ulva récoltée en estuaire au Maroc: *Ulva dangeardii.* nov. sp. *Botaniste* 42(1958):131-141.
Geitler, L. 1925. Cyanophyceae. *In* Pascher, A. Die Süsswasser-Flora Deutschlands, Österreichs und der Schweiz, No. 12. Jena.
Geitler, L. 1930-32. Cyanophyceae. *In* Rabenhorst, L. Kryptogamenflora, 2nd ed., vol. XIV, Leipzig.
Geitler, L. 1960. Schizophyzeen. *In* Handbuch der Pflanzenanatomie VI (1), 2nd ed. Berlin.
Gemeinhardt, K. 1939. Oedogoniales. *In* Rabenhorst, L. Kryptogamenflora, vol. XII. Leipzig.
Gerloff, J. 1962. Beiträge zur Kenntnis einiger Volvocales II. *Nova Hedwigia* 4:1-20.
Godward, M. 1942. The life cycle of *Stigeoclonium amoenum. New Phytol.* 41:293-301.
Herndon, W. 1958a. Studies on chlorosphaeracean algae from soil. *Am. J. Bot.* 45:298-308.
Herndon, W. 1958b. Some new species of chlorococcacean algae. *Am. J. Bot.* 45:308-323.
Hillis, L. W. 1959. A revision of the genus Halimeda (order Siphonales). *Publ. Inst. Mar. Sc. (Univ. Texas)* 6:321-403.
Hirn, K. E. Monographie und Iconographie der Oedogoniaceae. *Acta Soc. Sci. fenn.* 27(1):1900.
Hoek, C. van den. 1963. Sur la synonymie de trois Ulves d'eau saumâtre: *Ulva curvata. U. dangeardii* et *U. incurvata. Phycologia* 2:184-186.
Islam, A. K. M. N. 1961. The genus Cloniophora Tiffany. *Rev. Algol. n.s.* 6:7-32.
Köhler, K. 1956. Entwicklungsgeschichte, Geschlechtsbestimmung und Befruchtung bei Chaetomorpha. *Arch. Protistenk.* 101:223-268.
Kolkwitz, R., and Krieger, H. 1941-1944. Zygnemales. *In* Rabenhorst, L. Kryptogamen-Flora, vol. XIII. Leipzig.
Komárek, J. 1957. Das "Microcystis"-problem. *Taxon* 6:145-149.
Kornmann, P. 1959. Die heterogene Gattung Gomontia I. Der sporangiale Anteil, *Codiolum polyrhizum. Helgoländ. wiss. Meeresunters.* 6:229-238.
Kornmann, P. 1960. Die heterogene Gattung Gomontia II. Der fädige Anteil, *Eugomontia sacculata* nov. gen. nov. sp. *Helgoländ. wiss. Meeresunt.* 7:59-71.
Koster, J. Th. 1961. On the delineation of species of Cyanophyceae. *Bull. Res. Counc. Israel* 10D:90-93.
Kukk, E. G. 1959. K flore sinezelenych vodoroslej estonskoj U.S.S.R. *Bot. mat. otd. spor. rast.* 12:23-30.
Kuo Chi-fang. 1958. On the laboratory culture and life history of *Stigeoclonium subsecundum. Acta bot. sinica* 7:87-96.
Kützing, F. T. 1849. Species algarum. Leipzig.
Kützing, F. T. 1855. Tabulae phycologicae, vol. V. Nordhausen.
List, H. 1930. Die Entwicklungsgeschichte von *Cladophora glomerata* Kütz. *Arch. Protistenk.* 72:453-481.

Moewus, F. 1938. Die Sexualität und der Generationswechsel der Ulvaceen und Untersuchungen über die Parthenogenese der Gameten. *Arch. Protistenk.* 91:357-411.

Moruzi, C. 1960. Une nouvelle espèce de Cyanophycée de la flore algolique d'un lac à action thérapeutique: *Anabaenopsis teodorescui* Moruzi sp. nova *Rev. algol. n.s.* 5:193-197.

Muzafarov, A. M. 1959. Novye vidy i formy vodoroslej obnaružennye v vodoemach bassejna Amu-Dary. *Bot. mat. otd. spor. rast.* 12:30-37.

Novičkova, L. N. 1960. Novye i interesnye sinezelenye vodorosli takyrov. *Bot. mat. otd. spor. rast.* 13:30-34.

Obuchova, V. M. 1959. Novy vid roda Draparnaldiopsis. *Bot. mat. otd. spor. rast.* 12:129-132.

Papenfuss, G. F. 1960. On the genera of the Ulvales and the status of the order. *J. Linn. Soc. (Bot.)* 56:303-318.

Parriaud, H. 1954. Sur un hybride expérimental: *Fucus vesiculosus* Linn. × *Fucus chalonii* J. Feldm. *C. R. Acad. Sci., Paris* 238:832-834.

Parriaud, H. 1959. Sur deux Ulvacées récemment découvertes dans le bassin d'Arcachon: "*Ulva incurvata*" nov. sp. et "*Enteromorpha ahlneriana* Bliding." *P. V. Soc. linn. Bordeaux* 97(1958):1-11.

Petrov, Ju. E. 1961. Novye morskie sinezelenye vodorosli dlja Murmana. *Bot. mat. otd. spor. rast.* 14:107-111.

Pignatti, S. 1958. Secondo contributo alla flora algologica del Paveze. *Atti Ist. bot. Univ. Pavia, Crittog.*, ser. 5, 15:21-34.

Pocock, M. A. 1959a. Haematococcus in Southern Africa. *Trans. Roy. Soc. S. Afr.* 36:5-56.

Pocock, M. A. 1959b. *Letterstedtia insignis* Aresch. *Hydrobiologia* 14:1-71.

Pocock, M. A. 1960. Hydrodictyon: a comparative biological study. *J. S. Afr. Bot.* 26:167-319.

Powell, H. T., and Powell, H. G. 1957. Footnote *in* List of marine algae collected in the district around Dale Fort, Pembrokeshire: September 19-26, 1956. *Brit. Phycol. Bull.* 1(5):21-31.

Randhawa, M. 1959. Zygnemaceae. New Delhi.

Scagel, R. F. 1960. Life history of the Pacific coast marine alga, *Collinsiella tuberculata* Setchel and Gardner. *Can. J. Bot.* 38:969-985.

Schreiber, E. 1942. Die geschlechtliche Fortpflanzung von *Monostroma grevillei* und *Cladophora rupestris*. *Planta* 32:414-417.

Schussnig, B. 1928. Die Reduktionsteilung bei *Cladophora glomerata*. *Öst, bot. Z.* 77:62-67.

Schussnig, B. 1960. Handbuch der Protophytenkunde II. Jena.

Schwabe, G. H. 1960a. Zur Morphologie und Ökologie einiger Plectonema-Arten (Blaualgen und Lebensraum III). *Nova Hedwigia* II:243-268.

Schwabe, G. H. 1960b. Zur autotrophen Vegetation in ariden Böden (Blaualgen und Lebensraum IV). *Öst. bot. Z.* 107:281-309.

Segi, T. 1951. Systematic study of the genus Polysiphonia from Japan and its vicinity. *Journ. Fac. Fish., Univ. Mie* 1:169-272.

Segi, T. 1956. On the development of Monostroma in the sea. *Rep. Fac. Fish., Univ. Mie* 2:312-316.

Segi, T. Further study of Polysiphonia from Japan I. *Rep. Fac. Fish., Univ. Mie* 3:257-266.

Segi, T. 1960. Further study of Polysiphonia from Japan II. *Rep. Fac. Fish., Univ. Mie* 3:608-626.

Segi, T., and Gotô, W. 1955. On the planospores of *Monostroma latissimum* (Kütz.) Wittr. *Bull. Japan. Soc. Phycol.* 3:1-5.

Segi, T., and Gotô, W. 1956. On Monostroma and its culture I. *Bull. Japan. Soc. Phycol.* 4:55-60.

Silva, P. C. 1951. The genus Codium in California, with observations on the structure of the walls of the utricles. *Univ. Calif. Publ. Bot.* 25:79-114.
Silva, P. C. 1955. The dichotomous species of Codium in Britain. *J. Mar. Biol. Ass. U. K.* 34:565-577.
Silva, P. C. 1957. Codium in Scandinavian waters. *Svensk Bot. Tidskr,* 51:117-134.
Silva, P. C. 1959. The genus Codium (Chlorophyta) in South Africa. *J. S. Afr. Bot.* 25:103-165.
Silva, P. C. 1960. Codium (Chlorophyta) of the tropical western Atlantic. *Nova Hedwigia* 1:497-536.
Silva, P. C., and Womersley, H. B. S. 1956. The genus Codium (Chlorophyta) in Southern Australia. *Aust. J. Bot.* 4:261-289.
Simpson, G. G. 1961. Principles of animal taxonomy. New York.
Skuja, H. 1956. Review of Drouet, F., and Daily, W. A. Revision of the coccoid Myxophyceae. *Svensk Bot. Tidskr.* 50:550-556.
Starr, R. C. 1955. A comparative study of *Chlorococcum Meneghini* and other spherical, zoospore producing genera of the Chlorococcales. *Ind. Univ. Publ.* 20.
Stein, J. R. 1958a. A morphological study of *Astrephomene gubernaculifera* and *Volvulina steinii. Am. J. Bot.* 45:388-397.
Stein, J. R. 1958b. A morphologic and genetic study of *Gonium pectorale. Am. J. Bot.* 45:664-672.
Stein, J. R. 1959. The four-celled species of Gonium. *Am. J. Bot.* 46:366-371.
Tiffany, L. H. 1930. The Oedogoniaceae. Columbus, Ohio.
Transeau, E. N. 1951. The Zygnemataceae. Ohio.
Uspenskaja, W. J. 1929. Über die Physiologie der Ernährung und die Formen von *Draparnaldia glomerata* Agardh. *Z. Bot.* 22:337-394.
Welsh, H. 1961. Some new and interesting Cyanophyta from the Transvaal, South Africa. *Nova Hedwigia* 3:399-404.

The Gross Classification of Algae

Tyge Christensen

Institut for Sporeplanter
University of Copenhagen
Copenhagen, Denmark

While algal taxonomy as a whole has been subject to much change in later years, the system of Pascher (1914) is still rather closely followed by most authors as far as the classes go. The way of subdividing them, as well as that of grouping them together, may vary. But the very concept of an algal class is unchanged, demanding considerable uniformity with regard to chemical and cytological features and to the type of monadoid cells, while allowing for unlimited variation in thallus construction, and disregarding whether the prevailing vegetative condition is that of motile or of nonmotile cells.

The common acceptance of this class concept, of course, does not prevent authors from lumping or splitting single classes of Pascher's in accordance with increasing knowledge of algal chemistry and cytology. Personally, the author thinks one should be very strict both in not regarding differences in thallus construction and, on the other hand, in separating forms with different flagellar structure. Thus, the writer finds little basis, for example, for keeping the Zygnematales and the Charales in separate classes, as some authors do, but thinks the class of the Chrysophyceae must be restricted to comprise forms with one smooth flagellum and one with mastigonemes (the former of which may be a short internal structure; cf. Rouiller and Fauré-Fremiet, 1958). Likewise, he thinks the Chlorophyceae should comprise only forms with two or more smooth flagella.

Accordingly, in a recent survey the author (1962) has followed the suggestion by Parke (1961) that algae with golden-brown chromatophores, two smooth flagella, and often a haptonema between them, should be placed in a separate class. As no name has been given by Parke, the writer has introduced the designation Haptophyceae, hereby referring to the most striking feature of the group, not considering whether this feature is found in all its representatives or not (cf. the presence of monocarpic forms in the Polycarpicae and of "one-flowered umbels" in the Umbelliferae, both order names being maintained in spite of their irregular endings because they are found to be good names). As to the various genera to be removed from the Chlorophyceae because of different types of flagellation, the author has no personal opinion as to whether they belong together or not. The differences between their types of flagellation may perhaps be explained in the way of reduction and duplication (as possibly in the Euglenophyceae), but may at least as well express a considerable distance in relationship. As an easy common heading to be used for them until a more satisfactory classification can be attained, the author (1962) has introduced the designation Loxophyceae, referring to their lack of radial symmetry in the monadoid state.

Including these two new designations, the following list of algal classes can be set up:

1. Cyanophyceae
2. Rhodophyceae
3. Cryptophyceae
4. Dinophyceae
5. Raphidophyceae
6. Chrysophyceae
7. Haptophyceae
8. Craspedophyceae
9. Bacillariophyceae
10. Phaeophyceae
11. Xanthophyceae
12. Euglenophyceae
13. Loxophyceae
14. Prasinophyceae
15. Chlorophyceae

As for the arrangement of the classes in divisions, complete confusion appears on comparing the systems used by different modern authors. The only point of relative agreement is that the Cyanophyceae are usually considered to form a division of their own, probably more closely related to the first algae in evolution than are any of the algae provided with a true nucleus. Most authors also agree that the Rhodophyceae cannot be placed in a common division with any other of the classes listed. But there is little agreement as to the relations between the Rhodophyceae, the Cyanophyceae, and the rest of the algae. Some think that the red algae are primarily devoid of flagella, while all other nonflagellate algae with plastids and proper nuclei are supposed to have lost their motile stages secondarily. Such a view gives the red algae an isolated position somehow intermediate between the source of the present-day blue-green algae and that of algae with flagella. The view is supported by the presence of similar photosynthetic pigments in blue-green and red algae and has been held by several authors in the course of time, being particularly advocated by Kylin (1943). Still, many workers dealing with systematics prefer the sequence blue-green, green, brown, and red algae. In part they may do so because they find some practical advantages in sticking to this long-established sequence, or because they prefer not to interfere with the discussion of a subject so far from ordinary taxonomy and find they achieve this end best by sticking to the old. In any case no basis appears for assuming that flagella ever existed in the ancestry of the red algae, and Kylin's view seems to be gaining ground. Thus, it has recently been shared by Dougherty (1955) and by Chadefaud (1960).

Accepting this view and assuming that such complex features as flagella with eleven internal strands, nuclei and plastids, and a photosynthetic mechanism based on chlorophyll *a* have all arisen successfully only once in evolution, one may illustrate the phylogenetic relationships between the largest groups of organisms by a combination of the phylogenetic tree with a system of circles (Fig. 1). In this figure each circle represents one of the features mentioned and, accordingly, is entered only once by the tree but may be left by several of its branches independently.

Such an arrangement leads to a distribution of the algal classes in three remotely related groups of organisms, two of which comprise only one class each (or one comprises two, in case Cyanidium is regarded as being primarily devoid of flagella and still is not placed among the Rhodophyceae). The third large group, which contains all other algal classes in addition to the Embryophyta, the animals, and all or some of the fungi, is subdivided on the divisional level by all phycologists, but in very different ways:

Pascher (1914) distributed the algae with flagella in six divisions: Chrysophyta, Phaeophyta, Pyrrophyta, Chlorophyta, Eugleninae, and Chloromonadinae. This arrangement, slightly modified by its author in 1931, is now being gradually abandoned. Improved knowledge of pigments, reserve substances, and flagellar structure

FIG. 1. Diagram showing the supposed relationships between the major groups of organisms as related to the distribution of some fundamental chemical and cytological characters. (After Christensen, 1962.)

leaves very little doubt that the Phaeophyceae are quite closely related to the Chrysophyceae, which means that Pascher's two first divisions, comprising the classes numbered 6-11 in the above list, may be united into one division, though with some doubt regarding the Xanthophyceae. Most Phycomycetes undoubtedly belong in the same group, comprised by Copeland (1956) under the name of Phaeophyta. Pascher's third division, Pyrrophyta, was established for the Cryptophyceae and the Dinophyceae only, but was extended by Skuja (1948) to include the Raphidophyceae, and next by Copeland (1956) to comprise also the Euglenophyceae. Other authors have gone the opposite way, restricting the Pyrrophyta to comprise the Dinophyceae only and, with Graham (1951), leaving the Cryptophyceae as a class of uncertain affinity. The largest division concept is that of Chadefaud (1950), who refers all the algal classes mentioned, nos. 1-12 in the list on page 60, to a single division, *les Chromophycées*; later (1960) he adjusts the ending and at the same time lengthens the word to *Chromophycophytes*, following a hint by Seybold, Egle, and Hülsbruch (1941), as opposed to *Chlorophycophytes*, which comprise nos. 13-15.

Dougherty (1955) distributes the algae with flagella on the basis of presence or absence of chlorophyll *b*. This leads to a grouping similar to that used by Chadefaud except that the Euglenophyceae, despite their similarity to the Cryptophyceae and the Raphidophyceae, are moved from the *Chromophycées* to the *Chlorophycées*. Such shifting on a chemical basis may appear unnatural from a cytological point of view. But if the features common to the three classes mentioned (and the bilichromoprotein photosynthesis of the first mentioned) are assumed to have been taken over from common ancestors of all organisms with flagella, it makes little difference in principle whether the Euglenophyceae are referred to one or the other of such two large divisions. All together the group of algae with flagella does not seem to be divided by any gulfs comparable to that separating them from the Rhodophyceae, or these from the Cyanophyceae. One natural group of classes collects around the Chrysophyceae, and another around the Chlorophyceae, but there is little basis today for locating the classes 3-5, 11, and 12, except either in an equal number of divisions, or linked with one of the two larger groups, on the criteria used by Chadefaud or that used by Dougherty.

Future investigations may give a better foundation for an arrangement that accounts in detail for the relationships between the various classes. For the present, however, the author feels most inclined to support the simple bipartition based on the presence or absence of chlorophyll *b*.

LITERATURE CITED

Chadefaud, M. 1950. Les cellules nageuses des Algues dans l'embranchement des Chromophycées. *C. R. Acad. Sci., Paris* 231:788-790.

Chadefaud, M. 1960. Les végétaux non vasculaires (Cryptogamie). *In* Chadefaud, M., and Emberger, L. Traité de Botanique. Systematique. T. I. Paris.

Christensen, T. 1962. Alger. *In* Böcher T. W., Lange, M., and Sørensen, T. Botanik. vol. 2, no. 2, Copenhagen. (In Danish. An English edition is in preparation.)

Copeland, H. F. 1956. The classification of lower organisms. Palo Alto, California.

Dougherty, E. C. 1955. Comparative evolution and the origin of sexuality. *Syst. Zool.* 4:145-169, 190.

Graham, H. W. 1951. Pyrrophyta. *In* Smith, G. M. Manual of Phycology. Chapter 6. Waltham, Massachusetts.

Kylin, H. 1943. Verwandschaftliche Beziehungen zwischen den Cyanophyceen und den Rhodophyceen. *K. Fysiogr. Sällsk. Lund Förh.* 13(17):1-7.

Parke, M. 1961. Some remarks concerning the class Chrysophyceae. *Brit. Phycol. Bull.* 2(2):47-55.

Pascher, A. 1914. Über Flagellaten und Algen. *Ber. Deutsch. Bot. Ges.* 32:136-160.

Pascher, A. 1931. Systematische Übersicht über die mit Flagellaten in Zusammenhang stehenden Algenreihen und Versuch einer Einreihung dieser Algenstämme in die Stämme des Pflanzenreiches. *Beih. Bot. Centralbl.* 48, 2:317-332.

Rouiller, C. and Fauré-Fremiet, E. 1958. Structure fine d'un flagellé chrysomonadien: *Chromulina psammobia. Experimental Cell Res.* 14(1):47-67.

Seybold, A., Egle, K., and Hülsbruch, W. 1941. Chlorophyll- und Carotinoidbestimmungen von Süsswasseralgen. *Bot. Arch.* 42(2):239-253.

Skuja, H. 1948. Taxonomie des Phytoplanktons einiger Seen in Uppland, Schweden. *Symb. Bot. Upsal.* 9(3):1-399.

The Cytology of the Phaeophyta—A Review of Recent Developments, Current Problems, and Techniques

Margaret Roberts

Department of Botany
University of Hull
Hull, England

INTRODUCTION

The scope of this paper is restricted to a discussion of nuclear phenomena, a subject in which interest began at the turn of the century, and which has developed with the improvement of the microscope.

The pioneers in this field were Strasburger (1898) and Farmer and Williams (1896, 1898), who first investigated nuclear behavior. Naturally enough, they all chose large members of the group and all worked on species of Fucus or on other members of the Fucales. They described nuclear divisions in great detail, especially with regard to the achromatic figure, and the papers were illustrated in considerable detail. The chromosomes, however, proved small and intractable and these workers were unable to make accurate counts. Strasburger estimated the diploid number as 30, while Farmer and Williams obtained counts of 30 for the thallus and 14-15 for the oögonia.

The next notable contribution was that of Yamanouchi (1909), a detailed account of nuclear division, both mitotic and meiotic, in *Fucus vesiculosus,* illustrated by 79 drawings. He obtained counts of 64 on somatic cells and 32 at first divisions of the gametangia. This account has become accepted as "classic" and has never been challenged as to accuracy. It has not been repeated until very recently (Evans, 1962).

This was followed by numerous other accounts in the twenties and thirties when they were interrupted, presumably, by the war. Very little was published on the subject in the immediate postwar years, and it was 1954 before any considerable number of accounts were published, and since then there has been a fairly steady output. The majority of these investigations have been on members of the Fucales and Laminariales, but also include a number on Ectocarpales, Dictyosiphonales, Dictyotales, Cutleriales, Tilopteridales, and Sporochnales in descending order of popularity.* The majority, also, are the life history type of study, often incorporating an attempt to discover the place of reduction division. Very few include studies of nuclear behavior and structure (Giraud, 1956; Levan and Levring, 1942; Naylor, 1958b; Roy, 1938).

The results of these papers are very difficult to interpret. The general effect is one of confusion, probably due to inaccuracies arising from inadequate techniques—in this respect, it is perhaps significant that there is a tendency for counts to yield higher figures in recent years. A few points, however, are worth mentioning. Almost all accounts of the *Fucales* are Japanese in origin and give diploid and haploid values of 64 and 32 for almost all species investigated. Among the *Laminariales* most of the early papers gave counts of $n = 22$, or less. Naylor (1956) gave an estimate of 27-31 for three of the British species of Laminaria. A count of 31 has since been confirmed by Evans (unpublished) for *L. digitata,* while the three most recent Japanese papers give counts of about 30 for other members of the order, and Kemp and Cole (1961) give 31 for *Nereocystis lutkeana.* In the *Ectocarpales,* generally, the number is low, ranging from 8-12. It is also low in early papers on the *Dictyosiphonales*—viz. 8-10—but Abe (1940) gives a count of 18 for *Dictyosiphon foeniculaceus* and Naylor (1958a), about 26 for *Stictyosiphon tortilis.* For the *Sphacelariales* and for the *Dictyotales,* 16 is the usual figure, while Yamanouchi (1912, 1913) reports 24 for two members of the *Cutleriales.* It is thus impossible to draw any general conclusions, but figures suggest there may be a common basic number for each order.

The most recent paper on this subject (Evans, 1962) returns to the genus Fucus. Like Yamanouchi 53 years earlier, Evans describes both meiotic and mitotic divisions and has arrived at counts of 32

* For a complete list of these publications, see Naylor, 1958c and Roberts, 1962.

and 64 on haploid and diploid nuclei, respectively. Where then has progress occurred?

Progress has occurred in several fields, notably in technique and in recording results. Because of these improvements, more competent handling of small chromosomes is now possible and we may perhaps begin to see more consistent results, and also more detailed information is coming to light concerning nuclear structure and behavior.

Before considering these advances in detail, it is necessary to consider the problem of the small size of the chromosomes, and also another general point repeatedly stressed in the literature, that of the scarcity of dividing nuclei in preparations.

SIZE OF THE CHROMOSOMES

During prophase the contraction of the chromosomes is extreme so that by metaphase they are usually spherical in shape with a diameter of the order of 1μ, arranged on a plate often of the order of 5μ in diameter (larger in the diploid phase and at meiosis). It is thus necessary to employ a reliable technique and a first class optical system before detailed observations can be made. It may be possible to surmount this difficulty by employing some of the techniques now being developed in angiosperms to halt or delay prophase contraction of the chromosomes, but so far there is no record of any attempt being made in this direction.

SCARCITY OF DIVISIONS

The difficulty of finding dividing nuclei has been repeatedly emphasized, and various claims have been made as to the most favorable times for finding divisions. Yamanouchi (1909) claimed that figures were abundant in Fucus after one to two hours' immersion by the incoming tide. The same claim was made by Roy (1938), but has not been confirmed by other workers. I have found abundant figures in young and actively growing material of a variety of species if collected and fixed immediately after exposure by the ebbing tide. As far as totally submerged species are concerned, some workers have reported that meiotic divisions occurred in the early hours of the morning, but I was unable to confirm this in *Marginariella urvilliana* (Ach. Rich.) Tandy kept in tanks and fixed at intervals throughout the night, although the plants matured and discharged their oögonia during the period of observation. No estimates have been made as to the time required for the completion of division, and too few critical observations have been made to make any generalizations.

The natural outcome of such an unsatisfactory situation is that various attempts have been made to increase the number of divisions artificially. These may be summarized as follows:

i. *Application of colchicine.* This has been applied to actively growing regions or to cultures in solutions varying in concentration from 0.01 to 1%. Although reported by Levan and Levring (1942) to have little effect on the members of the Phaeophyceae they studied, other workers report that it certainly results in an increase of inhibited metaphases—but so contracted as to be quite useless. This has been my experience with a number of genera.

ii. *Wounding.* This was tried by Evans (1962) on Fucus, but he found the cells produced so full of dark contents as to be useless.

iii. *Indole-acetic acid.* This was reported by Davidson (1950) to stimulate thallus growth in Fucus, which suggests a possible application in this field, but it was found by Evans (1962) to have little effect.

iv. *Refrigeration.* This usually involves leaving material overnight in a refrigerator and then fixing after varying periods of exposure to temperatures of about 15 C and bright light. I have used this method with some success, but have not found it particularly reliable or spectacular in production of results.

Summing up, we can say that these results have been so little successful, or so inconsistent, that attention has tended to be concentrated on improving technique.

DEVELOPMENTS IN TECHNIQUE

Farmer and Williams, Strasburger, and Yamanouchi all used the classical method of embedding, sectioning and staining their material, usually with Heidenhain's iron alum hematoxylin. This is laborious and time-consuming, requiring about a week between the collection of the material and microscopic observation. Further, careful piecing together is required to ensure that all sections through a nucleus under consideration are taken into account. Although squash techniques were developed for the investigation of root tips, etc., of higher plants (Belling, 1926), they were not used to any degree in the algae until the acetocarmine technique was used by Godward (1948) on *Spirogyra,* followed by Cave and Pocock (1951) on the *Volvocales* and King (1953) on *Desmids.* Lewis (1956) and Naylor (1956) were the first to use this method on the Phaeophyceae, and since then it has been widely used (Kemp and Cole, 1961; Evans, 1962). The Feulgen technique was used by Mathias (1935) on *Stictyosiphon brachiata* and by Papenfuss (1935) on *Ectocarpus siliculosus.*

These methods, without significant modification, proved immediately successful in the Phaeophyta for filamentous species such as Ectocarpus, and for the adelophycean phases of heteromorphic genera such as Laminaria and Stictyosiphon. They are most successful when used on plants growing in culture, usually on slides or coverslips for easy handling. In the larger parenchymatous forms such as the Laminariales and the Fucales, the methods have had to include certain refinements.

The chief difficulty has been the nature of the cell walls, which almost invariably harden on fixation and so prevent successful squashing. A further difficulty is the presence in many tissues of abundant plastids which darken with chromic- or osmic-acid-containing fixatives, i.e., the traditional fixatives, and may mask the staining reaction.

Pigmentation

During recent years there has been an increase in the use of acetic alcohol or propionic alcohol fixatives. With this type of fixative the problem does not arise as the pigments are dissolved and the material decolorizes. This seems a satisfactory solution to the problem so far as the filamentous forms are concerned and for small portions of the parenchymatous species. In the latter, however, I have often had inconsistent results, even in neighboring cells, possibly because the alcohol in the fixative hardens the walls and prevents uniform penetration.

If a chromic or osmic fixative is used, the pigmentation can be removed by bleaching with H_2O_2. The fixative must be thoroughly washed out before the bleaching is carried out, and for this process I use 20% H_2O_2 (i.e., two parts 100 vol. solution to one part distilled water) for periods of up to 10 min. This in turn must be completely removed; otherwise the bleaching action continues and the material is ruined. Bleaching with H_2O_2 is often used by the Japanese workers in conjunction with Heidenhain's iron alum hematoxylin, but after embedding and before staining. Usually 10% solution is employed for periods of up to 48 hr.

Squashing

The treatment to soften the walls and aid spreading varies according to the type of staining to be used.

In the Feulgen technique, the hydrolysis with HCl usually softens the material sufficiently well to permit thorough squashing. Filamentous forms can be squashed directly, but with parenchymatous forms it is advisable to cut hand sections in the final bleach in SO_2 water before squashing.

With the acetocarmine technique, the hardening of the walls proved quite a problem, as this prevented both the penetration of the reagent as well as the flattening of the cells.

Lewis (1956) used two principal methods to overcome this difficulty in the Fucales. The first, which proved suitable for female gametangia, consisted of fixing receptacles using alcoholic fixation, removing the conceptacles and immersing them for 10-15 min in the light in a solution of equal volumes of saturated ammonium oxalate and 20 vol. H_2O_2, before washing in distilled water and staining with acetocarmine. Using this technique he obtained a count of $n = 32$ for *Ascophyllum nodosum* and demonstrated meiotic bivalents.

The second fixation maceration technique was used for preparations of antheridia. A 150-ml quantity of fixative was poured into a Waring blender, 30 g of material was added and macerated for 4 min and then poured into a cylinder to separate. From the middle layer preparations were made of antheridia, and in *Fucus vesiculosus* a count of $n = 32$ was made at the second postmeiotic mitosis.

Naylor (1957) used a method which depends on the extraction of alginate from the walls as sodium alginate, which is very soft and allows spreading. The material is fixed for 24 hr in an acid-containing fixative, washed and transferred to 6% sodium carbonate on a slide. It is then warmed gently until squashed by the pressure of the coverslip, washed and stained with acetocarmine. The limitations of this method are that it is rather drastic and difficult to control, and unless the sodium carbonate is completely removed it interferes with staining. However, it has been used quite successfully to obtain counts both on somatic cells in vegetative apices and on developing gametangia.

Evans (1962) has developed a technique consisting of fixation in acetic alcohol until bleached, followed by immersion in a 1 M solution of lithium chloride for 15 min. Salts of other monovalent metal ions of low atomic weight—viz., Na, K, Rb—were also successful, but with increasing atomic weight the spread lessened. This method is believed to depend on the replacement of divalent metal ions in the cell walls by monovalent ones, resulting in the breakdown of cross linkages into loose chains. This technique has been successfully applied to four species of Fucus, and has yielded both haploid and diploid counts.

The advantages of these two techniques—acetocarmine and Feulgen—include speed and the fact that the resulting preparations are of *entire* nuclei: the metaphase plates become flattened into the plane of the preparation, thus eliminating piecing together of sections through plates lying obliquely to the plane of the section. The specificity of the stain is a further advantage of the Feulgen reaction.

The acetocarmine technique has yet another advantage in that it swells the nuclei and chromosomes, thus slightly compensating for their small size. Thus, nuclei from the medulla of *Cystoseira tamariscifolia* treated by the Feulgen technique averaged (average of 50) 6.0μ in length, after acetocarmine treatment, 7.6μ (author; unpublished).

DETAILS OF NUCLEAR AND CHROMOSOME STRUCTURE

The outcome of advances in techniques is a more detailed knowledge of nuclei and chromosomes. The confusion that existed concerning the so-called "caryosome nuclei" is probably clearing, and the nuclei seem generally to resemble those of higher plants both in structure and behavior on division.

The resting nuclei are not normally characterized by the presence of chromatin and usually possess a conspicuous nucleolus, sometimes more than one. They may, however, contain Feulgen-positive bodies of quite considerable size—viz., of the order of 0.6μ in diameter. These chromocenters were first recognized by le Touzé (1912, p. 34, Pl. 9, Fig. 1) in *Halidrys siliquosa* (L.) Lyngb., and by Roy (1938, p. 162, Pl. 8) in this species and in several species of Cystoseira.

1. Chromocenters

***Halidrys siliquosa* (L.) Lyngb.** The observations of le Touzé and Roy on *Halidrys siliquosa* (L.) Lyngb. have been confirmed and extended by Naylor (1958b). The chromocenters are conspicuous in all somatic cells, whether stained with acetocarmine or by means of the Feulgen reaction. They vary in size and number in different tissues. Thus in the large medullary cell nuclei, which have a diameter of 10μ, they are larger and more numerous than in the small nuclei of the meristoderm cells, which are only 3μ in diameter. They vary in size from 0.75μ in diameter to 1.15μ, and in number from 2-9. Feulgen preparations of resting nuclei also reveal fine chromatin threads and many much smaller granules.

During prophase of somatic cell division, as the nucleus enlarges and the chromosomes differentiate, the chromocenters can be clearly seen to be arranged at intervals along the chromosomes. As prophase proceeds, the chromocenters become progressively less intensely stained, and, as the differentiating chromosomes shorten and thicken, become indistinguishable. Because of the extreme condensation it is impossible to say whether they are represented at metaphase by unstained portions of the chromosomes, as described by Darlington and la Cour in Trillium (1940). Divisions in the developing gametangia follow exactly the same course. In the antheridial initial, the arrangement of the chromocenters along the chromosomes has been most clearly demonstrated (Naylor, 1958b, Pl. 2d,e.). The chromocenters reappear during each interphase and are present in the mature sperms. The chromocenters show the same numerical variation in the developing nuclei of the antheridia, which are haploid, as in the diploid somatic cells. A similar lack of correlation between number of chromocenters and cytological phase was shown by Darlington and la Cour in Trillium (1940). In the oögonia the chromocenters are similarly present in the initial, disappear during prophase, and reappear during each interphase and in the mature egg.

Thus the chromocenters of *Halidrys siliquosa* (L.) Lyngb. in every way resemble those of angiosperms: in their distribution throughout the resting nucleus; in their arrangement at intervals along the chromosomes; in their variation in size and numbers and in their numerical independence of cytological phase.

Cystoseira. The observations of Roy (1938) on Cystoseira have been extended by the author (unpublished) and a few further points have emerged. Roy investigated four species of Cystoseira—*C. tamariscifolia* (Huds.) Papenfuss [sub. *C. ericoides* (L.) Ag.], *C. myriophylloides* Sauv., *C. granulata* (L.) Ag., and *C. foeniculacea* (L.) Grev. emend Sauv.—and found that *C. tamariscifolia* possessed chromocenters while the other three did not.

In *C. tamariscifolia* I have been able to confirm the presence of the chromocenters described by Roy, but find differences in the details of their distribution. In the cells of meristoderm and cortex I find between 6 and 12 chromocenters (Roy: 4-5), while in the large medullary cells there are fewer—in contrast to Halidrys where there are more in the medulla.

In *C. foeniculacea,* a species described by Roy as lacking chromocenters, I find chromocenters *in some tissues.* Thus, in the *meristoderm* the nuclei contain small and numerous granules (20 or more). In the *cortex* the granules are fewer and the larger—about 0.5μ in diameter—and resemble chromocenters. In the *medulla* the nuclei are larger, but contain fine granules as in the meristoderm. Again, as in *C. tamariscifolia,* there are fewer granules in the medulla than in the meristoderm.

In *C. granulata* I was unable to find chromocenters at any level, as described by Roy. The nuclei shows a finely granular structure throughout the plant. I have not been able to examine *C. myriophylloides,* but in seven other species of Cystoseira which I have examined, three have proved to possess chromocenters while four do not. [They are also present in *Bifurcaria rotunda* (Huds.) Papenfuss.]

Thus the presence of chromocenters is not a generic feature of *Cystoseira:* In fact, in *C. foeniculacea* the two types of nuclear structure occur within the individual.

2. The "Chromophilous Spherule"

A "staining body associated with the nucleolus" was described and figured by Yamanouchi (1909, Fig. 44a) in the oögonia of *Fucus vesiculosus,* and a similar body has been recorded occasionally since, associated with nuclei undergoing meiotic division, usually under the name of "chromophilous spherule." Little is known about this body. In *Halidrys siliquosa* (Naylor, 1958b) it is sometimes seen during prophase in the oögonial initial cell. It was not seen to divide, but a similar body appeared in each of the daughter nuclei during the second prophase and in each of the four resulting nuclei at the third prophase A similar body has been recorded in the oögonia of *Coccophora langsdorfii* (Tahara, 1929), *Sargassum horneri* (Okabe, 1929), and *Carpophyllum flexuosum* (Dawson, 1940), and in the antheridia of *Hizikia fusiformis* (Inoh and Hiroe, 1954a), *Sargassum piluliferum* (Inoh and Hiroe, 1954b), and *Sargassum tortile* (Inoh and Hiroe, 1956). It has also been recorded in early prophase of the first meiotic prophase of *Dictyopteris divaricata* (Yabu, 1958) and *Padina japonica* (Kumagae, Inoh and Nishibayashi, 1960; Kumagae and Inoh, 1960).

3. The Metaphase Chromosomes

In most of the 46 genera of the Phaeophyceae investigated, the metaphase chromosomes are so contracted as to be spherical in shape. One exception is *Himanthalia elongata* (L.) S. F. Gray. (Naylor, 1957, Fig. lb), where the chromosomes around the periphery of the plate do not condense so far. Usually there is no doubt as to shape as the chromosomes are seen scattered throughout the cell during the pro-metaphase, and in both side and polar view of the metaphase plate. This, of course, leads to special features at early anaphase separation into chromatids. At first the chromosomes become elongated in the direction of the poles, and a central constriction develops so that in side view they appear dumbbell shaped. The constriction becomes attenuated and eventually the chromosomes separate as two flat plates, and travel to the poles as flat plates—apart from occasional bridges and lagging chromosomes. This feature is perhaps of interest in relation to the parallel separation of long chromosomes seen in *Luzula* among angiosperms, and in *Cosmarium botrytis* (King, 1953) and species of *Spirogyra* (Godward, 1954) among the Chlorophyceae.

CONCLUSION

Although there has been a certain amount of progress in technique and consequently in information about nuclear structure in the Phaeophyceae, there is still need for a really reliable technique for handling small chromosomes.

LITERATURE CITED

Abe, K. 1940. Meiotische Teilung von *Dictyosiphon foeniculaceus. Sci. Rept. Tôhoku Univ.,* 4 Ser., 15:317.

Belling, J. 1926. The iron-aceto-carmine method of fixing and staining chromosomes. *Biol. Bull.* 50:160-162.

Cave, M. S., and Pocock, M. A. 1951. The acetocarmine technique applied to the colonial Volvocales. *Stain. Tech.* 26:173-174.

Darlington, C. D., and la Cour, L. F. 1940. Nucleic acid starvation of chromosomes in *Trillium. J. Genetics* 40:185.

Davidson, F. F. 1950. The effects of auxins on the growth of marine algae. *Am. J. Bot.* 37:502-510.

Dawson, A. E. E. 1940. Studies in the Fucales of New Zealand. 2. Observations on the female fround of *Carpophyllum flexuosum* (Esp.) Grev. *New Phytol.* 39:283.

Evans, L. V. 1962. Cytological studies in the genus Fucus. *Ann. Bot., N. S.* 26:345-360.

Farmer, J. B., and Williams, J. L. 1896. On fertilisation and segmentation of the spore in fucus. *Ann. Bot.* 10:479-487.

Farmer, J. B., and Williams, J. L. 1898. Contributions to our knowledge of the Fucaceae: their life-history and cytology. *Phil. Trans. Roy. Soc. (Lond.) B,* 140: 623-645.

Giraud, G. 1956. Recherches sur l'action de substances mitoclasiques sur quelques algues marines. *Rev. Gen. Bot.* 63:202-236.

Godward, M. B. E. 1948. The iron acetocarmine method for Algae. *Nature* 161:203.

Godward, M. B. E. 1954. The "diffuse" centromere or polycentric chromosomes in Spirogyra. *Ann. Bot., N. S.* 18:143-156.

Inoh, S., and Hiroe, M. 1954a. Cytological studies on the fucaceous plants. II. On the meiotic division in the antheridium of *Hizika fusiformis* Okamura. *La Kromosomo* 21:764.

Inoh, S., and Hiroe, M. 1954b. Cytological studies on the fucaceous plants. III. On the meiotic division in the antheridium of *Sargassum piluliferum* C. Ag. *La Kromosomo* 21:767.

Inoh, S., and Hiroe, M. 1956. Cytological studies on the fucaceous plants. VI. On the meiotic division in the antheridium of *Sargassum tortile* C. Ag. *La Kromosomo* 27-28:942.

Kemp, L., and Cole, K. 1961. Chromosomal alternation of generations in *Nereocystis luetkeana* (Mertens) Postels and Reprecht. *Can. J. Bot.* 39:1711-1724.

King, G. C. 1953. "Diffuse" centromere and other cytological observations on two Desmids. *Nature* 171:181.

Kumagae, N., and Inoh, S. 1960. Morphogenesis in Dictyotales. II. On the meiosis of the tetraspore mother cell in *Dictyota dichotoma* (Huds.) Lamour. and *Padina japonica.* Yamada. *La Kromosomo* 46-47:1521-1530.

Kumagae, N., Inoh, S., and Nishibayashi, T. 1960. Morphogenesis in Dictyotales. II. On the meiosis of the tetraspore mother cell in *Dictyota dichotoma* (Huds.) Lamour. and *Padina japonica* Yamada. *Biol. J. Okayama Univ.* 6:91-102.

Le Touzé, M. H. 1912. Contribution à l'étude histologique des Fucacees. *Rev. gen. Bot.* 24:33.

Levan, A., and Levring, T. 1942. Some experiments on c-mitotic reactions within Chlorophyceae and Phaeophyceae. *Hereditas* 28:400-408.

Lewis, K. R. 1956. A cytological study of some lower organisms with particular reference to the use of modern techniques (Ph. D. thesis). University of Wales.

Mathias, W. T. 1935. The life history and cytology of *Phloeospora brachiata. Publ. Hartley Bot. Labs.* 13:1-24.

Naylor, M. 1956. Cytological observations on three British species of Laminaria: a preliminary report. *Ann. Bot. N. S.* 20:431.

Naylor, M. 1957. An acetocarmine squash technique for the Fucales. *Nature* 180:46.

Naylor, M. 1958a. Some aspects of the life history and cytology of *Stictyosiphon tortilis* (Rupr.) Reinke. *Acta Adriatica* 8:3-22.

Naylor, M. 1958b. The cytology of *Halidrys siliquosa* (L.) Lyngb. *Ann. Bot. N. S.* 22:205-217.

Naylor, M. 1958c. Chromosome numbers in the algae, Phaeophyta. *Brit. Phycol. Bull.* 1(6):34-37.

Naylor, M. 1959. Feulgen reaction in the Fucales. *Nature* 183:627.

Okabe, S. 1929. Meiosis im Oogonium von *Sargassum horneri* (Turn.) Ag. *Sci. Rept., Tôhoku Imp. Univ.,* 4 Ser., 4:661.

Papenfuss, G. F. 1935. Alternation of generations in *Ectocarpus siliculosus. Bot. Gaz.* 96:421.

Roberts, M. 1962. Chromosome numbers in the algae, Phaeophyta II. *Brit. Phycol. Bull.* 2(3):165-166.

Roy, K. 1938. Recherches sur la structure du noyau quiescent et sur les mitoses somatiques de quelques Fucacees. *Rev. Algol.* 11:101.

Strasburger, E. 1898. Kernteilung und Befruchtung bei Fucus. *Jahrb. wiss. Bot.* 30:351-374.

Tahara, M. 1929. Ovogenesis in *Coccophora langsdorfii* (Turn.) Grev. *Sci. Rept., Tôhoku Imp. Univ.,* 4 Ser., 4:551.

Yabu, H. 1958. On the nuclear division in tetrasporangia of *Dictyopteris divaricata* (Okamura) Okamura and *Dictyota dichotoma* Lamour. *Bull. Fac. Fish. Hokkaido Univ.* 4:290-296.

Yamanouchi, S. 1909. Mitosis in Fucus. *Bot. Gaz.* 47:173-197.

Yamanouchi, S. 1912. The life history of *Cutleria. Bot. Gaz.* 54:441.

Yamanouchi, S. 1913. The life history of *Zanardinia. Bot. Gaz.* 56:1.

Environmental Conditions and the Pattern of Metabolism in Algae

G. E. Fogg

Department of Botany
Westfield College
London, England

A single strain of an algal species may show remarkable variation in the intensity and pattern of its metabolic activities according to the conditions to which it is exposed. This is sometimes obvious to the unaided eye—as, for example, when a green, actively growing culture of *Botryococcus braunii* is compared with a nitrogen-deficient one in which the cells have accumulated carotenoids and lipoids—and impressive evidence of it is given by many sets of analytical data for *Chlorella* and other unicellular algae. Nevertheless this variability is often overlooked, especially in biochemical studies on algae, and we are far from a complete understanding of how it occurs.

Much is known of the effects of conditions such as light intensity, temperature, and hydrogen ion concentration on the rates of individual metabolic processes and on their final expression in terms of growth in cell numbers; yet we are largely ignorant of the differential effects that these conditions may have on different processes. Sorokin (1960) discussed the effect on the over-all growth rate of *Chlorella* of the different effects of light and temperature on accumulation of cell material and on cell division but did not consider effects on metabolic patterns. Spoehr and Milner (1949), in an important paper, showed that the chemical composition of *Chlorella,* expressed in terms of carbohydrate, fat, and protein, varies according to the light intensity and temperature at which cultures are grown; yet, as they pointed out, the time factor must be taken into account in evaluating such results. If cultures are made in a limited volume of medium and the cell material is harvested after a relatively long period of growth it is difficult to distinguish between primary and secondary effects when assessing the effect of a given factor on the balance of metabolic processes. For example, temperature undoubtedly has direct effects; in cultures that have grown for some time, however, these effects may be masked by those produced by the different effective light intensities and nutrient deficiencies resulting from the different cell populations that have developed at the different temperatures. The method of continuous culture enables the direct effects of environmental conditions on metabolic pattern to be studied but, although Myers (1946) used this method to determine the effects of light intensity on various cellular characteristics of *Chlorella,* this technique has not been used, as far as I am aware, for any detailed studies of this sort.

The extent of the variation which may occur is perhaps best illustrated by some results of Fogg and Than-Tun (1960) for the nitrogen-fixing blue-green alga *Anabaena cylindrica,* which, although not obtained with continuous cultures, were obtained with cultures grown for a sufficiently short period (48 hr) for secondary effects arising from different rates of growth to be at a minimum. Cultures were grown at temperatures of 15, 20, 30 and 35C and light intensities of 2,000, 5,500, and 10,000 lux, in a purely mineral medium with elementary nitrogen as the only nitrogen source, and the changes in cell carbon and nitrogen determined. The assimilation of carbon showed the expected relationships to light intensity and temperature, with saturation at 5,500 lux, temperature limitation at 20C and below, and no inhibition at either the highest light intensity or the highest temperature. The assimilation of nitrogen showed a high temperature coefficient (Q_{10}= 5 for the range 20-30C) but was strongly inhibited at 35C; it showed much the same relationship to light intensity as did carbon assimilation—in agreement with the idea that nitrogen fixation in blue-green algae is closely dependent on photochemical reactions. Thus the ratio of carbon assimilated to nitrogen assimilated varied little with respect to light intensity but greatly with respect to temperature, being 40 or more at both 15 and 35C as compared with 10 at 30C, the optimum temperature for nitrogen assimilation. As a result the nitrogen content, expressed as a percentage of total dry weight, of cells which in all cases were actively growing varied from about 2.3 at 15C to 4.5 at 30C. These results are for a nitrogen-fixing alga; a parallel study with the organism growing on combined nitrogen was not carried out, but it is thought that the observed changes in rate of nitrogen assimilation result from effects on the later stages rather than on the fixation process. They may thus perhaps be taken as indicating that, in general, changes in temperature may produce marked alterations in the balance of the major anabolic processes in algae.

Changes in the supply or consumption of metabolites may have considerable effects on metabolic pattern, quite apart from those on the over-all rate of metabolism. The processes of intermediary metabolism are effected by reversible reactions and although we arbitrarily distinguish sequences of reactions such as glycolysis, the tricarboxylic acid cycle, and the carbon dioxide fixation cycle, these are so intermeshed through common intermediates as to form a single flexible system through which material can flow along various paths, largely according to supply and demand. The concepts of "overflow" and "shunt" metabolism which Foster (1949) put forward to explain organic acid production and other metabolic features of fungi are equally applicable to algae.

In an algal population growing exponentially, synthesis of proteins and other protoplasmic constituents predominates and directly utilizes intermediates of the photosynthetic carbon cycle. There is little accumulation of cell-wall materials or of reserve products, nitrogenous or nonnitrogenous. Transfer of such cells to a medium lacking a nitrogen source, as is commonly done for experiments on photosynthesis, does not have any immediate effect on the quantum efficiency or rate of photosynthesis but results in a drastic diversion of the intermediates produced to pathways other than those of protein synthesis. The enzyme system leading to carbohydrate synthesis accepts the major part of this overflow, as is shown both by the value of the photosynthetic quotient under such conditions and by direct analysis (for references see Fogg, 1959). More detailed evidence exists regarding the change in the pathway of carbon occurring when a nitrogen supply is restored to such cells. Fogg (1956) showed that the supply of ammonium nitrate to photosynthesizing nitrogen-deficient cells of the diatom *Navicula pelliculosa* brought about a dramatic change in the distribution among various cell fractions of radiocarbon supplied as bicarbonate, 87% of the carbon fixed then entering the fraction soluble in 80% ethanol but insoluble in benzene as compared with 24% in cells with no supplied nitrogen source. Holm-Hansen et al. (1959) carried out similar experiments with *Chlorella,* identifying individual intermediates by autoradiography. They found, for example, that in one experiment the supply of ammonium chloride increased the proportion of radiocarbon fixed in amino acids from 9.9 to 57% of the total while the radiocarbon incorporated in sugar phosphates correspondingly fell from 64 to 7%. That these switches should occur seems obvious now, yet until comparatively recently it was tacitly assumed that the pathway of carbon was the same in growing cells as in cells in a medium devoid of a nitrogen source. From this assumption arose the idea, still widely held, that carbohydrate is always the immediate product of photosynthesis.

A less obvious effect of the same sort is perhaps involved in the excretion of glycolic acid from photosynthesizing cells. Using radiocarbon as a tracer, Tolbert and Zill (1957) found that during short-term photosynthesis experiments *Chlorella pyrenoidosa* liberates glycolic acid as an extracellular product, its concentration in actively growing cultures reaching 3 to 8 mg/liter. Pritchard, Griffin, and Whittingham (1962), who have made a detailed study of the production of glycolate by *Chlorella,* conclude that it is derived from the carbon dioxide acceptor in photosynthesis, ribulose diphosphate, and that it accumulates only when photosynthesis is carbon dioxide limited. One would expect that glycolate would escape from the photosynthesizing cells until equilibrium was established between the intracellular and extracellular concentrations and it seems possible that under some circumstances glycolate might be the sole product of photosynthesis. Warburg and Krippahl (1960) have indeed reported a stoichiometric equivalence between glycolate excreted and carbon dioxide taken up by *Chlorella* such that one mole of glycolate is produced for every two moles of carbon dioxide assimilated. This would be expected to happen when cells begin photosynthesis in fresh medium. Only when steady diffusion gradients of glycolate have been established can the assimilatory power of the photochemical reaction become available for the synthesis of materials for the production of new protoplasm. Nalewajko, Chowdhuri, and Fogg (1963) have found that concentrations of glycolate of the order of 1 mg/liter do, in fact, abolish the lag phase in the growth of a planktonic strain of *Chlorella pyrenoidosa* under light-limited conditions. This addition has no appreciable effect on the relative growth rate, whereas additions of substances such as glucose, acetic acid, and pyruvic acid under similar conditions increase the relative growth rate but have no effect on the lag. Production of glycolate rather than carbohydrate may result in discrepant results in determinations of the quantum efficiency of photosynthesis. It may also have its ecological implications; phytoplankton may not be able to begin growth until steady diffusion gradients of glycolate have been established around the cells. This might partly explain the abruptness with which growth begins in spring in temperate waters and also its dependence on decrease in turbulence.

The availability of particular metabolites in the environment may thus have considerable immediate effects on metabolic pattern but nevertheless this pattern will be determined primarily by the relative activities of the various enzyme systems in the organism. The availability of a particular metabolite may, however, have longer term effects by producing alterations in the balance of these enzyme systems. This is most clearly illustrated by the effects of nitrogen deficiency. As we have seen, carbohydrate is the major product of metabolism immediately following the withdrawal of the nitrogen supply but it is clear that fat synthesis comes to predominate in most algae if the deficiency continues for several days.

The changes occurring during prolonged incubation in a medium with no nitrogen source, but otherwise complete, have been the subject of studies with *Chlorella* spp. (e.g., Bongers, 1956), *Navicula pelliculosa* (Fogg, 1956), and *Monodus subterraneus* (Fogg, 1959). The total amounts of protein and nucleic acids in such cultures remain constant or increase slightly at the expense of the soluble nitrogenous fraction of the cells and other constituents such as chlorophyll. The rate of photosynthesis declines and after a few days remains more or less constant at a value about 5% of that shown initially by the cell suspension. As a result of this continued photosynthesis the total dry weight of cell material increases and, cell division being limited, there is an increase in dry weight per cell. The proportion of fat synthesized shows no immediate alteration, evidently being limited by enzyme activity rather than by availability of intermediates. After about three days of nitrogen starvation the fat content on a percentage dry weight basis begins to increase. This change coincides with a climacteric in respiration and is evidently the result of a partial breakdown in organization of the cells. In *Navicula* the amount of fat expressed on a cell volume basis shows no increase and radiocarbon studies show a fall in the proportion of photosynthetically fixed carbon entering the fat fraction at this stage. It seems that the observed increase in fats on a dry weight basis is mainly the result of hydrolysis of other materials and their consequent loss by respiration or diffusion from the cells, the fats themselves remaining unchanged. In *Chlorella* and *Monodus*, however, there is a definite increase in the total amount of fat in cultures at this stage. The explanation of this is perhaps that a differential inactivation of enzymes has occurred with the result that the fat-synthesizing system is better able to compete for the available intermediates. Even in *Navicula* there appears to be some such change since nitrogen-starved cells of this diatom returned to a medium containing nitrate were found to continue to synthesize the same high proportion of fat for 24 hr although the rate of photosynthesis was nearly trebled. Only after two days in the nitrate medium, by which time, we may suppose, synthesis of fresh enzyme had altered the original balance, did it fall to that characteristic of cells grown with an adequate nitrogen supply.

Metabolic changes in the later stages of cultures in which growth is limited by the nitrogen supply are not necessarily of the same nature as those which occur when exponentially growing cells are transferred to a nitrogen-deficient medium. Among other things, the presence of metabolic products in the medium in the former situation and their absence in the latter may lead to important differences. Both direct analysis (e.g., Spoehr and Milner, 1949; Collyer and Fogg, 1955; Bongers, 1956) and tracer studies (Fogg, 1956) show that fat accumulation is characteristic of unicellular species belonging to the Chlorophyceae, Xanthophyceae, and Bacillariophyceae under such conditions. There does not seem to be any preliminary phase in which carbohydrate synthesis predominates, for this falls off concurrently with protein synthesis. The accumulation of fat is not directly determined by the concentration of available nitrogen (Fogg, 1959) and, again, seems to depend largely on the enzymic balance of the cells. The possibilities of alteration in enzymic balance appear to be rather different in exponentially growing cells transferred to a nitrogen-deficient medium from those in cells in a culture in which the nitrogen supply has been exhausted in the course of growth. As Tamiya and his collaborators have shown, the pattern of metabolism of *Chlorella* undergoes marked changes during the cycle of cell development and division. An exponentially growing population consists largely of the cell type distinguished by Tamiya et al. (1953) as D-cells, the transformation of which into L-cells is not dependent on a nitrogen source. The L-cells produced in the nitrogen-deficient medium can undergo division but the second generation of D-cells is unable to develop normally and the metabolic pattern becomes fixed at that characteristic of this developmental stage, with carbohydrate synthesis predominating (Nihei et al., 1954). Cells in a culture in which the nitrogen supply becomes exhausted as a result of growth, on the other hand, become arrested in the L-stage, perhaps because cell division is more sensitive than other processes to the by-products of metabolism which accumulate.

Nihei et al. (1954) found that whereas the photosynthetic quotient ($O_2/-CO_2$) of D-cells is about unity, that of L-cells may be as much as 3.3, indicating that the products of metabolism at this stage are more reduced than they are at other stages. If it is correct that fat accumulation in nitrogen-deficient cultures is the result of the halting of the developmental cycle at a point at which fat synthesis predominates, then other factors stopping the cycle at the same point should have a similar effect. It is not clear whether this is so. Phosphorus deficiency brings cell development to a standstill apparently at this stage but may not always induce fat accumulation (Spoehr and Milner, 1949).

Otsuka (1961), however, has found that sulphur-deficient cells of *Chlorella,* which become arrested at a stage when some cell extension has occurred and the nucleus has divided once, accumulate abnormally large amounts of fat. Also, von Denffer (1948) observed that fat accumulation in *Nitzschia palea* was the result of the accumulation of a staling product which blocked mitosis and was not necessarily confined to conditions of nitrogen deficiency. Probably, in addition to the effect of halting cyclic changes in enzymic balance at a particular point, other factors are involved in fat accumulation. Synthesis and degradation of enzymes in cultures of limited volume take place under the progressively modifying influence of falling nutrient concentrations and rising concentration of staling products. As in populations transferred abruptly to conditions of nitrogen deficiency, the activity of the fat-synthesizing system may be less susceptible to such adverse conditions than is that of other competing systems. Nitrogen deficiency evidently does not induce fat accumulation in Myxophyceae and Rhodophyceae (Collyer and Fogg, 1955), but it is not known what the basis of this difference from the other algal groups is.

Finally there are the changes in metabolic pattern resulting from adaptive enzyme formation. The classic instance of the appearance of a new pattern of metabolism in an alga following its adaptation to a new environment—the hydrogenase activity of hydrogen-adapted *Scenedesmus* (Gaffron, 1940)—is probably a matter of activation of an already existing enzyme rather than of the synthesis of new enzyme molecules, since similar adaptation in *Chlamydomonas moewusii* has an adaptation period of less than 10 min (Frenkel and Rieger, 1951). A few instances of adaptation by algae to carbohydrate substrates have been reported (for references see Fogg, 1953), and *Anabaena cylindrica* evidently needs to be adapted to nitrate before it can be utilized (Fogg and Wolfe, 1954), but in no case does any very thorough investigation of the adaptation appear to have been carried out. The indications are that algae are as well able to produce adaptive enzymes as other microorganisms; investigations of this question should be rewarding.

LITERATURE CITED

Bongers, L. H. J. 1956. Aspects of nitrogen assimilation by cultures of green algae. *Meded. Landb. Hoogesch., Wageningen* 56:1-52.

Collyer, D. M., and Fogg, G. E. 1955. Studies on fat accumulation by algae. *J. Exp. Bot.* 6:256-275.

Denffer, D. von. 1948. Über einen Wachstumshemmstoff in alternden Diatomeenkulturen. *Biol. Zentr.* 67:7-13.

Fogg, G. E. 1953. The metabolism of algae. Methuen and Co. Ltd., London.

Fogg, G. E. 1956. Photosynthesis and formation of fats in a diatom. *Ann. Bot.* 20: 265-285.

Fogg, G. E. 1959. Nitrogen nutrition and metabolic patterns in algae. *Symp. Soc. Exp. Biol.* 13:106-125.

Fogg, G. E., and Than-Tun. 1960. Interrelations of photosynthesis and assimilation of elementary nitrogen in a blue-green alga. *Proc. Roy. Soc.* 153*B*:111-127.

Fogg, G. E., and Wolfe, M. 1954. The nitrogen metabolism of the blue-green algae (Myxophyceae). *Symp. Soc. Gen. Microbiol.* 4:99-125.

Foster, J. W. 1949. Chemical activities of fungi. Academic Press, Inc., New York.

Frenkel, A. W., and Rieger, C. 1951. Photoreduction in algae. *Nature* 167:1030.

Gaffron, H. 1940. Carbon dioxide reduction with molecular hydrogen in green algae. *Am. J. Bot.* 27:273-283.

Holm-Hansen, O., et al. 1959. Effects of mineral salts on short-term incorporation of carbon dioxide in *Chlorella. J. Exp. Bot.* 10:109-124.

Myers, J. 1946. Culture conditions and the development of the photosynthetic mechanism. III. Influence of light intensity on cellular characteristics of *Chlorella. J. Gen. Physiol.* 29:419-427.

Nalewajko, C., Chowdhuri, N., and Fogg, G. E. 1963. Excretion of glycolic acid and growth of a planktonic *Chlorella*. pp. 171-183 *in* Studies on microalgae and photosynthetic bacteria. University of Tokyo Press.

Nihei, T., et al. 1954. Change of photosynthetic activity of *Chlorella* cells during the course of their normal life cycle. *Arch. Mikrobiol.* 21:155-164.

Otsuka, H. 1961. Changes of lipid and carbohydrate contents in *Chlorella* cells during the sulfur starvation, as studied by the technique of synchronous culture. *J. Gen. Appl. Microbiol.* 7:72-77.

Pritchard, G. G., Griffin, W. J., and Whittingham, C. P. 1962. The effect of carbon dioxide concentration, light intensity and *iso*nicotinyl hydrazide on the photosynthetic production of glycolic acid by *Chlorella*. *J. Exp. Bot.* 13:176-184.

Sorokin, C. 1960. Kinetic studies of temperature effects on the cellular level. *Biochim. Biophys. Acta* 38:197-204.

Spoehr, H. A., and Milner, H. W. 1949. The chemical composition of *Chlorella;* effect of environmental conditions. *Plant Physiol.* 24:120-149.

Tamiya, H., et al. 1953. Correlation between photosynthesis and light-independent metabolism in the growth of *Chlorella*. *Biochim. Biophys. Acta* 12:23-40.

Tolbert, N. E., and Zill, L. P. 1957. Excretion of glycolic acid by *Chlorella* during photosynthesis, pp. 228-231. *In* H. Gaffron (ed.). Research in Photosynthesis. Interscience Publishers, Inc., New York.

Warburg, O., and Krippahl, G. 1960. Glykolsäurebildung in *Chlorella*. *Z. Naturforschung*. 5b:197-199.

Micronutrient Requirements for Green Plants, Especially Algae

Clyde Eyster*

Charles F. Kettering Research Laboratory
Yellow Springs, Ohio

INTRODUCTION

The growth and normal biochemistry of green plants require the availability of from 15 to 20 elements. Eventually still others may be added to the list. Table I gives the list of these nutrient elements and indicates whether they are macronutrients or micronutrients. Some of the elements vary depending upon whether they are utilized by higher plants or by green or blue-green algae. Plants require mineral elements for various functions. The macronutrients are used generally as building materials, whereas the micronutrients are commonly metal constituents of enzymes which enter into biological reactions (McElroy and Nason, 1954).

TABLE I

Elements Required by Green Plants

Macronutrients (10^{-2}–10^{-4}M)	Micronutrients (10^{-5}M and less)
C, H, O, N	Fe, Mn, Cu, Zn
P, S, K, Mg	Mo, V, B
Ca*	Cl, Co, Si
Na†	

* Except for algae, where it is a micronutrient.
† For blue-green algae.

* Presently employed at Monsanto Research Corporation, Dayton, Ohio.

TABLE II

Mineral Requirements for *Chlorella pyrenoidosa* Based on Critical Concentrations Required in the Culture Medium

Mineral	Heterotrophic growth	Autotrophic growth
NO_3	2.5×10^{-3} M	2.5×10^{-2} M
Mg	2×10^{-2}	2×10^{-3}
K	4.3×10^{-3}	4.3×10^{-4}
P	1.8×10^{-4}	1.8×10^{-4}
S	2×10^{-4}	2×10^{-4}
Fe	1×10^{-9}	1.8×10^{-5}
Zn	0.77×10^{-10}	0.77×10^{-6}
Mn	1×10^{-9}	1×10^{-7}
Cl	3×10^{-6}	3.4×10^{-2}

A comparison between the mineral requirements for autotrophic and heterotrophic growth of *Chlorella* has shown some interesting differences (Table II). Except for Mg and NO_3 the critical quantities of macronutrients are quite similar for both autotrophic and heterotrophic conditions. The micronutrients such as manganese, iron, zinc, chlorine, and vanadium are required in much greater amounts for autotrophic growth than for heterotrophic growth. This suggests a comparatively high concentration requirement for photosynthesis and a much lower concentration requirement for heterotrophic growth.

Macronutrients, based on critical concentration in the culture medium, are required in the range of 10^{-2} M to 10^{-4} M. The quantitative requirements for micronutrients are less than 10^{-4} M. Of course, the amounts usually added to a culture medium may be in excess of the critical concentration.

MANGANESE

McHargue (1922) was the first to give definite evidence for the essentiality of manganese, and reported that plants deficient in manganese were stunted in their growth and produced no seeds. A mottled type of chlorosis accompanies manganese deficiency in higher plants. The essentiality of manganese for the growth of *Chlorella* was shown originally by Hopkins (1930). Additional studies which relate manganese to the growth of algae are Pirson and Bergmann (1955); Eyster et al. (1956a); Eyster, Brown, and Tanner (1956b); Reisner and Thompson (1956), Kessler (1957), and Eyster et al. (1958a). Manganese-deficient cells of *Chlorella* grown autotrophically are characterized by being abnormally clumped, by being about twice the volume of normal cells, and by having a reduction in chlorophyll (Brown, Eyster, and Tanner, 1958; Eyster, Brown, and Tanner, 1958b).

Manganese is one of the key elements in photosynthesis. The greater requirement for autotrophic growth than for heterotrophic growth shown for iron, zinc, chloride, and nitrate also applies to manganese (Eyster et al., 1958b). The critical concentration for autotrophic growth of *Chlorella pyrenoidosa* has been determined to be about 1×10^{-7} M compared to 1×10^{-9} M for heterotrophic growth. This greater requirement for autotrophic growth in the presence of carbon dioxide, bicarbonate, or carbonate is believed to be due to its role in photosynthesis.

Manganese-deficient *Chlorella* were shown (Pirson, 1937) to have a reduced level of photosynthesis which was restored to the normal level within a period of one or two hours upon the addition of manganese to the culture. Later these same results (Fig. 1) were obtained with manganese-deficient cultures of *Ankistrodesmus* (Pirson, Tichy, and Wilhelmi, 1952). Furthermore Arnon, Allen, and Whatley (1954, 1956) and Allen et al. (1955) have reported that manganese increased the capacity of isolated chloroplasts to fix carbon dioxide.

One contribution of the Charles F. Kettering Research Laboratory has been to present evidence that manganese affects growth and photosynthesis by being required for Hill reaction activity, which in the light in the presence of an oxidant causes the formation of gaseous oxygen from water (Eyster et al., 1956b, 1958a). The original experiments were based on the use of quinone as the Hill oxidant. More recently, it has been established that manganese is required for the reduction of TPN by chloroplast particles of *Chlorella* in the presence of photosynthetic pyridine nucleotide reductase also derived from *Chlorella* cells (Eyster, 1961). The manganese required for TPN reduction was shown to be associated with the chloroplast particles and did not seem to be a constituent of the reductase enzyme nor necessary for the formation of the reductase enzyme. Unpublished results (Eyster) show that manganese is a

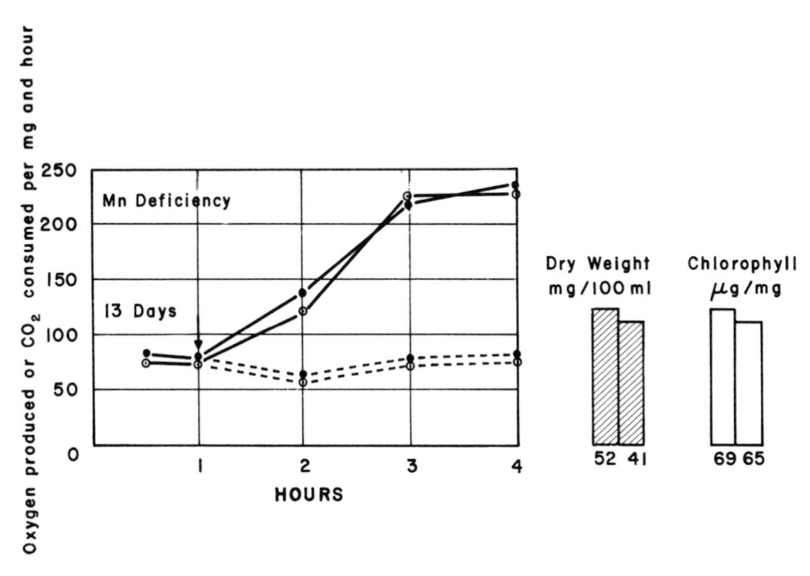

FIG. 1. Short-term recovery of photosynthesis after the addition of manganese to 13-day-old manganese-deficient cultures of *Ankistrodesmus*. (After Pirson et al., 1952.)

general requirement for oxygen evolution regardless of the Hill oxidant used. Quinone and nitrite are effective for *Chlorella* whole cells, whereas quinone, ferricyanide, TPN, and 2,6-dichlorophenol indophenol are effective Hill oxidants for sonicated *Chlorella* cells.

Growth, photosynthesis, and Hill reaction show a close parallel relationship at each concentration level of manganese added to the culture medium (Fig. 2). Most of the study involved the use of *Chlorella pyrenoidosa,* but was also extended to *Nostoc muscorum, Scenedesmus quadricauda, Porphyridium cruentum, Lemna minor* (duckweed), and sugar beets.

An interesting speculation has been that the size of the photosynthetic unit, which was reported to be about 2500 chlorophyll molecules (Emerson and Arnold, 1932-1933; Arnold and Kohn, 1934), is regulated by the role of manganese in photosynthesis. *Chlorella* grown with a minimum amount of manganese for maximum growth gave a chlorophyll-to-manganese ratio which tended to indicate such a relationship. Flashing light experiments (unpublished) of *Chlorella* cells successively subcultured to eliminate the reserve manganese and to obtain cells where essentially all the

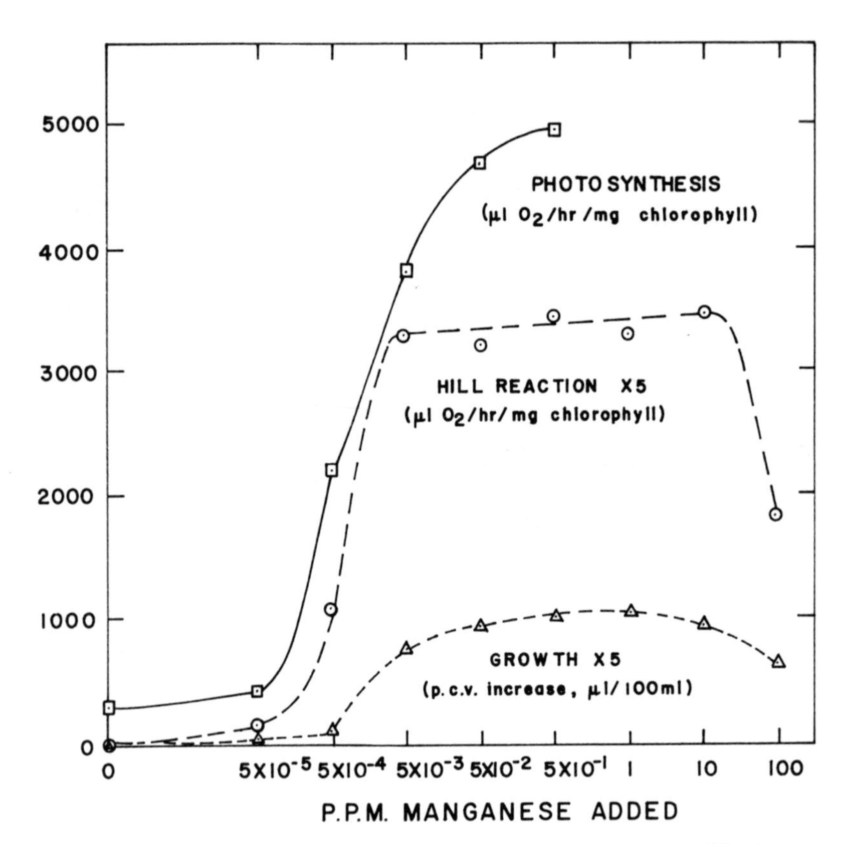

FIG. 2. Photosynthesis, Hill reaction, and growth of autotrophic *Chlorella pyrenoidosa* at various levels of manganese in the culture medium. (After Eyster et al., 1956b.)

manganese was simultaneously involved in photosynthesis showed a manganese-to-oxygen ratio which was gradually reduced and finally approached unity. The determinations were based on calculations of the moles of oxygen produced per flash and the moles of manganese (measured with Mn^{54}) contained in the cells.

IRON

The essentiality of iron for plants has been known for more than a century (see Gines, 1930; and Gilbert, 1957). Iron deficiency results in a severe chlorosis of the leaves, especially in the younger leaves, because iron is not readily transported within the plant.

More than 30 years ago Hopkins and Wann (1927) reported that iron was necessary for the culturing of *Chlorella.* Noack and Pirson (1939) and Albert-Dietert (1941) have studied the effects of iron deficiency in relation to nitrogen metabolism in *Chlorella.* Literature reviews dealing with iron nutrition include Hewitt (1951), Myers (1951), and Pirson (1955).

Eyster et al. (1958b) have shown that more iron was required for autotrophic growth of *Chlorella pyrenoidosa* than for its heterotrophic growth (Fig. 3). They reported that *Chlorella* had to have a minimum of 1.8×10^{-5} M iron for maximum growth and did not grow when the iron concentration was below 1.8×10^{-7} M. At high light intensities in very active dense cultures the optimum iron concentration may be 10-fold higher.

The biggest problem of iron nutrition is associated with its availability to plants. On standing, the dissolved iron changes to a colloidal form which reduces its availability. Fresh iron solutions need to be used for improved culturing conditions. Many prefer to chelate the iron with citrate or with ethylene diamine tetra acetic acid (EDTA). Although EDTA accentuates the need for more zinc, manganese, and calcium in the nutrient solution for *Chlorella,* it does not affect its iron requirement (Walker, 1954). The chelating effect of EDTA with iron must be very mild compared with the way

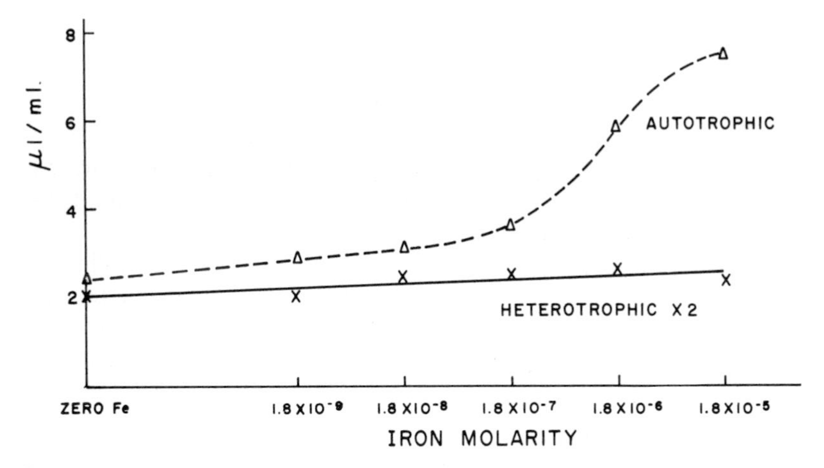

FIG. 3. Growth of *Chlorella pyrenoidosa* at different concentrations of iron. Autotrophic: 4 days growth from 1.4 μl washed cells. Heterotrophic: 21 days growth from 1 loopful cells.
(After Eyster *et al.*, 1958b.)

it binds zinc, manganese, and calcium. The application of iron as potassium ferricyanide has some advantage in certain media (Walker, 1954) because the entire molecule of potassium ferricyanide can be utilized to furnish four essential elements. In addition potassium ferricyanide solutions, if kept in a cold, dark place, do not deteriorate and aged solutions are as effective in nutrient media as fresh solutions.

The function of iron is associated with numerous iron-containing enzymes, e.g., peroxidase, catalase, cytochrome *c*, cytochrome oxidase, cytochrome *f*, cytochrome b_6, and photosynthetic pyridine nucleotide reductase. Review articles with special reference to plants and yeast dealing with iron compounds have been published by Scarisbrick (1947), Granick and Gilder (1947), and Lemberg and Legge (1949). Cytochrome *f* and cytochrome b_6 occur in chloroplasts of green leaves and are believed to be important in the transfer of electrons during processes involved in photosynthesis. The nature and possible function of chloroplast cytochromes has been reviewed by Hill and Bonner (1961). The study of cytochromes in algae and bacteria has been critically reviewed by Smith and Chance (1958). The iron content of photosynthetic pyridine nucleotide reductase was reported recently by Mower (1962), and an iron non-heme enzyme has been shown to be necessary for nitrogen fixation (Mower, 1962).

CHLORINE

Recently evidence has been reported to show that chlorine is a micronutrient element for higher plants (Broyer et al., 1954; Ozanne, Woolley, and Broyer, 1957; Johnson et al., 1957; and Martin and Lavollay, 1958). Tomato plants showed the following nutritional deficiency symptoms: wilting of leaflet blade tips, chlorosis, and bronzing and necrosis of the leaves proximal to the wilted areas. Compared with other micronutrients the chlorine requirement is not small. Bromine has been shown to be able to replace chlorine. However, neither fluorine nor iodine could substitute for chlorine.

Previously a chloride or anion effect had been reported for washed chloroplasts (Warburg and Lüttgens, 1946; Arnon and Whatley, 1949; and Gorham and Clendenning, 1952) and for lyophilized *Chlorella* cells (Schwartz, 1956). Warburg and Lüttgens designated chloride as a "coenzyme" of photosynthesis specifically concerned with oxygen evolution. This was later confirmed and more thoroughly investigated by Arnon (1959). He found that chloride was necessary for noncyclic photophosphorylation and for the riboflavin phosphate pathway of cyclic photophosphorylation. For the vitamin K pathway of cyclic photophosphorylation chloride was not required. Chloroplasts have all of these pathways for photophosphorylation whereas the chromatophores of photosynthetic bacteria can carry out only the vitamin K pathway of cyclic photophosphorylation.

Noncyclic photophosphorylation as it occurs in chloroplasts requires chloride and its biochemistry can be represented as follows:

$$2TPN + 2H_2O + 2ADP + 2H_3PO_4 \xrightarrow{PPNR} 2TPNH_2 + O_2 + 2ATP$$

If it is coupled with a ferricyanide Hill reaction the equation is as follows:

$$4Fe^{3+} + 2H_2O + 2ADP + 2H_3PO_4 \longrightarrow 4Fe^{2+} + O_2 + 2ATP + 4H^-$$

That two catalytic elements are known to be specifically concerned with oxygen evolution during photosynthesis should be emphasized. These two elements are manganese and chlorine.

When Eyster (1958a) formulated a special medium for *Chlorella* on the basis of critical concentrations of nutrients for autotrophic growth, he found that the special medium was seriously inadequate and could be corrected best by the addition of chloride. The amount of growth of *Chlorella* could be doubled with the addition of merely 0.02 mg NaCl/100 (3.4×10^{-6} M). However, the critical concentration for sodium chloride was found to be about 3.4×10^{-2} M (Fig. 4).

VANADIUM

Vanadium is one of the most recent additions to our list of micronutrients (Bertrand, 1942; Arnon and Wessel, 1953; and Arnon, 1954, 1958). *Scenedesmus obliquus* produced symptoms

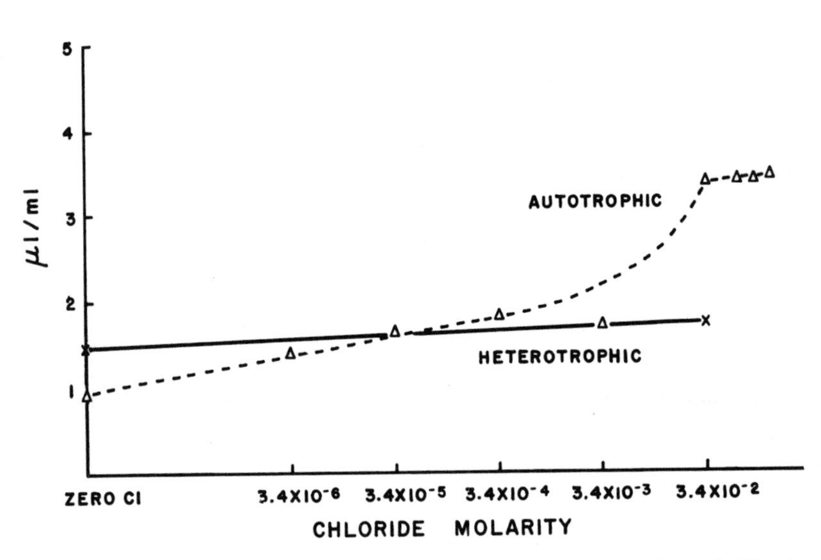

FIG. 4. Growth of *Chlorella pyrenoidosa* at different concentrations of chloride (NaCl). Autotrophic: 3 days growth from 0.65 μl washed cells. Heterotrophic: 20 days growth from 1 loopful cells.
(After Eyster *et al.*, 1958b.)

of chlorosis when grown without vanadium in the culture medium. Furthermore, photosynthesis based on a unit-chlorophyll basis at high light intensity of 20,000 lux was about twice as high in the cells with vanadium as in the cells deprived of vanadium. In weak light of 2,000 lux there was no significant difference (Fig. 5).

The work of Arnon dealing with vanadium as an essential micronutrient is quite convincing. Warburg and Krippahl (1954), and Warburg, Krippahl, and Buchholz (1955) extended the role of vanadium in photosynthesis to *Chlorella*. There has been no extension of this study to other algae or higher plants to indicate whether this essentiality for vanadium is general. Watching the literature on this subject during the next several years should prove very interesting.

It should be mentioned that vanadium can replace molybdenum as a catalyst in nitrogen fixation by *Azotobacter* (Bortels, 1930) and *Clostridium* (Burk, 1934).

ZINC

Zinc has definitely been established to be required for the growth of both higher green plants and algae. The earliest productive studies regarding the essentiality of zinc for plants center around the works of Raulin (1869), Voelcker (1913), and Maze (1914). However, it was not until 1926 that Sommer and Lipman presented convincing evidence of the necessity of zinc as a plant nutrient. Since then there have been numerous confirmations. Recently zinc-deficiency studies have been reported for cotton (Brown and Wilson, 1954), for bean (Viets, Boawn and Crawford, 1954), and for corn (Viets et al., 1953). Zinc-deficiency symptoms include leaf chlorosis, leaf necrosis, scant foliage and reduced leaf size, shortened internodes or rosette formation, and reduced fruit. "Mottled leaf" of citrus, pecan "rosette," and "little leaf" or "rosette" of fruit trees are diseases known to be caused by a deficiency of zinc. Early reports on the effect of zinc on plant growth

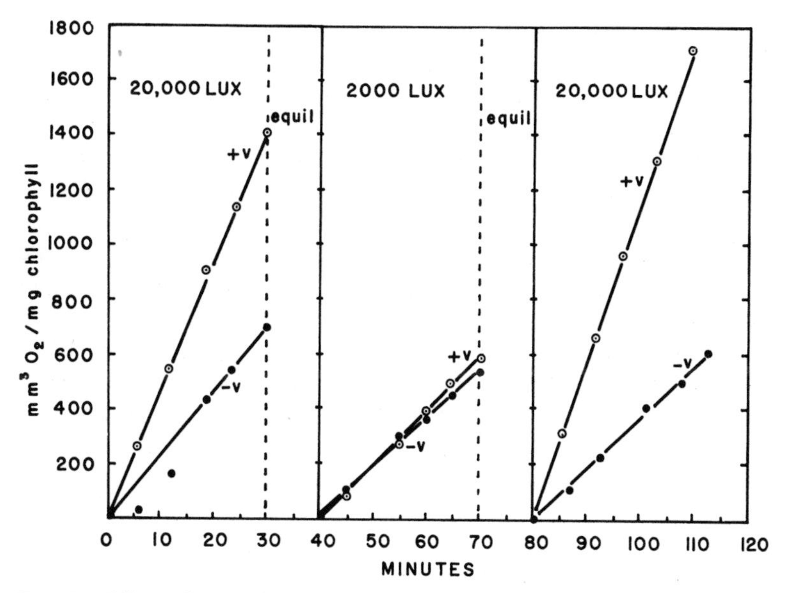

FIG. 5. Effect of vanadium on the photosynthesis of *Scenedesmus obliquus* at a high and a low light intensity. The same cells were exposed successively to high and to low and again to high light intensity with intervening equilibration periods. (After Arnon, 1958.)

are summarized by Brenchley (1927). Theories regarding the actions of zinc on plants are discussed by Chandler (1937) and Camp (1945). The metabolic role of zinc has been reviewed recently by Hoch and Vallee (1958).

The following zinc metalloproteins have been characterized: carbonic anhydrase, carboxypeptidase, alcohol dehydrogenase, glutamic dehydrogenase, and lactic dehydrogenase. The last three are pyridine nucleotide-dependent metallodehydrogenases. Partial evidence has been obtained for the existence of four more pyridine nucleotide-dependent zinc metalloproteins. These are glyceraldehyde-3-phosphate dehydrogenase, α-glycerophosphate dehydrogenase, malic dehydrogenase, and glucose-6-phosphate dehydrogenase (Vallee et al., 1956).

The existence of carbonic anhydrase in leaves was first reported by Neish (1939), who observed activity in both the crude chloroplast and cytoplasm fractions of four species. Evidence for the general occurrence of carbonic anhydrase in leaves was presented by Bradfield (1947). There have also been the published findings of Day and Franklin (1946), Steeman, Nielsen, and Kristainsen (1949), Osterlind (1950), Waygood and Clendenning (1950, 1951), Sibly and Wood (1951), and Brown and Eyster (1955). Carbonic anhydrase occurs in algae as well as in green leaves. Osterlind (1950) reported its presence in *Scenedesmus quadricauda* and *Chlorella pyrenoidosa.* Its presence in *Nostoc muscorum* was established by Brown and Eyster (1955). Hexokinase from *Neurospora crassa* (Medina and Nicholas, 1957a,b) may be a zinc metalloenzyme. Reed (1946) suggested a possible role of zinc in the hexokinase of tomato plant tissues. The synthesis of tryptophan seems to require zinc. There is an important connection with auxin biosynthesis inasmuch as tryptophan is a precursor (Skoog, 1940; Tsui, 1948). Zinc deficiency markedly reduced the ability of *Neurospora* (Nason, Kaplan, and Colowick, 1951), tomato plants (Possingham, 1956), and oat plant leaves (Wood and Sibly, 1952) to synthesize proteins. Zinc-deficient tomato plants contained about twice the total amount of amino acids and 10 times the amide content of normal zinc-supplemented controls.

Zinc-deficient *Rhizopus nigricans* had no pyruvic carboxylase activity (Foster and Denison, 1950). The conclusion reached is that zinc was not a constituent of the enzyme but that zinc was necessary for its synthesis. The aldolase activity of higher plants was decreased by a zinc deficiency (Quinlan-Watson, 1951, 1953).

Warburg and Lüttgens (1944, 1946) presented some evidence which caused them to speculate that zinc was a heavy metal intimately concerned with Hill reaction activity. They reported that the quantity of *o*-phenanthroline required to cause 100% inhibition was almost exactly equal to the amount required to completely chelate the zinc in the chloroplast fragments. The inhibitory action was corrected by the addition of zinc ions. Schwartz (1956), using lyophilized *Chlorella* cells, confirmed the reactivation of Hill reaction with zinc ions after removal of an unknown metal ion with phenanthroline. However, he could reactivate the phenanthroline inhibition by the addition of many divalent metal ions. Co^{++}, Cu^{++}, Mn^{++}, and Fe^{++} as well as Zn^{++} were effective, but Mg^{++} and Ca^{++} were completely without activity.

Additional studies which pertain to the requirement of zinc for the growth of algae include the effect of zinc deficiency on chlorophyll content of *Chlorella* (Noack, Pirson, and Stegmann, 1940; and Stegmann, 1940), the quantitative zinc requirement of *Chlorella* (Walker, 1954), and a comparison of the zinc requirements for *Chlorella pyrenoidosa* under autotrophic and heterotrophic conditions (Eyster et al., 1958b). Approximately 4.5 μg zinc is required for the growth of *Chlorella* in nitrate medium per gram dry weight and 6.5 μg zinc in urea medium (Walker, 1954). More zinc is required for autotrophic growth than for heterotrophic growth (Eyster et al., 1958b). Ondratschek (1941) has shown that zinc is a growth requirement for a number of flagellates.

Euglena gracilis has been shown by Price and Millar (1962) and by Price and Vallee (1962) to be an organism particularly suitable for physiological and biochemical studies involving zinc.

CALCIUM

Calcium is a macronutrient for higher plants. As calcium pectate, it is definitely known to be one of the components of the middle lamella of the cell wall. Furthermore, calcium is necessary for the continued growth of apical meristems and may enter into the composition of protoplasm and certain types of protein in the cell (Gilbert, 1957). Allen and Arnon (1955) have also shown that the nitrogen-fixing blue-green alga *Anabaena cylindrica* required macroquantities of calcium for growth regardless of whether the algae were given molecular nitrogen or nitrate nitrogen. The optimum amount appeared to be about 20 ppm calcium, with some indication that 20 ppm calcium was quite necessary for optimum fixation of nitrogen whereas it was somewhat more than adequate for growth with nitrate in the culture medium.

Green algae require only microquantities of calcium (Stegmann, 1940; Myers, 1951; and Walker, 1953) or do not require any calcium. The calcium requirement for *Chlorella* has not been too certain (Trelease and Selsam, 1939; Scott, 1943; and Hutner and Provasoli, 1951). Walker (1953) has given experimental evidence that *Chlorella* does become calcium-deficient when EDTA is present in the medium and that strontium was able to correct the deficiency better than calcium. A real calcium essentiality would not permit the substitution of any other element. Also one cannot be sure what preferential bindings occur when EDTA is in contact with so many different cations. The problem of calcium essentiality among the various algae is one which should be pursued vigorously. Ultimately one would then be certain about its requirement and also be able to determine its biochemistry.

Eyster (1959) has measured the amount of calcium required by *Nostoc muscorum* (a blue-green alga) to fix atmospheric nitrogen. The minimum amount in the modified Chu medium for maximum growth was reported to be 0.75×10^{-5} M or 0.3 ppm (a microquantity). Neither nitrate nor ammonium salt were included in the medium. The cultures were bubbled with 5% CO_2 in air and illuminated with about 1000 ft-c of light (Fig. 6).

Emission spectrographic analyses of crystalline α-amylases from human saliva, hog pancreas, *Bacillus subtilis,* and *Aspergillus oryzae* showed that these enzymes all contained at least 1 g-atom of very firmly bound calcium per mole of enzyme. The calcium content of amylase can be decreased to less than 1 g-atom/mole by dialysis against ethylenediaminetetraacetate. Incubation with this sequestrating agent resulted in inhibition of the amylases, and an addition of excess calcium completely reactivated the enzymes in all instances (Vallee, Stein, Sumerwell, and Fischer, 1959). A stabilizing effect of calcium salts on α-amylases was first reported more than 50 years ago by Wallerstein (1909), who patented their effect in connection with the brewing process.

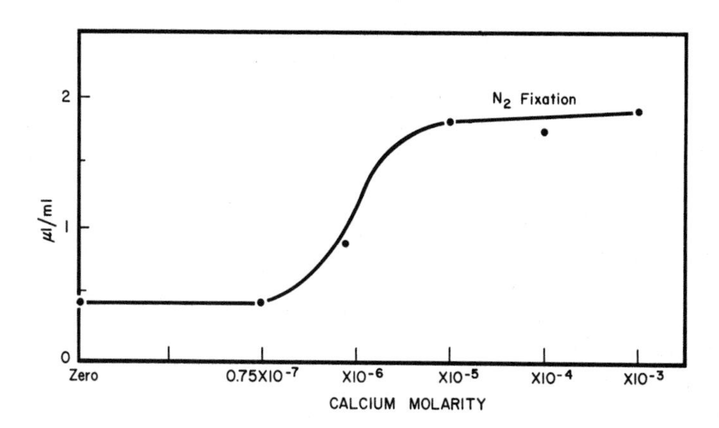

FIG. 6. Effect of different levels of calcium on the growth of *Nostoc muscorum* in the absence of nitrate, urea, and ammonium salts but in the presence of nitrogen gas. Cultures grown 6 days, while being shaken, bubbled with 5% CO_2 in air, and illuminated with 1500 ft-c cool white fluorescent light. Inoculum: 0.3 μl cells from calcium deficient culture.

BORON

Higher plants definitely require boron as a micronutrient. There is still some doubt about the essentiality of boron for algae except for its need by some algae for nitrogen fixation.

The importance of boron for plant growth was suggested by Agulhon (1910), Brenchley (1914), and Maze (1914). Conclusive evidence for its requirement was provided by Warington (1923) and Sommer and Lipman (1926). The following are some of the many reviews on the subject dealing with the requirement of boron for plant growth: Johnston (1928), Maze (1936), Dennis (1937), Greenhill, (1938), McMurtrey (1938), Shive (1945), Gauch and Dugger (1953), and Skok (1958). Boron deficiency produces the following symptoms: death of the apical meristems, failure of young roots to develop, hypertrophy of the cambium, disintegration and discoloration of the phloem, and breakdown of xylem tissue blocking the conducting system.

It has been reported that boron acts as a coenzyme for inositol synthesis (Rosenberg, 1946, 1948), stimulates nitrogen fixation by

Azotobacter (Jordan and Anderson, 1950), stimulates tyrosinase activity (MacVicar and Burris, 1948), increases oxidase activity, causes liberation of inorganic phosphate, glucose accumulation, lack of ATP, and accumulation of certain soluble nitrogen compounds, probably plays a part in the transportation and utilization of carbohydrates (Gauch and Dugger, 1953), and possibly is functional in protein metabolism. The severity and onset of boron-deficiency symptoms bears a relationship to the amount of calcium present such that the boron requirement increases with an increase in amount of calcium (Stiles, 1961; Berger, 1949; Gauch and Dugger, 1953; and Eyster, unpublished). In other words, calcium accentuates boron deficiency. The exact function of boron in plant metabolism is still obscure and there has been much speculation about its role in cellular functions (see review by Skok, 1958).

Except for its role in nitrogen fixation, the boron requirement for algae is uncertain. Geigel (1935) reported a marked stimulation in the growth of *Chlorella* in the presence of 10 ppm boron. McIlrath and Skok (1957) obtained increased growth of *Chlorella* with 0.5 ppm boron in the culture medium. There was a 60% increase in cell number and 137% increase in dry weight per cell. An essentiality

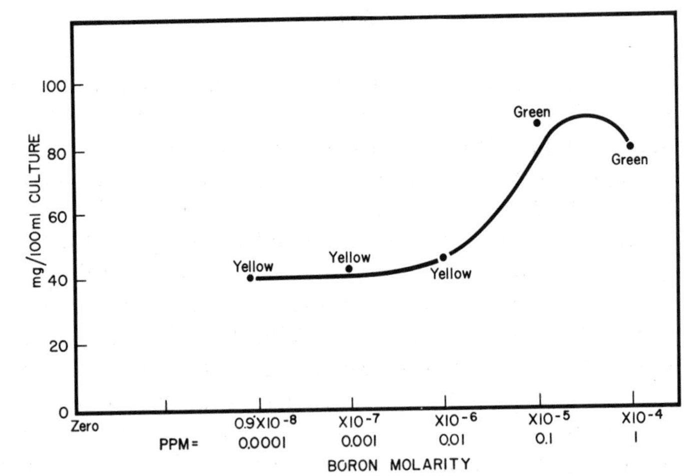

FIG. 7. Effect of different levels of boron on the growth of *Nostoc muscorum* in the absence of nitrate, urea, and ammonium salts but in the presence of nitrogen gas. Cultures grown in 5 weeks, while being shaken, bubbled with 5% CO_2 in air, and illuminated with 1500 ft-c cool white fluorescent light. Inoculum: 0.3 μl cells from boron-deficient culture.

for boron by algae which do not fix nitrogen may well exist, but much more needs to be done to substantiate such a claim.

Eyster's work (1952, 1958, 1959) has shown that *Nostoc muscorum* requires boron for growth. Cultures grown without added boron in Corning (boron-free) flasks resulted in a relatively small reduction (39%) in cell count compared with control cultures to which boron was added. However, there was a chlorosis which started slowly during the third week and the cells became completely white by the end of eight weeks. The boron-deficiency chlorosis was corrected by the addition of boron. The optimum boron concentration for nitrogen fixation appeared to be about 0.1 ppm or 0.9×10^{-5} M boron (Fig. 7). As yet no strong evidence has been reported that boron is required by algae for another function, but if it is, the amount would be far less than that required for nitrogen fixation.

MOLYBDENUM

Molybdenum is an important essential micronutrient for all green plants including algae. Its essentiality for higher plants was first reported by Arnon and Stout (1939) and Piper (1940). Previously it was reported that both molybdenum and vanadium promoted the growth and nitrogen fixation of *Azotobacter* (Bortels, 1930) and that molybdenum was necessary for the growth of *Aspergillus niger* and for nitrogen metabolism (Garner, 1935; and Steinberg, 1936). The following review articles are suggested: Hoagland (1945), Mulder (1950), Anderson (1956), and Nason (1958).

Symptoms of molybdenum deficiency include chlorosis followed by wilting, curling, and withering of the leaves (Hewitt, 1956; Stout, 1956).

Molybdenum is necessary for nitrate utilization and for nitrogen fixation. Considerably more molybdenum is required by *Azotobacter* for nitrogen fixation than for nitrate utilization (Bortels, 1930; Burema and Wieringa, 1942; and Mulder, 1948; Evans, 1956).

Regarding the essentiality of molybdenum for algae it has been shown that molybdenum-deficient cells of the green alga *Scenedesmus obliquus* fail to assimilate nitrate nitrogen (Arnon et al., 1955; and Ichioka and Arnon, 1955) (Fig. 8).

Eyster (1959) reported that the critical concentration of molybdenum for nitrogen fixation by *Nostoc muscorum* was 1×10^{-7} M and

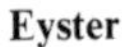

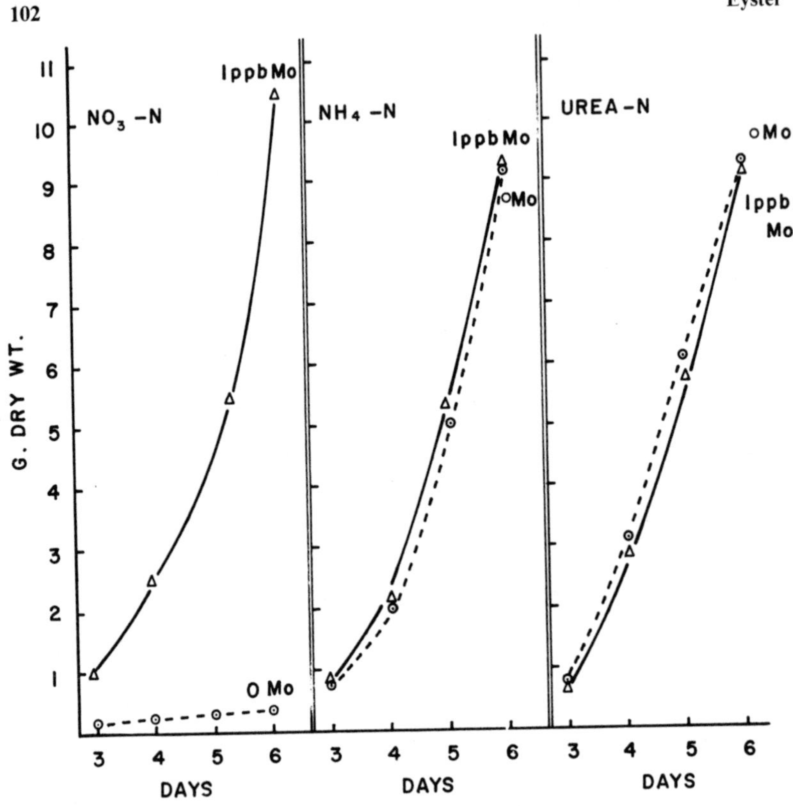

FIG. 8. Effect of molybdenum on the growth of *Scenedesmus obliquus* supplied with nitrate, ammonia, or urea nitrogen. Ordinate represents grams dry weight of cells per liter of nutrient solution. (After Arnon, 1958.)

that the critical concentration of molybdenum for the utilization of nitrate was 1×10^{-10} M. No molybdenum was required when ammonium ions furnished the nitrogen. Likewise it was found that *Chlorella pyrenoidosa* required a critical concentration of 1×10^{-10} M molybdenum whereas no molybdenum was necessary for growth with an ammonium salt. To determine the critical concentration of molybdenum for nitrate utilization it was necessary to highly purify the reagents for the culture medium. This was done by an electrolytic method (Eyster, unpublished). The electrolytic apparatus is described in Fig. 9. Ten-fold concentrations of nutrient medium were placed in this electrolytic cell which by means of the silver chloride reference electrode was provided with a voltage of -1.7 for 24 hr. At this voltage all of the cations were removed from the

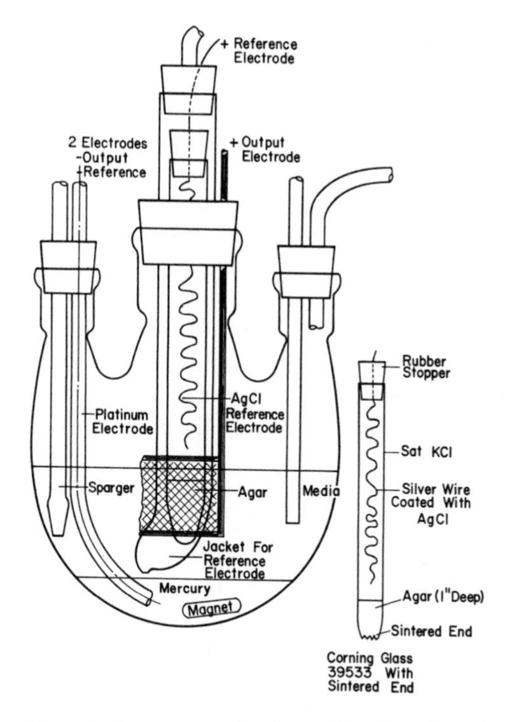

FIG. 9. Electrolytic apparatus for the purification of nutrient media.

medium except K, Na, Mg, and Ca. Special effort was observed to exclude oxygen from the system during electrolysis. The system was continuously bubbled with nitrogen gas which had been passed through oxysorbent to remove any oxygen. The mercury in the bottom of the flask served as the negative electrode, and being negative at a voltage of 1.7 it caused the migration of positive ions of manganese, molybdenum, iron, copper, and zinc from the medium into the mercury. The voltage was maintained at a constant value of -1.7 by means of a potentiostat (Lingane and Jones, 1950). At the end of about 24 hr, while the current was still on, half of the one liter of 10-fold nutrient medium in the flask was removed into a super clean flask. Doctor Paul Delahay of Louisiana State University, who was consultant on this project, emphasized that it would be unwise to remove any medium after the positive electrode lost contact with the medium, because the trapped cations which previously were contaminants in the medium would tend to migrate back into the medium. During the electrolysis, nitrate was reduced to nitrite. Nitrite is poisonous to algae. To surmount this problem, the macronutrient minus potassium nitrate was purified. Fisher Certified Reagent grade postassium nitrate was found to be pure enough so that it could be used without further purification to demonstrate molybdenum deficiency in *Chlorella.* Fisher Certified Reagent grade compounds were used where possible for all the macronutrients. Also shown was that the only salt usually somewhat contaminated with cations such as molybdenum, manganese, iron, etc., was the magnesium sulfate. Microelements added to the culture medium were the "Spec Pure" brand obtained from Johnson, Matthey and Co., Ltd.

COBALT

A cobalt requirement has been demonstrated by Holm-Hansen, Gerloff, and Skoog (1954) for bacteria-free cultures of *Calothrix parientina, Nostoc muscorum, Coccochloris Peniocystis,* and *Diplocystis aerugina.* The nutritional function of cobalt seemed to be associated with the role of vitamin B_{12}, and the cobalt requirement of *Nostoc muscorum* could be satisfied by the addition of relatively minute amounts of vitamin B_{12}. Forty parts per trillion of cobalt was sufficient to satisfy the needs of *Nostoc muscorum* for growth.

Buddhari (1960) was able to show a distinct cobalt requirement for *Anabaena cylindrica* and *Nostoc* sp. based on dry weight, chlorophyll, and phycocyanin determinations of cultures grown with and without added cobalt in the form of $Co(NO_3)_2$. He purified the nutrient salts by CuS coprecipitation (Arnon et al., 1955). There was almost no growth in the absence of added cobalt compared with good growth in the presence of added cobalt. A critical concentration of 0.5 μg Co/liter was sufficient to support full growth. Toxicity was observed at a concentration of 10 mg Co/liter.

Recently it has been established that cobalt is essential for soybean plants grown under symbiotic conditions for nitrogen fixation (Shaukat-Ahmed and Evans, 1960, 1961). Also, Reisenauer (1960) reported that a small amount of cobalt resulted in a 66%

increase in weight and a marked increase in the nitrogen content of symbiotically grown alfalfa plants. There is an earlier report by Bolle-Jones and Mallikarjuneswara (1957), who observed a beneficial effect of cobalt at a concentration of 0.005 ppm of the nutrient medium on the growth of rubber plant seedlings grown in purified sand culture.

Clethra barbinervis has been reported to be a cobalt accumulator plant (Yamagata and Murakami, 1958). It was found to contain 661 ppm Co whereas other plants in the same location had only 2 to 5.3 ppm Co.

The biological significance of cobalt has been reviewed by Marston (1952) and Young (1956). Cobalt has been definitely established to be an essential micronutrient element for animals in which certain physiological processes such as the syntheses of labile methyl groups, methionine, and purines are now known to be dependent on cobalt.

COPPER

Sommer (1931) was the first to clearly demonstrate that copper was an essential micronutrient for tomato, sunflower, and flax plants. Tomato plants grown in nutrient solutions under controlled conditions showed the following copper deficiency symptoms: stunted growth of shoots and roots, curling of leaves, failure to produce flowers and a bluish-green color of the leaves (Sommer, 1945). Lipman and MacKinney (1931) grew copper-deficient barley plants which were unable to produce seed. They concluded that copper was an essential element for plant growth. Splendid reviews dealing with the essentiality of copper for plants have been prepared by Sommer (1945), Arnon (1950), and Steinberg (1950).

The metabolic role of copper in plants is not completely understood. Evidence to date links copper with oxidations in the plant. Laccase, ascorbic acid oxidase, and polyphenol oxidase (tyrosinase) are plant enzymes which have been shown to contain copper as a metal constituent. The function of laccase in the metabolism of the plant has not yet been explained. The latex of several species of lacquer trees contains laccase.

The role of ascorbic acid oxidase in plant metabolism has been reviewed by James (1946), Waygood (1950), and Arnon (1950).

Some evidence exists that the enzyme acts as a terminal oxidase bringing about electron transfer between reduced ascorbic acid and atmospheric oxygen. The ascorbic acid system in plants is quite similar to the cytochrome system in animal tissues and in yeast in that ascorbic acid is reversibly oxidized and reduced in the former and cytochrome *c* in the latter.

The present status of polyphenol oxidase in plant metabolism can be evaluated by referring to the review papers of Nelson and Dawson (1944), Dawson and Mallette (1945), and Arnon (1950). The enzyme catalyzes the oxidation of a variety of phenols aerobically and is considered by Onslow (1931), Boswell and Whiting (1938), Baker and Nelson (1943), and Bonner and Wildman (1946) to be a terminal oxidase in respiration. There is evidence that the enzyme catalyzes the oxidation of $DPNH_2$ and $TPNH_2$ according to the following scheme:

$$\text{Substrates} \longrightarrow \text{dehydrogenases} \begin{matrix} \nearrow \text{DPN} \longrightarrow \text{DPNH}_2 \searrow \\ \searrow \text{TPN} \longrightarrow \text{TPNH}_2 \nearrow \end{matrix} \text{o-quinone} \longrightarrow$$

$$\text{o-phenol} \longrightarrow \text{Cu}^{++}\ \text{enzyme} \longrightarrow \text{Cu}^{+}\ \text{enzyme} \longrightarrow \text{O}_2$$

The aforementioned copper oxidases make it obvious that copper is essential for the respiration of plants. There is indirect evidence that copper has a function in photosynthesis. Clover leaves were found (Neish, 1939) to concentrate copper in the chloroplasts. Potassium ethyl xanthate, a strong inhibitor of the light reaction of photosynthesis and believed to be fairly specific for copper enzymes, was shown by Arnon (1950) to have its inhibition counteracted either by dialysis or by the addition of copper which was the only metal effective in reversing the inhibition. Arnon postulated that "copper may prove to be at least one of the metals concerned in the light reaction of photosynthesis in green plants." Twelve years have elapsed and still there is no direct evidence that copper plays a direct role in photosynthesis. During this period of time manganese has been shown to be essential to photosynthesis, being concerned with oxygen evolution.

Walker (1953) is the only one who has reported a copper requirement for the growth of green algae (Fig. 10). He has shown that *Chlorella* requires copper when grown in a glucose-urea (or

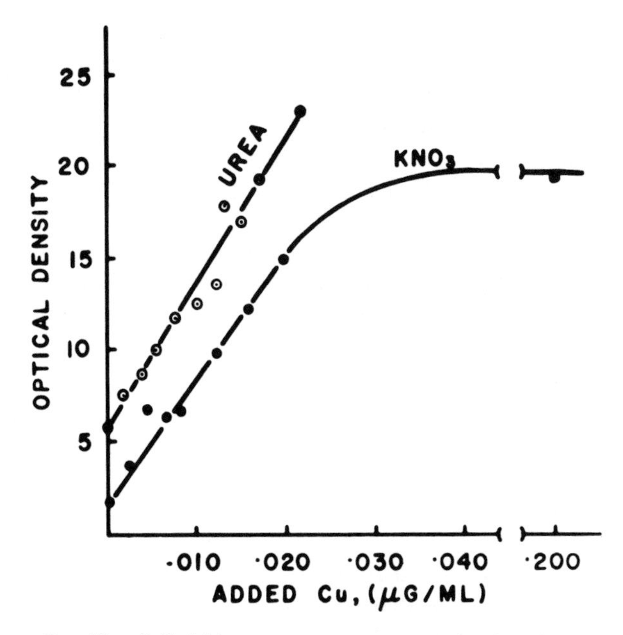

FIG. 10. Cell yield versus copper concentration for *Chlorella pyrenoidosa* grown in glucose-urea-EDTA-salts and glucose-nitrate-EDTA-salts media deficient only in copper. (After Walker, 1953.)

nitrate)-EDTA-salt medium deficient in copper. The copper requirement could not be satisfied by beryllium, magnesium, barium, cobalt, iron, zinc, manganese, molybdenum, nickel, thallium, gallium, aluminum, cadmium, vanadium, germanium, titanium, zirconium, arsenic, bismuth, tin, chromium, mercury, sodium, potassium, boron, silver, lead, or gold when ions of these elements were added singly to growth cultures lacking copper. He was unable to demonstrate any requirement for copper with the glucose-urea-salt medium; however, he found that copper was required for maximum growth in the glucose-nitrate-salt medium. This difference could have been due to a copper impurity in the urea or to an increased copper requirement for nitrate assimilation. The problem of copper essentiality for algae is by no means solved. Information on the copper requirement of several more commonly cultured algae as well as confirmations about the copper requirement for *Chlorella* would be quite desirable.

SILICON

Although silicon seems to be required by some plants in macroquantities, it will be discussed here in connection with micronutrients. Sommer (1926) was able to improve the growth of rice and millet by the addition of silicon to the culture medium. Ishibashi (1937) was able to increase the growth of rice by adding air-dried silicic acid to the soil. Sunflower and barley were reported by Lipman (1938) to be definitely benefited by the presence of silicon in the culture medium. Raleigh (1939) concluded that silicon was an indispensable chemical element for the growth of the beet plant. Wagner (1940) reported that rice, barley, cucumber, corn, tomato, and tobacco gave increased growth when silicon was added and that table beets did not respond to added silicon. Woolley (1957) concluded that if silicon were essential for the growth and development of the tomato plant, silicon would be required in amounts of less than 0.2 μg atoms/g dry weight of the plant.

Lewin (1954, 1957) and Jorgensen (1955a, b) have made studies on silicon metabolism in diatoms which are known to have silicon in the cell wall. Lewin grew *Navicula pelliculosa* in medium containing 1 ppm Si and found that silicate uptake was an aerobic process inhibited in the absence of oxygen. Washed cells lost the ability to utilize silicon which could be partially restored by the addition of sulfate (2×10^{-3} M). Na_2S (4×10^{-5} M), glutathione (2×10^{-4} M), l-cysteine (2×10^{-4} M), *dl*-methionine (2×10^{-3} M), and $Na_2S_2O_3$ (4×10^{-5} M) were found to give complete restoration of silicate uptake even after five washings. The fact that reduced sulfur compounds were more effective than sulfate suggested the role of a reducing agent in addition to a source of sulfur. Ascorbic acid alone (2×10^{-5} M) was found to induce only partial recovery in washed cells, as did sulfate, but ascorbic acid plus sulfate gave almost complete recovery.

Silicon-deficiency symptoms of higher plants (Raleigh, 1939) were necrosis of roots and wilting of leaves.

SODIUM

The essentiality of sodium, really a macronutrient, will be discussed here because blue-green algae definitely require sodium for their growth and development.

The blue-green alga *Anabaena cylindrica* (Allen and Arnon, 1955) and bladder saltbush, *Atriplex vesicaria* (Brownell and Wood, 1957), have been shown to require sodium for their growth. Hodgson (1955) investigated the possible essentiality of sodium in the nutrition of radish, barley, and celery. Although he found that the yield of celery was doubled when 60 ppm of sodium was added to the culture solutions, he concluded that sodium was not essential but that it served some beneficial function. The results of Hodgson were confirmed by Woolley (1957), who obtained a 12% increase in the dry weight of tomato plants with additions of sodium at the rate of 1 mM NaCl/liter of solution.

There are experimental results which indicate that sodium is quite important under certain conditions of plant nutrition. This is particularly true of sugar beet (Lill, Byall, and Hurst, 1938), cotton (Matthews, 1941), flax (Lehr and Wybenga, 1955), barley (Lehr and Wybenga, 1958), and red table beet (Wybenga and Lehr, 1958).

Lehr (1941a,b) states that for sugar beets sodium may almost be deemed an indispensable nutrient element, approaching potassium in importance. A rather comprehensive review of the literature in this field is given by Lehr (1941a) and by Mullison and Mullison (1942).

A sodium concentration of 5 ppm or higher is required for optimal growth of *Anabaena cylindrica* (Allen and Arnon, 1955). Furthermore, they found that potassium, lithium, rubidium, and caesium could not substitute for sodium. Sodium-deficient cultures were shown to contain less phycocyanin than those with adequate sodium. Their chlorophyll content was not affected. In order to maintain a steady high rate of photosynthesis in resting cells it was necessary to suspend the cells in a buffer containing sodium salts. This effect of sodium, first observed on *Synechococcus cedrorum* (Allen, 1952), has been noted for *Anabaena* as well.

Benecke (1898) described an *Oscillatoria* which grew in a medium in which all of the potassium salts had been replaced by sodium compounds. Emerson and Lewis (1942) reported that *Chroococcus* did not grow when sodium was omitted from the medium. Recently Allen (1952) found that 23 cultures of various blue-green algae grew in sodium salt media without added potassium. The growth of *Microcystis aeruginosa* was reported by Gerloff, Fitzgerald, and Skoog (1952) to be benefited by sodium carbonate or sodium silicate, the effect of which could have been due to a more suitable pH rather than to their sodium content. Kratz and Myers (1955) showed that sodium had a marked effect on the growth rate of *Anacystis nidulans*.

Unpublished data by Eyster have shown that *Nostoc muscorum* required from 3 to 5 ppm sodium; that approximately the same amount of sodium was required regardless of whether the nitrogen source was NO_3, NH_4, or N_2; and that the sodium requirement was influenced by the amount of potassium in the medium. At higher levels of potassium more sodium was required.

DISCUSSION

Presently only five micronutrient elements (Eyster, 1962) seem to be directly concerned with photosynthesis of green plants including algae (Fig. 11). Manganese and chlorine each have a role in oxygen evolution. Vanadium accelerates photosynthesis under high light intensity but as yet this effect has been reported only

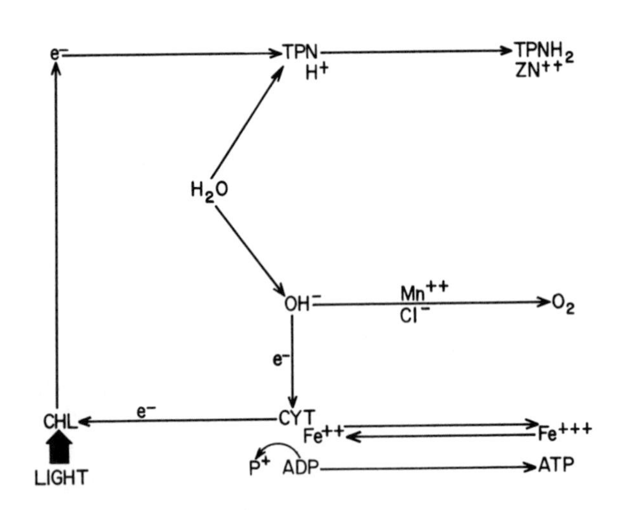

FIG. 11. An attempt to indicate the function of micronutrient elements concerned with photosynthesis using Arnon's (1961) scheme for noncyclic photophosphorylation in chloroplasts. The function of vanadium is still not sufficiently clarified to be included in this figure.

for *Scenedesmus obliquus* and *Chlorella.* Iron is the metal part of at least two plant cytochromes which occur in the chloroplasts of green cells and which very likely facilitate the transfer of electrons during photosynthesis. Although more zinc is required in light-grown than in dark-grown cultures of *Chlorella,* its role in photosynthesis is uncertain. One might suspect zinc to aid in hydrogen transfer since zinc is definitely known to be the metal constituent of various dehydrogenases.

Nitrogen fixation in algae which have the capacity to fix gaseous nitrogen requires the following elements: boron, calcium, iron, molybdenum, and perhaps cobalt. Each of these elements may be required for one or more other functions. The critical concentration of an element for its two or more functions is usually always much higher for nitrogen fixation. Algae which are forced to fix nitrogen have very high boron and calcium requirements. These same algae when furnished with fixed forms of nitrogen either require no boron and calcium or the need is so meager that it is extremely difficult to measure. *Nostoc muscorum* demands 10-fold more iron when it fixes nitrogen than when it is not forced to do so. This same blue-green alga requires about 1000-fold more molybdenum than for nitrate utilization. Holm-Hansen et al. (1954) discuss the nutritional function of cobalt with reference to nitrogen fixation. They reported that cobalt was primarily required for normal growth in the presence of fixed nitrogen but that cobalt may have an additional function in nitrogen fixation. However, Buddhari (1960) found that blue-green algae always needed essentially the same amount of cobalt for growth regardless of the form of nitrogen supplied to them. He concluded that cobalt is not directly involved in the process of nitrogen fixation.

There is also some evidence that small amounts of manganese, calcium, boron, cobalt, and copper are required for other growth functions. Silicon is required by diatoms for cell wall formation. Blue-green algae definitely require sodium.

SUMMARY

1. The following five elements are required by algae and green plants generally for photosynthesis: manganese, iron, chlorine, zinc, and vanadium.

2. The following five elements are required for nitrogen fixation by those algae which possess the capacity to fix nitrogen gas: iron, calcium, boron, molybdenum, and cobalt. We are still not too sure about cobalt.

3. Small concentrations of the following six elements are or may be required for other metabolic functions: manganese, calcium, boron, cobalt, copper, and silicon.

LITERATURE CITED

Agulhon, H. 1910. Use of boron as a catalytic fertilizer. Thése de l'Université de Paris, 158 pp. *Compt. rend.* 150:288-91.

Albert-Dietert, F. 1941. Die Wirkung von Eisen und Mangan auf die Stickstoff-assimilation von *Chlorella. Planta* 32:88-117.

Allen, M. B. 1952. The cultivation of Myxophyceae. *Archif. Mikrobiol.* 17:34.

Allen, M. B., and Arnon, D. I. 1955. Studies on nitrogen-fixing blue-green algae. I. Growth and nitrogen fixation by *Anabaena cylindrica* Lemm. *Plant Physiol.* 30:366-372.

Allen, M. B., et al. 1955. Photosynthesis by isolated chloroplasts. III. Evidence for complete photosynthesis. *J. Am. Chem. Soc.* 77:4149-4155.

Anderson, A. J. 1956. Molybdenum deficiencies in legumes of Australia. *Soil Sci.* 81:173-182.

Arnold, W., and Kohn, H. I. 1934. The chlorophyll unit in photosynthesis. *J. Gen. Physiol.* 18:109-112.

Arnon, D. I. 1950. Functional aspects of copper in plants. *in* McElroy, W. D., and Glass, B. (eds.). A symposium on copper metabolism. pp. 89-114. The Johns Hopkins Press, Baltimore, Maryland.

Arnon, D. I. 1954. Some recent advances in the study of essential micronutrients for green plants. 8eme Congr. intern. botan., 8eme Congr. Paris. Sect. 11:73-80.

Arnon, D. I. 1958. The role of micronutrients in plant nutrition with special reference to photosynthesis and nitrogen assimilation. Chapt. I. *In* Lamb, C. A., Bentley, O. G., and Beattie, J. M. (eds.). Trace elements. Academic Press, Inc., New York, New York.

Arnon, D. I. 1959. Conversion of light into chemical energy in photosynthesis. *Nature* 184:10-21.

Arnon. D. I. 1961. Cell-free photosynthesis and the energy conversion process. *In* McElroy and Glass (eds.). Light and life. The Johns Hopkins Press, Baltimore, Maryland.

Arnon, D. I., Allen, M. B., and Whatley, F. R. 1954. Photosynthesis by isolated chloroplasts. *Nature* 174:394-396.

Arnon, D. I., Allen, M. B., and Whatley, F. R. 1956. Photosynthesis by isolated chloroplasts. IV. General concept and comparison of three photochemical reactions. *Biochim. Biophys. Acta* 20:449-461.

Arnon, D. I., et al. 1955. Molybdenum in relation to nitrogen metabolism. I. Assimilation of nitrate nitrogen by *Scenedesmus. Physiol. Plantarum* 8:538-551.

Arnon, D. I., and Stout, P. R. 1939. Molybdenum as an essential element for higher plants. *Plant Physiol.* 14:599-602.

Arnon, D. I., and Wessel, G. 1953. Vanadium as an essential element for green plants. *Nature* 172:1039-1040.

Arnon, D. I., and Whatley, F. R. 1949. Is chloride a coenzyme of photosynthesis? *Science* 110:554-556.
Baker, D., and Nelson, J. M. 1943. Tyrosinase and plant respiration. *J. Gen. Physiol.* 26:269-276.
Benecke, W. 1898. Über Culturbedingungen einiger Algen. *Bot. Zeit.* 56:83.
Berger, K. C. 1949. Boron in soils and crops. *Advances in Agron.* 1:321-351.
Bertrand, D. 1942. Vanadium, an oligosynergic element for *Aspergillus niger. Ann. Inst. Past.* 68:226-44.
Bolle-Jones, E. W., and Mallikarjuneswara, V. R. 1957. A beneficial effect of cobalt on the growth of the rubber plant *(Hevea brasilienses). Nature* 179:738-739.
Bonner, J., and Wildman, S. G. 1946. Enzymatic mechanism in the respiration of spinach leaves. *Arch. Biochem.* 10:497-518.
Bortels, H. 1930. Molybdans als Katalysator bei der biologischen Stickstoffbindung. *Arch. Mikrobiol.* 1:333-342.
Boswell, J. G., and Whiting, G. C. 1938. A study of the polyphenol oxidase system in potato tubers. *Ann. Botany, N. S.* 2:847-864.
Bradfield, J. R. G. 1947. Plant carbonic anhydrase. *Nature* 159:467-468.
Brenchley, W. E. 1914. On the action of certain compounds of zinc, arsenic, and boron on the growth of plants. *Ann. Botany.* 28:283-301.
Brenchley, W. E. 1927. Inorganic plant poisons and stimulants. 2nd ed. Cambridge Univ. Press, London.
Brown, L. C., and Wilson, C. C. 1952. Some effects of zinc on several species of *Gossypium* L. *Plant Physiol.* 27:812-817.
Brown, T. E., and Eyster, C. 1955. Carbonic anhydrase in certain species of plants. *The Ohio J. Sci.* 55:257-262.
Brown, T. E., Eyster, H. C., and Tanner, H. A. 1958. Physiological effects of manganese deficiency. Chapt. 10, pp. 135-155. *In* Lamb, Bentley, and Beattie (eds.). Trace elements. Academic Press, Inc., New York.
Brownell, P. F., and Wood, S. G. 1957. Sodium as an essential micronutrient element for *Atriplex vesicaria* Heward. *Nature* 179:635-636.
Broyer, T. C., et al. 1954. Chlorine—a micronutrient element for higher plants. *Plant Physiol.* 29:526-532.
Buddhari, W. 1960. Cobalt as an essential element for blue-green algae (Doctor's thesis). University of California, Berkeley, California.
Burema, S. J., and Wieringa, K. T. 1942. Molybdenum as a growth factor for *Azotobacter chroococcum. Antonie van Leeuwenhoek. J. Microbiol. Serol.* 8:123-133.
Burk, D. 1934. Azotase and nitrogenase in *Azotobacter. Ergeb. Enzymf.* 3:23-56.
Camp, A. F. 1945. Zinc as a nutrient in plant growth. *Soil Sci.* 60:157-164.
Chandler, W. H. 1937. Zinc as a nutrient for plants. *Bot. Gaz.* 98:625-646.
Dawson, C. R., and Mallette, M. F. 1945. The copper proteins. *Adv. Protein Chem.* 2:179-248.
Day, R., and Franklin, J. 1946. Plant carbonic anhydrase. *Science* 104:363-365.
Dennis, R. W. G. 1937. Relation of boron to plant growth. *Science Progress* 32:58-69.
Emerson, R., and Arnold, W. 1932-1933. The photochemical reaction in photosynthesis. *J. Gen. Physiol.* 16:191-205.
Emerson, R., and Lewis, C. M. 1942. The photosynthetic efficiency of phycocyanin in *Chroococcus* and the problem of carotenoid participation in photosynthesis. *J. Gen. Physiol.* 25:579.
Evans, H. J. 1956. Role of molybdenum in plant nutrition. *Soil Sci.* 81:199-208.
Eyster, C. 1952. Necessity of boron for *Nostoc muscorum. Nature* 170:755.
Eyster, C. 1958a Chloride effect on the growth of *Chlorella pyrenoidosa. Nature* 181:1141-1142.

Eyster, C. 1958b. The microelement nutrition of *Nostoc muscorum. The Ohio J. Sci.* 58:25-33.
Eyster, C. 1959. Mineral requirements of *Nostoc muscorum* for nitrogen fixation. p. 109. Proceedings IX International Botanical Congress. Montreal. Volume II Abstracts.
Eyster, C. 1961. Relation of manganese to photosynthetic reduction of TPN. *Plant Physiol.* 36:suppl. iii.
Eyster, C. 1962. The requirements and functions of micronutrients by green plants with respect to photosynthesis. Biologistics for Space Systems. Papers presented at Wright-Patterson Air Force Base Symposium Workshop, May, 1962.
Eyster, C., et al. 1956a. The role of manganese in growth, photosynthesis, respiration and Hill reaction using *Chlorella pyrenoidosa* and spinach chloroplasts. *Plant Physiol.* 31:suppl. xvii.
Eyster, H. C., Brown, T. E., and Tanner, H. A. 1956b. Manganese requirement with respect to respiration and Hill reaction in *Chlorella pyrenoidosa. Arch. Biochem. Biophys.* 64:240-241.
Eyster, C., et al 1958a. Manganese requirement with respect to growth, Hill reaction, and photosynthesis. *Plant Physiol.* 33:235-241.
Eyster, H. C., Brown, T. E., and Tanner, H. A. 1958b. Mineral requirements for *Chlorella pyrenoidosa* under autotrophic and heterotrophic conditions. Chapt. 11, pp. 157-174. *In* Lamb, Bentley, and Beattie (eds.). Trace elements. Academic Press, Inc., New York.
Foster, J. N., and Denison, F. W., Jr. 1940. Role of zinc in metabolism. *Nature* 166:833-834.
Garner, W. W. 1935. Chemical elements essential for nutrition of plants. p. 25. U.S. Dept. Agr. Bur. Plant Indus. Rept.
Gauch, H. G., and Dugger, W. M., Jr. 1953. The role of boron in the translocation of sucrose. *Plant Physiol.* 28:457-466.
Geigel, A. R. 1935. Effect of boron on the growth of certain green plants. *J. Agr. Univ., Puerto Rico* 19:5-28.
Gerloff, G. C., Fitzgerald, G. P., and Skoog, F. 1952. The mineral nutrition of *Microcystis aeruginosa. Am. J. Botany* 39:26.
Gilbert, F. A. 1957. Mineral nutrition and the balance of life. University of Oklahoma Press, Norman, Oklahoma. 350 pp.
Gines, F. G. 1930. Relative effects of different iron salts upon growth and development of young rice plants. *Philippine Agr.* 19:43-52. *Experiment Sta. Rec.* 67:23.
Gorham, P. R., and Clendenning, K. A. 1952. Anionic stimulation of the Hill reaction in isolated chloroplasts. *Arch. Biochem. Biophys.* 37:199-223.
Granick, S., and Gilder, H. 1947. Distribution, structure, and properties of the tetrapyrroles. *Adv. in Enzymol.* 7:305-368.
Greenhill, A. W. 1938. Boron deficiency in horticultural crops: Recent developments. *Sci. Hort.* 6:191-198.
Hewitt, E. J. 1951. The role of the mineral elements in plant nutrition. *Ann. Rev. Plant Physiol.* 2:25-52.
Hewitt,. E. J. 1956. Symptoms of molybdenum deficiency in plants. *Soil Sci.* 81:159-171.
Hill, R., and Bonner, W. D., Jr. 1961. The nature and possible function of chloroplast cytochromes. pp. 424-435. *In* McElroy, W. D., and Glass, B., (eds.). A symposium on light and life. The Johns Hopkins Press, Baltimore, Maryland.
Hoagland, D. R. 1945. Molybdenum in relation to plant growth. *Soil Sci.* 60:119-123.

Hoch, F. L., and Vallee, B. L. 1958. The metabolic role of zinc. Chapt. 22. *In* Lamb, C. A., Bentley, O. G., and Beattie, J. M. (eds.) Trace elements. Academic Press Inc., New York.

Hodgson, J. F. 1955. I. The essentiality of sodium for plant growth. II. Quantitative determination of sodium with the flame photometer (Doctoral dissertation). University of Wisconsin.

Holm-Hansen, O., Gerloff, G. C., and Skoog, F. 1954. Cobalt as an essential element for blue-green algae. *Physiol. Plantarum* 7:665-675.

Hopkins, E. F. 1930. The necessity and function of manganese in the growth of *Chlorella* sp. *Science* 72:609-610.

Hopkins, E. F., and Wann, F. B. 1927. Iron requirement for *Chlorella. Bot. Gaz.* 84:407-427.

Hutner, S. H., and Provasoli, I. 1951. The phytoflagellates. Vol. I. pp. 27-128. *In* Lwoff (ed.). Biochemistry and physiology of protozoa. Academic Press, Inc., New York.

Ichioka, P. S., and Arnon, D. I. 1955. Molybdenum in relation to nitrogen metabolism. II. Assimilation of ammonia and urea without molybdenum by *Scenedesmus. Physiol. Plantarum* 8:552-560.

Ishibashi, H. 1937. The effect of silica on the growth of cultivated plants. V. The effect of silica on the growth of rice plants growing on soils of various depth. *J. Sci. Soil and Manure (Japan)* 11:535-549.

James, W. O. 1956. The respiration of plants. *Ann. Rev. Biochem.* 15:417-434.

Johnson, C. M., et al. 1957. Comparative chlorine requirements of different plant species. *Plant and Soil* 8:337-353.

Johnston, E. S. 1928. Boron: its importance in plant growth. *J. Chem. Educ.* 5:1235-1242.

Jordan, J. V., and Anderson, G. R. 1950. Effect of boron on nitrogen fixation by *Azotobacter. Soil Sci.* 69:311-319.

Jorgensen, E. G. 1955a. Variations in the silica content of diatoms. *Physiol. Plantarum* 8:840-845.

Jorgensen, E. G. 1955b. Solubility of the silica in diatoms. *Physiol. Plantarum* 8:846-851.

Kessler, E. 1957. Stoffwechselphysiologische Untersuchungen an Hydrogenase enthaltenden Grünalgen I. Über die Rolle des Mangans bei Photoreduktion und Photosynthese. *Planta* 49:435-454.

Kratz, W., and Myers, J. 1955. Nutrition and growth of several blue-green algae. *Am. J. Botany* 42:282.

Lehr, J. J. 1941a. The importance of sodium for plant nutrition. I. *Soil Sci.* 52:237-244.

Lehr, J. J. 1941b. The importance of sodium for plant nutrition. II. Effect on beets of the secondary ions in nitrate fertilizers. *Soil Sci.* 52:373-379.

Lehr, J. J., and Wybenga, J. M. 1955. Exploratory pot experiments on sensitiveness of different crops to sodium. C. Flax. *Plant and Soil* 6:251-261.

Lehr, J. J., and Wybenga, J. M. 1958. Exploratory pot experiments on sensitiveness of different crops to sodium. D. Barley. *Plant and Soil* 9:237-253.

Lemberg, R., and Legge, J. W. 1949. Hematin compounds and bile pigments. Interscience Publishers, Inc., New York.

Lewin, J. C. 1954. Silicon metabolism in diatoms. I. Evidence for the role of reduced sulfur compounds in silicon utilization. *J. Gen. Physiol.* 37:589-599.

Lewin, J. C. 1957. Silicon metabolism in diatoms. IV. Growth and frustule formation in *Navicula pelliculosa. Can. J. Microbiol.* 3:427-433.

Lill. J. G., Byall, S., and Hurst, L. A. 1938. The effect of applications of common salt upon the yield and quality of sugar beets and upon the composition of the ash. *J. Am. Soc. Agron.* 30:97-106.

Lingane, J. J., and Jones, S. L. 1950. Improved potentiostat for controlled potential electrolysis. *Anal. Chem.* 22:1169-1172.

Lipman, C. B. 1938. Importance of silicon, aluminum and chlorine for higher plants. *Soil Sci.* 45:189-198.

Lipman, C. B., and MacKinney, G. 1931. Proof of the essential nature of copper deficiency. *Plant Physiol.* 6:593-599.

MacVicar, R., and Burris, R. H. 1948. Relation of boron to certain plant oxidases. *Arch. Biochem.* 17:31-39.

Marston, H. R. 1952. Cobalt, copper and molybdenum in the nutrition of animals and plants. *Physiol. Rev.* 32:66-121.

Martin, G., and Lavollay, J. 1958. Chlorine, an indispensable trace element for *Lemna minor. Experientia (Basel)* 14:333-4.

Matthews, E. D. 1941. Evidence of the value of the sodium ion in cotton fertilizers. Georgia Agr. Exp. Sta. Circ. 127.

Maze, P. 1914. Influences respectives des elements mineral sur le developpement du mais. *Ann. Inst. Pasteur* 28:1-48.

Maze, P. 1936. The role of special elements (boron, copper, zinc, manganese, etc.) in plant nutrition. *Ann. Rev. Biochem.* 5:525-538.

McElroy, W. D., and Nason, A. 1954. Mechanism of action of micronutrient elements in enzyme systems. *Ann. Rev. Plant Physiol.* 5:1-30.

McHargue, J. S. 1922. The role of manganese in plants. *J. Am. Chem. Soc.* 44:1592-1598.

McIlrath, W. J., and Skok, J. 1957. Influence of boron on the growth of *Chlorella. Plant Physiol.* 32:suppl. xxiii.

McMurtrey, J. E., Jr. 1938. Distinctive plant symptoms caused by deficiency of any one of the chemical elements essential for normal development. *Bot. Rev.* 4:183-203.

Medina, A., and Nicholas, D. J. D. 1957a. A zinc-dependent hexokinase from *Neurospora crassa. Nature* 179:87-88.

Medina, A., and Nicholas, D. J. D. 1957b. Some properties of a zinc-dependent hexokinase from *Neurospora crassa. Biochem. J.* 66:573-578.

Mower, H. F. 1962. Central Research Department, E. I. du Pont de Nemours and Company, Wilmington, Delaware. Personal communication.

Mulder, E. G. 1948. Importance of molybdenum in the nitrogen metabolism of microorganisms and higher plants. *Plant and Soil* 1:94-119.

Mulder, E. G. 1950. Mineral nutrition of plants. *Ann. Rev. Plant Physiol.* 1:1-24.

Mullison, W. R., and Mullison, E. 1942. Growth responses of barley seedlings in relation to potassium and sodium nutrition. *Plant Physiol.* 17:632-644.

Myers, J. 1951. Physiology of the algae. *Ann. Rev. Microbiol.* 5:157-180.

Nason, A. 1958. The metabolic role of vanadium and molybdenum in plants and animals. Chapt. 19. *In* Lamb, Bentley, and Beattie (eds.). Trace elements. Academic Press, Inc., New York.

Nason, A., Kaplan, N. O., and Colowick, S. P. 1951. Changes in enzymatic constitution in zinc-deficient *Neurospora. J. Biol. Chem.* 188:397-406.

Neish, A. C. 1939. Sudies on chloroplasts. II. Their chemical composition and the distribution of certain metabolites between the chloroplasts and the remainder of the leaf. *Biochem. J.* 33:300-308.

Nelson, J. M., and Dawson, C. R. 1944. Tyrosinase. *Adv. Enzymol.* 4:99-152.

Noack, K., and Pirson, A. 1939. Effect of iron and manganese on nitrogen assimilation of *Chlorella. Ber. deut. botan. Ges.* 57:442-452.

Noack, K., Pirson, A., and Stegmann, G. 1940. The need of *Chlorella* for traces of elements. *Naturwiss.* 28:172-173.

Ondratschek, K. 1941. Mineral requirements of heterotrophic flagellates. *Arch. Mikrobiol.* 12:241-253.

Onslow, Muriel W. 1931. The principles of plant biochemistry. Cambridge University Press, London—New York.

Osterlind, S. 1950. Inorganic carbon sources of green algae. II. Carbonic anhydrase in *Scenedesmus quadricauda* and *Chlorella pyrenoidosa. Physiol. Plantarum* 3:430-434.

Ozanne, P. G., Woolley, J. T., and Broyer, T. C. 1957. Chlorine and bromine in the nutrition of higher plants. *Australian J. Biol. Sci.* 10:66-79.

Piper, C. S. 1940. Molybdenum as an essential for plant growth. *J. Australian Inst. Agr. Sci.* 6:162-164.

Pirson, A. 1937. Ernährungs- und Stoffwechselphysiologische Untersuchungen an *Fontinalis* und *Chlorella. Z. Bot.* 31:193-267.

Pirson, A. 1955. Functional aspects in mineral nutrition of green plants. *Ann. Rev. Plant Physiol.* 6:71-114.

Pirson, A., and Bergmann, L. 1955. Manganese requirement and carbon source in *Chlorella. Nature* 176:209-210.

Pirson, A., Tichy, C., and Wilhelmi, G. 1952. Stoffwechsel und Mineralsalzernährung einzelliger Grünalgen. I. Vergleichende Untersuchungen an Mangelkulturen von *Ankistrodesmus. Planta* 40:199-253.

Possingham, T. V. 1956. The effect of mineral nutrition on the content of free amino acids and amides in tomato plants. *Australian J. Biol. Sci.* 9:539-551.

Price, C. A., and Millar, E. 1962. Zinc, growth and respiration in *Euglena. Plant Physiol.* 37:423-427.

Price, C. A., and Vallee, B. L. 1962. *Euglena gracilis:* A test organism for study of zinc. *Plant Physiol.* 37:428-433.

Quinlan-Watson, T. A. F. 1951. Aldolase activity in zinc deficient plants. *Nature* 167:1033-1034.

Quinlan-Watson, T. A. F. 1953. The effect of zinc deficiency on the aldolase activity in the leaves of oats and clover. *Biochem. J.* 53:457-460.

Raleigh, G. J. 1939. Evidence for the essentiality of silicon for growth of the beet plant. *Plant Physiol.* 14:823-828.

Raulin, J. 1869. Etudes chimiques sur la vegetation. *Ann. sci. nat. Bot. 5 Ser.* 11:93-299.

Reed, H. S. 1946. Effects of zinc deficiency on phosphate metabolism of the tomato plant. *Am. J. Botany* 33:778-784.

Reisenauer, H. M. 1960. Cobalt in nitrogen fixation by a legume. *Nature* 186:375-376.

Reisner, G. S., and Thompson, J. F. 1956. The large scale laboratory culture of *Chlorella* under conditions of micronutrient element deficiency. *Plant Physiol.* 31:181-185.

Rosenberg, A. J. 1946. Action of boron and of m-inositol on *Clostridium saccharobutyricum. Compt. rend.* 222:1310-1311.

Rosenberg, A. J. 1948. Influence of *l*- and *d*-inositol and some pyrone derivatives on cultures of *Clostridium saccharobutyricum* inhibited by malonate. *Compt. Rend. Soc. Biol.* 142:443-444.

Scarisbrick, R. 1947. Hematin compounds in plants. *Ann. Rept. Chem. Soc.* 44:226-236.

Schwartz, M. 1956. The photochemical reduction of quinone and ferricyanide by lyophilized *Chlorella* cells. *Biochim. Biophys. Acta* 22:463-470.

Scott, G. T. 1943. The mineral composition of *Chlorella pyrenoidosa* grown in culture medium containing varying concentrations of calcium, magnesium, potassium and sodium. *J. Cellular Comp. Physiol.* 21:327-338.

Shaukat-Ahmed and Evans, H. J. 1960. Cobalt: a micronutrient element for the growth of soybean plants under symbiotic conditions. *Soil Sci.* 90:205-210.

Shaukat-Ahmed and Evans, H. J. 1961. The essentiality of cobalt for soybean plants grown under symbiotic conditions. *Proc. Nat. Acad. Sci.* 47:24-36.

Shive, J. M. 1945. Boron in plant life—a brief historical survey. *Soil Sci.* 60:41-51.

Sibly, P. M., and Wood, J. G. 1951. The nature of carbonic anhydrase from plant sources. *Australian J. Sci. Res. B.* 4:500-510.

Skok, J. 1958. The role of boron in the plant cells. Chapt. 15. *In* Lamb, C. A., Bentley, O. G., and Beattie, J. M. (eds.). Trace elements. Academic Press, Inc., New York.

Skoog, F. 1940. Relationships between zinc and auxin in the growth of higher-plants. *Am. J. Botany* 27:939-951.

Smith, L., and Chance, B. 1958. Cytochromes in plants. *Ann. Rev. Plant Physiol.* 9:449-482.

Sommer, A. 1926. Studies concerning essential nature of aluminum and silicon for plant growth. *Univ. Calif. Publ. Agr. Sci.* 5:2.

Sommer, Anna L. 1931. Copper as an essential for plant growth. *Plant Physiol.* 6:339-345.

Sommer, Anna L. 1945. Copper and plant growth. *Soil Sci.* 60:71-79.

Sommer, Anna L., and Lipman, C. B. 1926. Evidence on the indispensable nature of zinc and boron for higher green plants. *Plant Physiol.* 1:231-249.

Steeman, N. E., Nielsen, E., and Kristainsen, J. 1949. Carbonic anhydrase in submerged autotrophic plants. *Physiol. Plantarum* 2:325-331.

Stegmann, G. 1940. The significance of the trace elements for *Chlorella. Z. Botanik* 35:385-422.

Steinberg, R. A. 1936. Relation of accessory growth substances to heavy metals including molybdenum, in the nutrition of *Aspergillus niger. J. Agr. Res.* 52:438-448.

Steinberg, R. A. 1950. The copper nutrition of green plants and fungi. pp. 115-140. *In* McElroy, W. D., and Glass, B. (eds.) A symposium on copper metabolism. The Johns Hopkins Press, Baltimore, Maryland.

Stiles, W. 1961. Trace elements in plants. Cambridge Univ. Press, Cambridge, England.

Stout, P. R., and Johnson, C. M. 1956. Molybdenum deficiency in horticultural and field crops. *Soil Sci.* 81:183-190.

Trelease, S. F., and Selsam, M. E. 1939. Influence of calcium and magnesium on the growth of *Chlorella. Am. J. Botany* 26:339-341.

Tsui, C. 1948. The role of zinc in auxin synthesis in the tomato plant. *Am. J. Botany* 35:172-179.

Vallee, B. L., et al. 1956. Pyridine nucleotide dependent metallodehydrogenases. *J. Am. Chem. Soc.* 78:5879-5883.

Vallee, B. L., Stein, E. A., Sumerwell, W. N., and Fischer, H. 1959. Metal content of α-amylases of various origins. *J. Bio. Chem.* 234:2901-2905.

Viets, F. G., Jr., Boawn, L. C., and Crawford, C. L. 1954. Zinc content of bean plants in relation to deficiency symptoms and yield. *Plant Physiol.* 29:76-79.

Viets, F. G., Jr., et al. 1953. Zinc deficiency in corn in Central Washington. *Agron. J.* 45:559-565.

Voelcker, J. A. 1913. Pot culture experiments. *J. Roy. Agr. Soc., England* 74:411-422.

Wagner, F. 1940. Die Bedeutung der Kieselsäure für das Wachstum einiger Kulturpflanzen, ihren Nährstoffhaushalt, und ihre Anfälligkeit gegen echte Mehltaupilze. *Phytopath. Z.* 12:427-479.

Walker, J. B. 1953. Inorganic micronutrient requirements of *Chlorella*. I. Requirements for calcium (or strontium), copper, and molybdenum. *Arch. Biochem. Biophys.* 46:1-11.

Walker, J. B. 1954. Inorganic micronutrient requirements of *Chlorella*. II. Quantitative requirements for iron, manganese, and zinc. *Arch. Biochem. Biophys.* 53:1-8.

Wallerstein, M. 1909. U.S. Patent no. 905,029.

Warburg, O., and Krippahl, G. 1954. Photosynthesis enzymes. *Angew. Chem.* 66:493-496.

Warburg, O., Krippahl, G., and Buchholz, W. 1955. Action of vanadium on the photosynthesis. *Z. Naturforsch.* 10b:422.

Warburg, O., and Lüttgens, W. 1944. The assimilation of carbon dioxide. *Naturwiss.* 32:161.

Warburg, O., and Lüttgens, W. 1946. Photochemische Reduktion des Chinons in grünen Zellen und Granula. *Biokhimiya* 11:303-322.

Warington, K. 1923. The effect of boric acid and borax on the broad bean and certain other plants. *Ann. Botany* 37:630-670.

Waygood, E. R. 1950. Physiological and biochemical studies in plant metabolism. II. Respiratory enzymes in wheat. *Can J. Res., C.* 28:7-62.

Waygood, E. R., and Clendenning, K. A. 1950. Carbonic anhydrase in green plants, *Can J. Res., C.* 28:673-689.

Waygood, E. R., and Clendenning, K. A. 1951. Intracellular localization and distribution of carbonic anhydrase in plants. *Science* 113:117-179.

Wood, J. G., and Sibly, P. M. 1952. Carbonic anhydrase activity in plants in relation to zinc content. *Australian J. Sci. Res.* B5:244-255.

Woolley, J. T. 1957. Sodium and silicon as nutrients for the tomato plant. *Plant Physiol.* 32:317-321.

Wybenga, J. M., and Lehr, J. J. 1958. Exploratory pot experiments on sensitiveness of different crops to sodium. E. Red table beet. *Plant and Soil* 9:385-394.

Yamagata, N., and Murakami, Y. 1958. A cobalt-accumulator plant, *Clethra barbinervis* Sieb. and Succ. *Nature* 181:1808.

Young, R. S. 1956. Cobalt in biology and biochemistry. *Science Progress* 44:16-37.

Some Problems Remaining in Algae Culturing*

A. G. Wurtz

Director, Station d'Hydrobiologie Appliquée
Le Paraclet par Boves, France

The subject of algae culturing is one so large that even a book would not exhaust our knowledge about the results so far reached. Therefore, I shall point out certain modern developments which either have not received a satisfactory answer, or are still creating some confusion. Among such problems are: the estimation of growth or yield; different methods of purification to obtain bacteria-free cultures; the influence and the size of the inoculum on the growth of the culture; more or less pure cultures of algae used as indicators for ecological water properties or value of the water supply at all; algae considered as purifiers of industrial wastes and sewage; and algae and pollution control in general.

1. METHODS OF EVALUATING THE GROWTH OF ALGAE CULTURES. ESTIMATION OF THE YIELD

In order to understand the evolution of the methods of measurement in this field of work, let us only remember that in the early years of algal culture authors worried above all about the speed at which the obtained culture was working, and what the "energy cost" of the whole operation was. Their aims being only systematic or morphological, they were not concerned about the purity of the cultures. Even ecological data valuable in the relationships between organisms could be obtained with these non-bacteria-free cultures. Then came the time when bacteriologists introduced their techniques into algology. The great names of the so-called "pioneers" appeared—Beijerinck, R. Chodat, Molisch,

*This paper was presented at the NATO Advanced Studies Institute by Dr. T. T. Macan, Deputy Director, Freshwater Biological Association, Ambleside, England.

Benecke, Klebs, Pringsheim. The last cited wrote (in 1945): "Progress in the study of algae will largely depend on the use of the existing methods of culture and their successful modification for special purposes. . . ." In recent years, we have seen authors giving up the pure cultures and returning to the ecological multi-species culture. If we find the time, among the opinions we shall have to give, we shall at least say what can be expected from both these old and new methods in the near future

As could be expected, the easiest method at hand was to count the cells of the algae. By taking care to have the rate of numeration a little less than the ordinary rate of division of the cells, the results proved valuable, in spite of a climate of deficiency in which the first algae numerators lived. It could be rather well established that many algae showed sigmoid, exponential (plotted against time) growth curves; even better, the data could show that the curve was represented by the equation characteristic of auto-catalyzed, monomolecular reactions, that have described the growth of population and of individual organisms.

The differential form of this equation expresses the rate of growth and should be written:

$$\frac{dcs}{dt} = kx(A - x)$$

Upon integration it becomes:

$$\log \frac{x}{A - x} = K(t - t_1)$$

where x represents population per cubic millimeter at time t; A represents the maximum number of cells attained in the same unit volume; t_1 is the time when $x = A/2$; and $K = kA/2.3$. K is inversely proportional to the duration of the growth period. To give an example: By using this technique, the regular decrease of the growth rate during nearly the entire period of growth allowed Pratt to suggest the production of a growth-inhibiting factor by the cell of *Chlorella vulgaris,* which was later (in 1940) called chlorellin.

Counting of cells by the indirect way of the growth rate has also been successfully used in continuously run, large-scale cultures, especially with *Chlorella.* More generally, growth rate refers to the amount of growth in a culture per unit of time. In a batch culture, growth-rate is measured in terms of increase in cell-count per unit volume per unit time. In the given case, in a continuous run system where population density and other variables are held constant, growth rate, expressed in liters of overflow of culture per liter per day, is measured by the amount of dilution (or the amount of overflow of culture obtained due to the dilution) necessary to maintain constant population density for a constant volume of culture. In a continuous system which maintains culture conditions, the growth rate is directly proportional to the rate of reproduction and can be defined as:

$$K_d = \frac{V_o}{V_c}\frac{1}{\Delta t_D}$$

where K_d = growth rate (for one day);
V_o = volume of overflow in liters;
V_c = volume of culture in liters;
Δt_D = time in days (24 hr);
Y_d = yield (grams dry weight per liter per day);
D_c = population density (grams dry weight per liter).

At the same time the definition of the yield can be given. The yield (grams per liter per day) is the product of the growth rate (liters per liter per day) and the population density (grams per liter). In the formula, yield is defined as:

$$Y_d = \frac{V_o}{V_c}\frac{D_c}{\Delta t_D} \qquad \text{or} \qquad Y_d = K_d D_c$$

The volume of overflow is constant with time and the slope of the line is the growth rate on an hourly basis. Growth rate (K_d) is equal to 1.33. The maximum production obtained from such a system is 0.48 g dry weight/liter each day.

Having given these short definitions, I shall briefly discuss two additional methods used for determining algal growth in cultures. The first of these is the measurement of the turbidity of the culture—or optical density (percentage of light transmitted)—method, which has been very largely used by many authors (I myself considered it as my main method); the second is the colorimetric determination of the chlorophyll content of the cells in culture (determined generally in a given wavelength, mostly red). Since these two methods give numbers as measurement, or a biochemical value of the chlorophyll, they are seemingly very accurate and reliable. However, they suffer somewhat from the fact that no clear and constant relationship has ever been demonstrated between the growth rate and the growth, on one hand, and the optical density or the accumulation of chlorophyll, on the other. I do not mean to advise abandoning either one of these two last methods of measurement; I mean only to recommend care in interpreting the results obtained. Besides, a dry weight terminating a growth curve (measured in turbidity) would always remain a food measurement, and may be kept until other methods of measurement are discovered, or at least standardized.

The conclusion to be drawn from the foregoing is that we are unfortunately still lacking good standarized methods of measuring the growth of algae in cultures. No doubt if the progress is the same in the next 50 years as it has been in the last 50 years, our efforts may be successful.

2. DIFFERENT METHODS OF PURIFICATION IN ORDER TO OBTAIN BACTERIA-FREE ALGAE CULTURES. HOW SOME OF THESE METHODS ALLOW AN UNDERSTANDING OF THE ECOLOGY OF ALGAE

Every phycologist is aware of the difficulties of establishing pure cultures. Therefore, once obtaining one culture of algae does not mean anything if the results are not worked out in order to understand the ecology. It is not necessary to insist again on the need for bacteria-free algae cultures if physiological work is to be undertaken, and if one does not want, at the very beginning of the experimental cultures, to work with many unknown substances.

The achievement of pure cultures of algae must not simply be a question of agility and nimbleness of the fingertips of the operator. Most often, however, such is the case. I personally have nothing against the pipetting and washing methods, and I admire an operator who obtains a pure culture by washing ten times under the binocular or the microscope a few cells at the beginning and one cell at the end of the purification, as well as another who obtains the same result by irradiation of the mass culture with ultraviolet light. The significance of obtaining a pure culture, however, depends on what is done afterward with that culture. The most important point I would like to make in this section is that it seems to me that work becomes useless if one stops at the pure culture itself, and if achieving a pure culture is the only aim. In order to make things clear, there should be two specialized aspects of the question: On one side, there should be the laboratories who obtain the pure cultures by their own work and the creation of vast centers of pure culture-collections of algae; on the other side, there should be the workers who have ideas and make their experiments with the material they have at hand. It is promising to see that, despite the difficulties inherent in such questions and despite the nonexistence of an international "brain-trust" which would coordinate all the work done in the world, we have already reached a certain stage where the work can be carried on in this way.

If this NATO Conference is to reach conclusions about the increasing interest in the knowledge of the relations between Algae and Man, perhaps we, all of us, might come to the conclusion that there are not yet enough people working on algae in the world. That is at least the impression I have from what I know of Europe, and especially of France.

Another example, illustrating how methods of isolation may improve ecological knowledge, is the story of three very important blue-green algae, planktonic ones, which seemed impossible to grow in culture. They are *Microcystis aeruginosa, Gloetrichia echinulata* and *Aphanizomenon flos-aquae*, obtained by Gerloff, Fitzgerald, and Skoog in the years 1948-1949. These algae are "top algae" in fish farming because they produce water-blooms in fish ponds and were in lakes where the waters are considered as highly productive not on account of the direct consumption of these algae by fish or other animals, but on account of the indirect utilization of the products of their decomposition. Until they came into pure culture (which seemed very difficult), it was suggested that perhaps they needed high concentrations of organic nitrogen, or perhaps mineral nitrogen or phosphates, or unknown organic compounds. I was always interested in these three algae and I knew how to produce water-blooms of at least the first two mentioned by introducing mineral manure of phosphates into ponds. Here, however, was where the discrepancy began. When Gerloff, Fitzgerald, and Skoog obtained these three algae in bacteria-free cultures, it was in one of the well-balanced solutions published by Chu, entirely inorganic, with a special amount of inorganic nitrogen, and without any organic compound. The unexpected result was that it is the inorganic nitrogen which is important in cultures; and in ponds, fish farmers obtain the water-blooms of these three planktonic algae by introducing only phosphates. We must admit that the cultures, despite the wonderful results obtained, do not give a completely satisfactory answer to the question of the ecological or environmental requirements of these algae.

I do not want to leave the subject without adding that I might have found the answer by another approach. As a matter of fact, after I became a specialist in the mud and water exchanges in shallow waters, I often had the opportunity to confirm that the introduction of phosphates into the water first raised the pH of the mud, accelerated the reactions of decomposition in the mud, and liberated from the mud not only inorganic but also organic nitrogen. When I was on an F.A.O. United Nations expert assignment in East Africa about three years ago, I used to produce water-blooms of *Microcystis flos-aquae,* in order to increase the production and found in the open water an increase of phosphates coming from the mineral manure, but also an increase of dissolved organic nitrogen. I think that the answer is here, since it has been confirmed even in tropical ponds. We must admit that the three algae are able to grow well in pure mineral nutrient solutions, but the answer to the ecological background is not completely satisfactory if we consider the inorganic nitrogen as the only important factor and if we try to transpose the results from cultures into natural conditions. The interpretation of the results is even more subtle. There are factors of replacement possible, e.g., dissolved organic nitrogen in place of the inorganic. That is why the mud is able to release so many substances, giving the impression that phosphates are able to produce nitrogen and, consequently, the water-blooms in question. And there remains still more work to be done on these algae, in cultures and by taking more account of the ecological environmental conditions.

It is appropriate, before concluding this section, and speaking a little more of the mud, to raise the question of the "soil-water technique" extensively used by Pringsheim for growing sensitive algae which would not have grown otherwise. I am much indebted to Professor Pringsheim for all I learned from him. Perhaps the existence of these soil-water cultures is one of the reasons why I wanted to specialize in the exchange processes between mud and water, trying to explain environmental requirements of algae, which could not have been explained otherwise.

The soil-water technique, which is not considered as a substitute for pure cultures, is therefore a unialgal culture for algae which need growth factors, micronutrients, vitamins, and perhaps other organic substances which are not destroyed by the tyndallization (heating in running stream). As the technique does not require bacteriological sterility, it is mainly described as a preparatory method before final purification. The principal interest of the method lies in the fact (as Pringsheim writes) of the imitation of natural conditions and the provision in the test tube of a miniature artificial pond. That is why the cultures contain the organic substances which keep indispensable elements, especially heavy metals such as iron and manganese, in solution. In the soil-water cultures the mud phase serves as an accumulator and a place of reduction and synthesis. More than 500 strains, especially fragile Euglenophyta, Dinophyceae, Cryptophyceae, etc., have been isolated and kept by Pringsheim, with great success. Even to the physiologist, and as already pointed out to the ecologist, these cultures offer new problems but also a new road of approach.

A new series of problems has been posed by my efforts to find a simple method of isolation of algae, especially the method of growing cultures in deep agar, a technique inspired from bacteriology, i.e., growing and isolation of anaerobic bacteria in deep agar. It consists of boiling the organic medium for half an hour, in order to expel all the oxygen present, and of inoculating the bacteria by dipping a glass wire (closed at one end) into twelve successive tubes, without reloading the glass wire between each operation. Since the loop is washed in the semifluid agar in the first tubes (from 1 to 6, let us say), there is a good chance that only a few cells will remain in the last tubes. These cells are allowed to grow and to form small colonies which are isolated with the help of a very fine glass pipette by cutting the tubes and the agar just below them. In this way it is possible to obtain a pure culture of anaerobic bacteria within three or four days.

Initially I used the same technique for algae with inorganic media, but taking great care not to boil the agar, even letting the tubes rest a few days after sterilization in order to obtain the maximum of oxygen diffusion into the depth of the agar. This technique worked very well and I was able to purify or isolate clonic strains (coming from one cell) of several cultures of algae. The idea then occurred to me of boiling the agar, preparing in this way an anaerobic medium. Surprisingly, a few algae grew in the medium, supposed to be without oxygen. What had happened? Did the algae really grow in an anaerobic medium, did the operator introduce a little oxygen by the inoculation, or did the first alga cell produce enough oxygen to allow the start of growth? Now the problem was seriously put; I could not get rid of it, and as yet I have not found a satisfactory explanation. On the contrary, this strange technique resulted in my becoming more and more closely engaged in mud problems and trying to understand better the relations between algae and mud. How deep could algae penetrate into the mud? And what about the darkness? Everybody knows that many algae are able to grow in the darkness, when they have at their disposal an organic source of carbon. But what about the oxygen? In my cores of mud and water samples, kept in the laboratory in the daylight for study, I found blue-green algae (such as Oscillatoria) at a depth of 12 to 15 cm in the mud, together with sulphur-reducing bacteria (purple like Chromatium) at a potential of oxydoreduction measured as low as -250 mv (at pH 7.0). Again we have seen how an isolation technique served to raise a multitude of particular questions—on the relations between algae living on the top layers of mud, those living a little in the mud and the water immediately above, either in ponds or lakes or in flowing waters, clear or polluted.

Before turning from this section on methods of purification and their possible relations to ecology, I want to mention three final methods which seem of great interest, although I have never practiced them and am therefore not competent to discuss them in any detail. They are: the method of purification by ultraviolet light (in which dilute suspension of algae is placed in a quartz-windowed chamber and irradiated for 20 to 30 min with 2750 A ultraviolet light from quartz-jacketed mercury vapor lamp; generally the algae are more resistent than the bacteria to the ultraviolet); the method of washing the algae with dilute mineral solutions (again, everything could be tried which would be more toxic to bacteria than to the algae); and the method of trying to kill the bacteria by antibiotic or bacteriostatic substances.

3. INFLUENCE OF THE SIZE OF THE INOCULUM ON THE FUTURE GROWTH OF THE CULTURE

One thing is certain, the amount of inoculum does not influence the growth of the culture, or the final number of cells obtained, or the yield, all conditions remaining the same. And it is well that this is so, as expected; otherwise it would never have been possible to deduce a mathematical formula for growth. Many experiments have been done in order to prove this fact. The only thing on which the final number of cells depends (all other conditions being the same, I repeat) is the amount of food, or mineral salts, which is available to the algae. What varies is the slope of the curve, i.e., the rate of multiplication throughout the growth period, indicated by the velocity constant.

For example, a *Chlorella* growth curve which I observed showed the following figures: With an initial inoculum of 0.1, 1, or 100 cells/mm^3, the maximum of cells (very nearly 100,000 cells/mm^3) was reached, respectively, after 330 hr (roughly 13 days), 240 hr (10 days), and 160 (roughly 7 days). The necessary quantity of inoculum to be used depends on what we want to know: If we want a quick answer, we shall inoculate a big inoculum from the mother culture. If we have time and want to follow the growth curve, we shall inoculate with a small amount of inoculum. Whatever happens, if the environmental conditions do not vary, the total number of cells reached at the end of the culture will be the same.

This technique could be used for toxicologic research on algae, i.e., the effect of wastes, pollutants and all sorts of products coming from man's industry or agriculture on the algae of a river, where they are considered the best self-purifying agents in a stream. It occurs to me that mother cultures of the algae living in the given conditions could be kept somewhere in the main laboratories beside the main rivers. And then, in order to hold a pollution control, cultures could be inoculated, composed half of the medium suitable for algae and half of the stream water before pollution, and then diluted with wastes in different ways: before treatment, after treatment, etc. The question is to know whether the cultures have to be kept bacteria-free, pure, or unialgal. For the first step, even though the stream water could be sterilized it would not be necessary, and I would not recommend it because the waste, which may contain toxic organic and other substances, is not to be sterilized since it is not in nature. The initial objective would be to make a sort of survey of what the algae can tolerate, and what they can purify. Afterward, going deeper into research work, the cultures could be completely sterile. One must not forget that algae are the best helpers to men in the self-purification of a stream. And that wonderful power must not be lost. I wanted just to go a few steps forward, and indicate the problems and possibilities. We have not yet started this research work in France, despite the fact that pollution is becoming more and more severe.

4. CULTURING OF ALGAE USED FOR ECOLOGICAL PURPOSES

Before offering any thoughts about the need for bacteria-free (pure) cultures or not in ecological research, I would like to point out that what I am going to say is only my personal opinion. I hope, however, I shall agree with the majority, in the actual needs of our researches. If we do algae culturing in a laboratory, it is difficult to imagine them not bacteria-free, pure, and the algae being tested successively; otherwise the experiments could never be reproduced and the results could not be multiplied. Only then are pure cultures possible. That is one of the reasons why algal culture started suddenly and progressed, when workers with bacteriological formation brought their techniques. One can say that algae culturing really started around the year 1900 with the bacteriological background, i.e., by eliminating the *Bacteria*. Now, more than 60 years later, many authors are thinking that we cannot make any more progress, at least in ecology, without joining up the bacteria-free cultures, which put algae in conditions which are far from natural. Can we really do that after having worked properly for 70 years? These two points of view demand discussion.

a. When an alga is removed from its natural environment to start a pure culture, it is put in an absolutely new way of life. If some conditions such as temperature and light can be easily reproduced, others such as nutrition interrelations with other organisms, or space, cannot. Then why insist? There are conditions in a laboratory which are quite different from those prevailing in ponds, lakes, rivers, and streams, and we shall perhaps never be able to make the transition from one to the other. Of course we feel that it is a nice success when an author succeeds in getting pure cultures of planktonic algae, which just mean a question of space. In fact, in nature, it is not so much the space which is important as the easy replacement of food requirements, and space allows just such an easy replacement. That is why it is astonishing that so many authors have been able to isolate and grow true planktonic algae. As new media are found, the growth of benthic algae should also become increasingly easier. That is why, until we have made more progress on the interreactions of organisms in the water, we must continue to use pure bacteria-free algae cultures.

b. By starting the argument in a different way, what is our aim? We want to study a whole by splitting it into its elements, and we do not know how big these elements may be. There is no discussion even possible: They must be the smallest possible. They must be the species of alga in which we are interested, not influenced by the organisms living at the same level as algae; by organisms living on a level below, as bacteria, but which may be the strongest competitors; nor by the animals living on a level above. The only studies we can make in this field are those on the influence of the chemical composition of the medium on one alga. By all experimental means, we may change the composition of the medium, and other physical or chemical factors which may affect the growth, but we may do it with only one alga. Difficult though this may be, one alga must be tested after another, even if we feel that this will take more than 50 years.

Then later, when all the elements are at hand, we can try to re-create something that resembles a natural water by putting one well-known alga with another, then three, four, five, and then introducing one or two bacteria. That is the program. Of course we are well aware of the difficulties: The bacteria are strong competitors and develop generally much more quickly if the medium contains the smallest amount of organic substances, frequently overgrowing the algae. We must learn to work slowly and accurately. Has there ever been any worker who would say that culturing of algae was not a matter of patience? When the moment has come, we should call this new science "constructive synecology."

An approach to this work was undertaken at Windermere in the English Lake District by D. Cannon, J. W. Lund, and J. Sieminska in 1961. The project dealt with the growth of *Tabellaria flocculosa* var. *flocculosa* (Roth) Knuds under natural conditions of light and temperature. I would like to emphasize a few points of this work, not to summarize it, but to demonstrate the potential value of an ecological work based on algae cultures realized in glass-stoppered bottles, which, after starting growth in the laboratory under artificial light, are suspended in deep lake water.

The number of cell divisions expresses the growth rate. The best medium, after many attempts to grow the diatoms (which are, as everybody knows, mostly very difficult to grow), was found to be Chu's medium No. 10 with 0.2‰ extract of Windermere mud taken from a depth of 30-60 m. After suspending the cultures of *Tabellaria flocculosa* in the lake, the influence of light, temperature, and depth (the mineral composition of the nutrient medium remaining the same) was studied in detail. Thus, for example, growth took place between limits of temperature of 19C (which was the highest reached in mid-September) and 4C (below which no growth or cell division took place). During the entire year, the maximum growth was not always attained at the highest level of exposure (0.5 in. below the surface), but depth profiles showed that a maximum was always attained in the upper 2 m. No growth took place at any time at depths of more than 8 m. Growth rates of six cell divisions carried out in the population were considered good. Two cell divisions were considered poor. Differences in growth equivalent to less than half a cell division by the population considered as a whole could not be reliably distinguished.

The vast interest of experimental work based on such a pattern is evident, putting as it does algal cultures produced out of a natural medium (here a lake) at the disposal of ecology, in order to explain the ecological phenomena. Therefore let us only remember the method of suspended bottles of pure cultures of algae, in order to measure the growth of plankton algae, or even to do physiological work, such as measuring photosynthesis at different depths, at different temperatures. This method should be applied to the study of streams polluted by industrial wastes where evidence should be gathered on how turbidity and large temperature changes affect (generally adversely) the growth of useful algae. Methods of taking account of the changes in chemical composition in a stream, after the inflow of pollutants, have not yet been worked out because these changes may vary in myriad unknown ways; but their use should be kept in mind, despite the difficulties.

5. ALGAE CONSIDERED AS PURIFIERS OF INDUSTRIAL WASTES AND SEWAGE WITH THE HELP OF CULTURES

One of the modern developments in the utilization of algal cultures is the consideration of algae as a help for man in the purification of wastes and sewage before these are released into streams. The work thus started would be achieved in streams by the power of self-purification of the streams, which is also the final work of algae. Of course it could only be a question of inorganic solutions, containing only small amounts of high organic solutions, in the presence of which algae could not live.

As an example, may I cite one case not far from my Hydrobiological Station in the northern part of France: the pollution of the Somme River by the domestic and industrial wastes of the town of Amiens. Between the exit of the Somme River from the town and a point a few miles downstream there is a zone of dangerous pollution in which the water contains more than 200,000 bacteria/ml, besides *Escherichia coli* and *Enterococcaceae,* which render the water dangerous to man. There are no algae. Then, after the ordinary decompositions, the dangerous pathogenic bacteria disappear and are replaced by others unwelcome in rivers such as the filamentous *Sphaerotilus natans* and the filamentous sulfur bacteria *Beggiatoa alba,* the presence of which shows that the water contained traces of H_2S. Then algae appear and the struggle begins. I count them by suspending small plastic slides in the water in an effort to determine the relationships among them, and which ones are dominant. After about 10 miles the algae have performed their work; they are now dominant and have eliminated the bacteria which, encountering more organic substances, have disappeared. By doing more and more photosynthesis in the stream, they have purified the water, even producing supersaturation.

On the basis of this example, I can imagine some very large open-air algae-culturing device, a sort of purification station in which the organic wastes would first be decomposed or reduced by some means, and would then flow, after being supplied with oxygen, through long open-air water pipes, be enriched in the adequate algae cultures, and achieve their purification by becoming only algae cultures. I have no idea of the cost of such an operation; I do not even know if it would work or if anybody has ever tried to put it into practice with a pilot plant. It is still a theoretical idea of mine, supported by the well-known purification effect of algae, or the self-purification power of streams due to the work of algae.

The other idea, in the same category, has already entered the experimental algae-culturing phase—the purification of inorganic wastes. We have been asked at my station by the Atomic Energy Commission to initiate a project on the purification of wastes coming from the extraction of uranium metal from ore. Modern industries are concerned with extraction of the greatest amounts of ammonium nitrates from their wastes, and we had to find the best algae which would be able to do this properly. Since I knew that algae might be able to utilize ammonium nitrate, because I had cultures growing in media having this mineral salt as a nitrogen source, I accepted the challenge, and the experiments began. Many algae have been tested: *Volvocales* (mostly *Chlamydomonas*), *Chlorococcales* (mostly *Ankistrodesmus*), *Scenedesmus* and *Pediastrum* and many diatoms. *Pediastrum* and all the diatoms tested were unable to utilize even the smallest amounts of ammonium nitrate. *Ankistrodesmus* utilized it slightly, *Scenedesmus* a little more, but not enough to be of interest in technical work. So far, *Chlamydomonas* has been found to be the best utilizer, being able to utilize up to 1 g ammonium nitrate liter during a three-week period. This was very promising and an unusually high amount for an alga.

It was easy, after this experiment, to imagine vast culture tanks in which, by means of appropriate illumination, *Chlamydomonas* would destroy enormous amounts of ammonium nitrate. When we learned the exact amount of ammonium nitrate which had to be destroyed (6 tons dry weight daily), however, we were less sure of the result and a little disappointed. Despite these difficulties, the work goes on and we hope to have satisfactory results shortly. Why should we not be able to start and control cultures of algae on a very large scale, such as 10,000-liter tanks, and more? In any case, despite the difficulties due to the fact that they need illumination, a start has been made in finding algae cultures able to effect an appropriate waste treatment, at least in some wastes which have only inorganic composition.

6. ALGAE AS INDICATORS OF WATER QUALITY AND PRODUCTION

There is one area where our knowledge about the environmental requirements of algae must be completed by culturing—algae as indicators of water quality, qualitatively (but then we must know exactly what their presence represents among the ecological conditions) and quantitatively. Too, since the amount of algae is important for the final production of proteins such as fish, we must know the place of the given algae as a link in the food chains. This second problem will be treated in the last section of this paper.

For the first problem, allow me to describe an example from my experience which, even if the necessary explanations are somewhat complicated and lengthy, I believe you will find as striking as it was for me the first time I encountered it.

When I arrived in East Africa and first saw the fish farm where I was to study the ecology of fish ponds, fish food, water chemistry, etc., I was struck by the red color of many of the small, shallow fish ponds. It was water-bloom, and the fisheries-officers and people concerned with fisheries who were present already knew that the ponds in question were "doing well." As soon as I got a microscope, I put a name to that water-bloom. It was the well-known *Euglena sanguinea*. The red water-bloom appeared regularly every two or three weeks, remained three or four days, then disappeared. These phenomena meant that a whole chemical and biological life cycle was accomplished in a fortnight, i.e., bacteriological transformations and decompositions, mineralization, development and growth of the algae. *Euglena* came to the surface only when the time had come to produce a water-bloom or, better to say, to conduct photosynthesis. The speed of these transformations was astonishing; we could only guess what happened, and could not explain the complicated biochemical reactions which took place in the mud or in the water.

I was even more surprised when a few months later we had a visit from the Fisheries Adviser, who knew about techniques in the Far East. When I told him about the curious problems of the water-blooms, he told us (without knowing the name of the alga) that the Chinese fish farmers had known from early times 2000 years ago that the ponds were "red flowering" twice a month or a moon. They could expect at these times a very good production of Tilapia, a fish that eats higher plants and has no relationship with the water-blooms at all. Without knowing even that algae existed the Chinese knew how to utilize water-blooms as indicators, and to make empirical estimates of the production of fish.

We do not know much more now. *Euglena sanguinea* remains a good indicator of high production, but we do not know why. Only cultures of *Euglena sanguinea* could solve the problem. And that is why the need of working together with the same aim, but with three different disciplines, appeared: Nobody could have sufficient knowledge of the three disciplines—1) algae culturing; 2) chemistry and especially biochemistry of the mud and the water explaining the ecological needs of an algae to grow in nature or in culture; 3) bacteriology of the mud and the water, and chemistry of the fermentations and transformations—necessary to explain why and how given chemical, mineral, or organic substances appeared. That is teamwork which could not be easily started anywhere. But, in order to improve our knowledge of algae, that teamwork could perhaps be suggested for future work.

Let us imagine for a moment, as in a dream, that there exists somewhere an electronic calculating machine into which all the data—taxonomic, ecological, and from cultures all the characteristics such as optimums of temperature, of oxygen for development, production of oxygen, environmental requirements, needs for nutrients, needs for every chemical substance entering in the life cycle, optimums of everything—could be fed to be digested and filed on perforated cards, and that it would only be necessary to press a button to have at once all these characteristics before our eyes. Then it would be easy to explain why *Euglena sanguinea* comes and goes, what is the factor responsible for its appearance, death, etc., and why the fish ponds in which it produces water-blooms are very productive, which is the only thing of which we are sure. Unfortunately it is no more than a dream. Our knowledge is not yet so far advanced; such a machine does not exist; and even if it did we have not enough data to feed it. In reality, how many algae exist whose characteristics are *completely* known and could be utilized this way by an electronic synthesizing machine? Not even 10, and there should be 10,000. What has been done until now in algal culturing and ecology is nothing compared to what still remains to be done.

Another field of research where an approach should be made to a better understanding of algal ecology is the culture of algae which could be indicators of pure water supply, such as diatoms living in pure, cold running waters in the mountains (*Ceratoneis arcus, Meridion circulare, Diatoma hiemale, Eunotia* sp. plur., *Rhoicosphenia curvata*) and blue-green algae such as the group of *Chamaesiphonales* living in pure running waters, mostly rich in calcium carbonate. A good knowledge of the nutrient and environmental requirements of these algae, obtained from cultures, would allow us to fix levels and limits of all sorts of water supplies, for all purposes, industry and even drinking water. Everybody knows that it must be very difficult to grow such cultures in pure, running, cold water and no author speaks of having had success in such an attempt. Nobody even speaks of having tried or failed. Nowadays, technical difficulties should not stop us. Since an important organism living in polluted, running waters, the filamentous bacteria *Sphaerotilus natans*, has been cultured in continuous running (pure, bacteria-free) cultures, there should be no major difficulty in obtaining cultures of diatoms and blue-green algae of running, pure waters. And what an approach this would be to new problems such as pure water supply and control, in relation to the needs of man.

7. ALGAE CULTURING IN ORDER TO UNDERSTAND HOW ALGAE ARE LINKS IN THE FOOD CHAINS

There are many algae which are a nuisance in natural waters. Others are useless: Though they produce some oxygen during their life, they become dangerous after their death by the oxygen absorption they may suddenly occasion, and the production of mud by their mass development. Most, however, are useful, being at the beginning of life in the waters, as food for every kind of animals living in the waters. Let us explain.

Algae are primary producers, and all the animals live and feed on them, or sometimes on their detritus. Which are the animals which are supposed to feed on algae? There are *Infusoria,* then higher, but still small and transparent, animals such as *Rotatoria, Cladocera, Copepoda,* etc., small worms living in the mud, and then fish fry and even full-grown vegetarian fish, grazing on microscopic, filamentous algae. The best way to make sure if algae are food for these animals is to have bacteria-free cultures of algae ready and put them in an as sterile a way as possible in the presence of the organisms. Let us suppose, for instance, that we want to test a pure, bacteria-free culture of *Chlamydomonas* against a Rotatorian such as *Brachionus.* Into a fully grown, bright green culture of *Chlamydomonas* we introduce about 10 to 20 sterile washed *Brachionus,* or grown in sterile cultures from the egg, and wait. If the *Rotatoria* behave well and if they multiply, that means that the algae are a good food for them; this can be confirmed under the microscope by observing whether the green algae fill the digestive tract.

This should be repeated with all sorts of algae, and all sorts of animals possible, in order to understand the food relationships between them. Algae are the first step in the food chain; they are the first link between inorganic substances and organic life; they are the primary producers, while the animals, which are the consumers, depend upon them. It is unnecessary to repeat how important it is to operate with pure cultures of algae, and even more, we would like to add that bacteria-free cultures of algae are the only way to solve the problems of food-chain relationships between producers and consumers.

Things are a little different with bigger animals which cannot be grown in pure cultures, and which live in running waters, such as insect larvae (*Ephemeroptera, Plecoptera, Trichoptera,* etc.) grazing on algae covering the stones where they live. Here it would be necessary to get the algae from the digestive tract by dissection and to put them in the adequate culture solutions, in order to see which algae will grow and to distinguish: 1) those which have not been ingested; 2) those which have been ingested but not digested (and should grow); and 3) those which have been digested (the cell walls only remain or they have disappeared between the ingestion and the rectum). It is easy to understand that working this way is still hard work, rendered even more difficult by the fact that some of the diatoms which have to be studied are known to be hard to culture. For this reason very few workers have successfully ventured to attack this problem of food relations.

When we realize that algae may be food for *Infusoria,* then for nearly all the aquatic animals from zooplankton to small benthic animals, for small fish fry, for bigger full grown fish, and even for men, we see the amazing world algae open to us, and the huge possibilities they offer. And whatever the problems are, in order to understand that world, or to get at least a better approach, algae cultures are necessary in most cases. The time is past when cultures serve as an end in themselves, allowing the taxonomist to amuse himself by preserving for a short time the strain he likes. Now, cultures of algae have become a fundamental need and an excellent tool in all fields of modern research on waters, whatever they may be.

The Ecology of Benthic Algae

F. E. Round

Department of Botany
The University
Bristol, England

From the earliest years of the study of algal ecology, the community which has received most attention is the phytoplankton. A considerable number of descriptive studies have resulted in a mass of data relating to the taxonomy and distribution of species, seasonal cycles and to water chemistry; now experimental studies are leading to a better understanding of the growth characteristics of the species and community. Yet, even after more than a century of work, new species are still to be found, while few detailed studies of the biology, physiology, or biochemistry of individual species have been undertaken. This preponderance of effort on the one community is hardly surprising and not unwarranted, since the phytoplankton often forms the bulk of the biomass in at least the lotic environment, and is readily sampled, cultured, and amenable to experimentation.

The benthic flora, on the other hand, has been subjected to very few systematic studies, although many of the taxa are well known. Little is known of the geographical distribution of species (except from taxonomic records), seasonal cycles, flora of the microhabitats, relation to flow, or relation to water chemistry, and experimental studies on the communities or species are rare. Although there are still many problems associated with the sampling of phytoplankton and particularly with attempts to determine productivity, these are slight compared with the general lack of knowledge concerning sampling of the benthos, where separation

This investigation was supported in part by funds from Public Health Service Grant WP – 68.

of the algal cells and substrata is necessary. Therefore, quantitative studies are scarce and for some associations, completely lacking, although the taxonomic position of most species is at least as well known as for the plankton. Yet the benthic algal flora is distributed in at least three quite separate habitats, is as highly developed in running as in standing waters, is an important community of all small bodies of water (in small streams and ponds the benthic algal flora often forms the major algal biomass, since plankton is poorly developed in these waters), and is probably the habitat in which over nine tenths of all algal species grow, e.g., practically all the species of the pennate diatoms, while the majority of Conjugales, Cyanophyta, Euglenophyta, Xanthophyceae, and Chrysophyceae are benthic organisms.

Thus it may be appropriate that this series of lectures on "Algae and Man" is commencing with an attempt to review the ecology of the benthic community and direct attention to this aspect of phycology much neglected by man. No attempt will be made to review the whole field from a literature standpoint, but merely to point out some of the features of the benthic algal flora and some of the approaches and problems involved. The benthic algal flora includes the coastal macrophytic seaweeds but these will not be considered here.

The simplest definition of the benthic community is the assemblage of organisms living on the bottom of fresh-water or brackish ponds, lakes, rivers, and the sea bed. The term *benthic* is from the Greek *βενθοσ (bottom)*, but the term has been extended to include not only the flora on sediments *(epipelic)* and on rock or stone surfaces *(epilithic),* but also to that on the macrophytic benthic plants *(epiphytic).* These words are descriptive of the substrata on which the algae grow in natural habitats and are preferable to such terms as *periphyton* or *Aufwuchs.* The latter are collective terms, and obscure the fact that separate and well-defined floras are to be found on the three substrata beneath the water surface. The term *periphyton* (= *Bewuchs* of some German authors) has also been used for the growth on artificial substrata such as glass slides (e.g., in Sládečková, 1960), which primarily attract a mixture of species from the epiphytic and epilithic habitats, which then secondarily trap species from the epipelic and planktonic habitats. Fuller reviews of the terminology are in Cooke (1956) and Sládečková-Vinnikova (1956).

A simplified diagram of the distribution of the three

communities is given in Fig. 1. Other categories have been designated for different substrata but they are of doubtful usefulness at the present stage of studies. The algal flora in these three subcommunities are quite distinct, although casual species from one may creep into another or one may be obliterated and be replaced by another, e.g., as silt covers rock surfaces. Between the epipelic subcommunity and the others, there is little interchange of species, but the epilithic and epiphytic share many species, although few detailed studies have been made of the composition and seasonal sequence of the latter two subcommunities. The epipelic is essentially a subcommunity of motile forms since motility is necessary to enable species to move to the surface after disturbance of the sediment.

The exceptions to this "rule of motility" are the genera *Scenedesmus* and *Pediastrum* among the Chlorococcales; *Melosira varians, Fragilaria construens, F. intermedia, F. virescens,* and other rarer *Fragilaria* species among the diatoms; and *Microcystis* spp. and *Aphanothecae stagnina* among the Cyanophyta. Other apparently nonmotile genera such as *Merismopedia* and *Holopedium* are in fact slightly motile. On the sediments of small ponds, e.g., in moorlands, a relatively nonmotile flora occurs, made up of mucilage-forming genera such as *Frustulia, Chroococcus,* and desmids, both unicellular (*Euastrum, Cosmarium, Micrasterias, Xanthidium,* etc.) and filamentous (*Hyalotheca, Spondylosium, Sphaerozosma, Desmidium,* etc.); these sediments, however, are not greatly disturbed by wave action or macroscopic animals, as are the sediments of the littoral zone of lakes, where in fact these genera are scarce or completely lacking. Investigation of the flora of the surface sand at La Jolla from the beach region down to 35 m below sea level yielded no nonmotile species at all, a feature undoubtedly correlated with the great movement of the surface material in this surf-beaten marine habitat (Round, unpublished). A nonmotile flora of minute diatoms attached to the sand grains also occurs, forming an as yet undescribed association (Round, unpublished). This flora has also been recognized on the sand grains of Danish fiords by Grøntved (1960).

On the sediments of streams and rivers with moderate to fast flow, the flora is composed almost exclusively of motile species, and only in regions of extremely low rate of water flow can the nonmotile species survive. Thus, in an eighteen-month survey of the epipelic algal flora of small streams the dominant species were

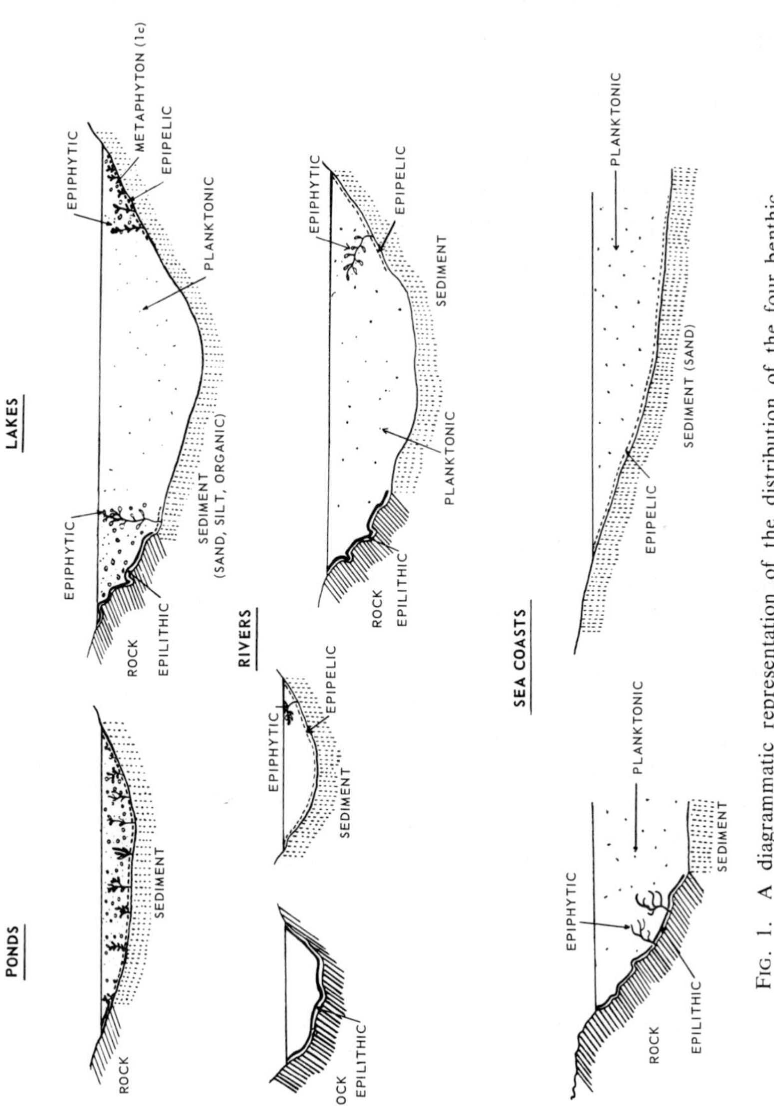

FIG. 1. A diagrammatic representation of the distribution of the four benthic subcommunities in ponds, lakes, rivers (3 different types of channel), and sea coasts (rocky and sandy shore).

all motile, viz., *Nitzschia dissipata, N. palea, N. angustata* var. *acuta, N. tryblionella, Navicula viridula, N. cuspidata, Caloneis amphisbaena, Gyrosigma acuminatum, Surirella ovata,* and *Cymatopleura solea* (Round, unpublished).

The epilithic and epiphytic subcommunities are essentially nonmotile, although there may be admixture of some motile forms as the flora becomes dense. The species comprising this flora are characteristically forms with mucilage attachment pads or stalks, e.g., *Achnanthes* spp., *Cymbella, Gomphonema, Chamaesiphon*; or in the case of some diatoms the mucilage sticks the cell down like a postage stamp, e.g., *Cocconeis* and *Epithemia,* and the prostrate discs of *Coleochaete, Stigeoclonium, Dermocarpa,* etc. Here also the larger forms with modified basal holdfast cells occur, e.g., *Oedogonium* and *Bulbochaete.*

The epipelic flora is in essence an extension of the soil or beach flora down beneath the water surface and indeed shares a few species with the subaerial flora, but there is little connection between the subaerial epiphytic and aquatic epiphytic flora. The growth form of the algae in the epipelic and soil flora is very similar, e.g., filamentous Cyanophyta are common in both, as are unicells such as *Euglena, Nitzschia, Navicula, Caloneis, Pinnularia,* while mucilaginous colonies such as those of *Aphanothecae* occur in the epipelic flora and *Stigonema, Nostoc,* and *Cylindrocystis* on the soil. The beach flora in marine habitats tends to be a reduced version of the marine epipelic flora with some contaminants from the adjacent soil flora, e.g., *Hantzschia amphioxys,* and with the same life forms predominating, i.e., motile dinoflagellates and diatoms. Any derivation of one flora from the other is extremely difficult to detect, although it would seem most likely that the soil flora is in part at least a remnant of the epipelic flora adapted to an aerial environment. The terrestrial epiphytic flora is very poorly developed in temperate regions and hence is not a source of species for the aquatic epiphytic associations.

The extent of the benthic flora in a lake basin, river course, or in the littoral of the sea is greatly influenced by the morphometry of the habitat, a factor which is of much less consequence to the planktonic community. Thus, whereas a lake basin of moderate size often contains a single planktonic community which varies little from place to place, the benthic flora may be developed to varying extent in different regions of the basin, on differing substrates, affected by inflow and outflow and other currents, by prevailing wind direction, by depth, etc. Similar but often more drastic variation is found along the course of a river or along the seacoast.

The idea has crept into the literature that the benthos gives rise to the plankton; this concept goes back to the early naturalists who assumed that resting spores must be formed for overwintering of the plankton and that these fell onto the sediment and there germinated when conditions were favorable. From this the idea has developed that there is a steady build-up of plankton on the sediments which then floats off into the water when conditions become favorable. Both Wesenberg-Lund (1908) and Nauman (1927) perpetuated this idea. Much evidence can be produced to show that this is an entirely erroneous concept, at least in moderate to large lakes and in the oceans. Never in the author's experience is there any conspicuous build-up of plankton on the sediments before a lake plankton bloom, but only afterward when the dead plankton settles out onto the sediment.

This is not invalidated by Lund's (1954-5) finding that *Melosira* settles out of the plankton when turbulence is insufficient to maintain a floating population and remains in a resting state on the sediments until disturbance of these brings the filaments into the plankton once more. Here there is no evidence for any degree of cell division while the *Melosira* is on the sediments and in all probability the main mass of filaments are on the deep sediments below the photic zone. Possible exceptions, and I believe they are few, might be the colonial Cyanophyta, which occur on littoral sediments as small almost unidentifiable colonies of *Microcystis-Aphanocapsa* type. These are rarely seen to form gas vacuoles, but it is possible that they are juvenile stages of planktonic forms, which by the formation of gas vacuoles rise into the open water. This, however, does not apply to common colonial forms such as *Aphanothecae stagnina,* which produces macroscopic globules on littoral sediments. The prerequisite of motility for the epipelic flora means that few planktonic species could survive on the sediments. They may, however, find a niche in the metaphyton (see below), although it is doubtful whether or not they can do more than exist there and grow when conditions are generally favorable in the whole water mass. When species have apparently disappeared from the plankton, extensive sampling may still show isolated cells from which new populations can develop, e.g., the demonstration that cells of *Asterionella* were present in the plankton of Windermere at all times, albeit in extremely low numbers (Lund, 1949). In small ponds it is more difficult to isolate the plankton from the bottom community and each is contaminated from the other, especially during stormy periods. Even in large lakes it is possible to find large littoral forms in the plankton after autumn gales, e.g., in a reservoir in Wales the author found *Pleurotaenium trabecula* dominant in the phytoplankton only in September-October during the autumn overturn coincident with gales; however, at the same time it was present in large numbers on the sediments, where it is to be found throughout most of the year, while it is absent from the plankton (Round, 1956). A similar explanation is possible for the occurrence of other "large" genera in the plankton, e.g., *Surirella*, although in some tropical lakes these appear to be truly planktonic; however, the epipelic flora in tropical lakes has not been examined in detail.

In fast-flowing streams there is little plankton and this is derived from sediments, plants, rocks, ponds, etc. in the drainage basin. In slow-flowing rivers it seems that species of *Melosira, Cyclotella, Oscillatoria, Pediastrum* and *Coelastrum*, etc. are found and may be able to grow on the sediments and in the plankton, although no studies have been made to show the relationship between the two communities. In the Weser around Bremen, Hustedt (1959) lists 31 species, mainly from the Centrales, which he considers as euplanktonic; here the flow is greatly reduced by tidal pressure leading to more lake-like conditions and the forms are undoubtedly not well adapted for a benthic existence and would seem to be synergistic.

There is in some waters, particularly in small ponds, an unattached, nonmotile benthic flora composed of algae lying loose on the sediments and often extending several centimeters or more into the water mass. Behre (1956) termed the mixture of algae found between water plants but not attached to them the *metaphyton* (probably synonymous with the *pseudoperiphyton* of Sládečková, 1960); this term may be used more widely to indicate all the algae which are neither clearly associated with a substrate nor free-floating. Species from this association do occur also on the sediments, but it is not true to imply as Behre does that the bottom flora is composed of species derived from the metaphyton. Typical components of this flora are the "ball-like" masses of *Cladophora* and the collections of *Spirogyra, Zygnema, Mougeotia,* and *Ulothrix* lying loose on the sediments of lakes and ponds. In sluggish streams *Hydrodictyon* often occurs in this habitat. Around macroscopic plants and between the filaments of *Spirogyra* etc. a large number of unattached desmids, diatoms, Chlorococcales, Chroococcales, and flagellates are found, some completely free-living throughout their growth cycle and others broken away from the plant substrata, e.g., *Tabellaria flocculosa, Tolypothrix,* and *Oedogonium* spp. As Behre emphasizes, the metaphyton is characterized by nonmotile species and he believes that this community is probably the richest of the littoral zone. It is the habitat of numerous desmid species of the genera *Closterium, Arthrodesmus, Cosmarium, Euastrum,* and *Staurastrum* and is present around macrophytic plants in rivers as well as lakes and ponds, e.g., in a Finnish river system many desmids were present among samples of plant material (Round, 1960c), and it was from this metaphyton and from that between epiplithic algae that the unusual abundance of desmids was recorded. In small ponds and bog pools the metaphyton is a very rich source of desmids, e.g., in between *Sphagnum.*

THE FLORA

Details of the flora of the benthic habitats are given in very abridged form in Table I. Discussion of floras tends to degenerate into long lists of species, yet this is the primary essential before any attempt can be made to separate the associations of waters of different chemical composition, age, geographical position, etc. The terminology relating to communities, associations, etc. is confused, but certain broad principles seem to be accepted and may be summarized in the followed scheme. The term community is used for the assemblage of organisms living in a defined habitat; the organisms will tend to exhibit a characteristic life form(s) and recur in various proportions throughout the range of the habitat. It is not easy and probably not desirable to use generic names in a description of the community, since it tends to be an abstraction only definable from a habitat basis; thus, in the aquatic environment there are two main communities, planktonic and benthic, i.e., communities of the open water and of the bottom and its associated plants.

In the former the life form tends to be adapted to flotation,

TABLE I

Distribution of Some Common Benthic Algae in Subcommunities of Fresh-Water and Marine Habitats

	EPIPHYTIC	EPILITHIC	METAPHYTON(IC)	EPIPELIC
SPRINGS — COLD	Achnanthes minutissima A. lanceolata A. affinis Cocconeis placentula Denticula tenuis	Meridion circulare Achnanthes	Absent?	Meridion circulare Fragilaria leptostauron F. construens and vars. Navicula cryptocephala Nitzschia dissipata Campylodiscus noricus
SPRINGS — HOT	Absent?	Synechococcus Synechocystis Phormidium Mastigocladus Plectonema Scytonema Cyanidium caldarium	Absent?	Absent?
STREAMS AND RIVERS	Chamaesiphon Oncobyrsa Dermocarpa Rivularia Aphanochaete Chaetophora Oedogonium Bulbochaete Cocconeis Achnanthes Synedra Cymbella Gomphonema	Hildenbrandia rivularis Lithoderma fluviatilis Chamaesiphon Rivularia Meridion circulare Diatoma hiemale Cocconeis placentula Achnanthes Synedra Gomphonema Cladophora Vaucheria Lemanea	Scenedesmus and other Chlorococcales Euglena Phacus Desmids {Staurastrum, Cosmarium, Euastrum, etc.} Spirogyra Mougeotia Zygnema	Melosira varians Fragilaria intermedia Frustulia Gyrosigma Caloneis Neidium Diploneis Stauroneis Navicula Amphiprora Amphora Cymbella (motile spp.) Bacillaria Nitzschia Cymatopleura Surirella Scenedesmus Pediastrum Oscillatoria Spirulina

PONDS AND LAKES	Characium	Gloeocapsa	Mougeotia	Chroococcus
	Characiopsis	Nostoc	Spirogyra	Aphanocapsa
	Ophiocytium	Calothrix	Zygnema	Aphanothecae
	Coleochaete	Scytonema	Binuclearia	Merismopedia
	Chaetophora	Tolypothrix	Ulothrix	Oscillatoria
	Stigeoclonium	Schizothrix	Microspora	Phormidium
	Bulbochaete	Dichothrix	Oedogonium	Lyngbya
	Oedogonium	Achnanthes	+ mixture of diatoms,	Fragilaria
	Gloeotrichia	Eunotia	flagellates, etc.	Frustulia
	Synedra	Cymbella		Anomoeoneis
	Tabellaria	Tabellaria		Stauroneis
	Eunotia	Frustulia		Caloneis
	Achnanthes	Cladophora		Neidium
	Cocconeis			Gyrosigma
	Cymbella			Navicula
	Gomphonema			Mastogloia
	Epithemia			Diploneis
	Rhopalodia			Amphora
				Pinnularia
				Nitzschia
				Cymatopleura
				Surirella
				Closterium
				Euastrum
				Synura
				Cryptomonas
				Euglena
				Phacus
				Trachelomonas
SEA SHORE	Dermocarpa	Pleurocapsa	Navicula (in tubes)	Merismopedia
	Lyngbya	Calothrix		Navicula
	Licmophora	Rivularia		Diploneis
	Grammatophora	Phormidium		Amphiprora
	Synedra	Cocconeis		Amphora
	Rhabdonema	Achnanthes		Pleurosigma
	Striatella	Navicula		Caloneis
	Cocconeis			Nitzschia
	Achnanthes			Surirella
	Isthmia			

albeit in an unknown way; and in the latter the organisms are adapted to living on relatively solid surfaces. Both these communities need to be broken down into smaller units before a floristic study is undertaken and for want of a better term, these units may be termed subcommunities; these are assemblages of organisms associated with particular physical facies of the habitat occupied by the community, e.g., the surface film of water, the stone surfaces of the benthos, etc. (see Table II for details).

The flora of the subcommunities can be and often is listed without regard to the degree of dominance, constancy, or fidelity of the species, but it is possible and desirable to distinguish different species groupings within a subcommunity; these are a reflection of differences in water chemistry, flow, substrate, etc. The term association has generally been used for these and they can best be typified by the construction of an association table after the manner of higher plant sociologists of the Braun-Blanquet school, e.g., this system has been used by Margalef (1953), Symoens (1957), and Schlüter (1961a,b). At this level there is confusion between what higher plant sociologists term alliances (i.e., groupings of closely related associations), the association itself and also subassociations. I believe this is due in the aquatic field to insufficient attention to sampling of well-defined habitats.

At the present stage of algal sociology the association is the most appropriate unit, together with subassociations which should have some reality in terms of microhabitats, e.g., the assemblage of epiphytic algae on *Oedogonium* or *Bulbochaete* may be termed a subassociation while the filamentous algae themselves are components of an association in the epiphytic subcommunity. Finer distinctions or larger groupings may be necessary as knowledge increases. These concepts are summarized in Table II. Some of these associations have long been recognized, e.g., Nauman (1920) noted the *Aphanothecetum* on lake sediments; Margalef (1953), the *Melosiretum arenariae* of flowing waters; Symoens (1957), the *Diatometo hiemalis-Meridionetum circularis*, while Schlüter (1961b) has a system of alliances, e.g., *Meridio-Naviculion gregariae* with subordinate associations, e.g., *Naviculetum rhyncocephalae-viridulae* and subassociations, e.g., *Synedretrum minusculae;* the latter is not synonymous with a subassociation as defined above. In Table II some of the genera are given which, either alone or in combination, may form the dominants in the associations, but no association

TABLE II

Some Characteristic Species of Fresh-Water and Marine Benthic Associations and Subassociations

COMMUNITIES	PLANKTON	BENTHOS		
SUBCOMMUNITIES	Neuston Euplankton	Epipelic	Epiphytic	Epilithic
ASSOCIATIONS	LAKES	1. *Navicula radiosa* 2. *N. oblonga* 3. *N. pupula* 4. *Pinnularia viridis* 5. *Nitzschia* spp. 6. *Closterium* 7. *Aphanothecae stagnina* 8. *Oscillatoria limosa*	1. *Achnanthes minutissima* 2. *Gomphonema* 3. *Epithemia turgida* 4. *Stigeoclonium* 5. *Aphanochaete repens* 6. *Gloeotrichia pisum* 7. *Tolypothrix distorta*	1. *Rivularia haematites* 2. *Tolypothrix distorta* 3. *Calothrix parietina* 4. *Cladophora glomerata* 5. *Frustulia rhomboides* 6. *Eunotia* 7. *Epithemia*
	SPRINGS	1. *Campylodiscus noricus* 2. *Fragilaria leptostauron* 3. *Amphora ovalis* v. *pediculus*	1. *Melosira arenaria* 2. *Achnanthese lanceolata* 3. *Gomphonema intricatum* v. *pumila*	1. *Meridion circulare* 2. *Achnanthes minutissima*
	STREAMS OR RIVERS	1. *Caloneis amphisbaena* 2. *Nitzschia sigmoidea* 3. *Navicula graciloides* 4. *Cymatopleura solea* 5. *Surirella ovata* 6. *Oscillatoria limosa*	1. *Eunotia pectinalis* 2. *Tabellaria flocculosa* 3. *Synedra ulna* 4. *Oedogonium* 5. *Bulbochaete*	1. *Achnanthes lanceolata* 2. *Meridion circulare* 3. *Diatoma hiemale* 4. *Ceratoneis arcus* 5. *Cladophora glomerata* 6. *Chamaesiphon* 7. *Hildenbrandia*
SUBASSOCIATIONS	LAKES AND STREAMS	1. *Synedra parasitica* 2. *Amphora ovalis* v. *pediculus*	1. *Achnanthes minutissima*	
ASSOCIATIONS	MARINE LITTORAL	1. *Pleurosigma* 2. *Amphora* 3. *Navicula* 4. *Nitzschia*	1. *Grammatophora* 2. *Licmophora* 3. *Synedra* 4. *Cocconeis*	1. *Calothrix scopulorum* 2. *Rivularia atra* 3. *Cocconeis scutellum*
	MARINE SALT MARSH	1. *Pleurosigma angulatum* 2. *Navicula gregaria* 3. *N. pygmaea* 4. *N. cincta* v. *heufleri* 5. *Nitzschia obtusa* v. *scalpeliformis* 6. *Caloneis amphisbaena* v. *subsalina*	1. *Cocconeis scutellum* 2. *Synedra affinis*	

names (i.e., by use of the generic name with the suffix *-etum* and with the species name in the genitive case) are used, although I think it is possible to define a limited number of associations in this way.

The compilation of such tables pinpoints many problems, such as the interaction of species and floras, and enables the range of tolerance of species to be recognized, which may indicate unsuspected conditions for growth, reproduction, competition, etc. As examples one might cite the tolerance of *Trachelomonas* species to hydrogen sulfide (Round, 1955), the growth of certain epiphytes only in the furrows of leaves (Düringer, 1958), the occurrence of *Navicula hungarica* var. *capitata* on silted sediments but not on sand in the English Lake District (Round, 1957c), the restriction of species groups to a single drainage system, e.g., *Pinnularia cardinaliculis* to the lakes draining into Windermere (Round, 1957c), and the loss of motility of certain diatom genera in saline streams (Round, unpublished). Only from detailed floristic analyses will the individual peculiarities be determined and from these the experimental approaches designed.

But this raises the difficult question of sampling each flora and separating the associations. Often the algologist is in a position which would be comparable with that of the land ecologist if he had to sample with a grab from a helicopter flying above the clouds and if he then had to sort out the forest trees, from the shrubs, from the herbs, from the epiphytic Bryophytes, and from the "planktonic" insects flying among them all. This type of analogy has been used with great force by submarine geologists and sedimentologists but is no less applicable to much of the benthic flora. Three basic requirements are necessary to give the floristic groundwork a firm basis: 1) the recognition of the various habitats and their associations; 2) the quantitative sampling of these habitats so that direct comparisons can be made; and 3) the separation of casual species from descriptions of the associations.

The construction of association tables depends on the sampling of sites of comparable nature. Hence it is necessary to select carefully the sites in each subcommunity and to avoid combining samples, e.g., from epipelic and epilithic sources in streams since clean, i.e., unsilted, rock surfaces, have a flora distinct from that on the sediment of a stream. Nevertheless it may not always be possible to find ideal situations and in fact many regions in streams are of silted rock surfaces and these may have to be recognized as another subcommunity. Assuming that the habitat of a subcommunity is recognized, e.g., the sediment of a stream or the moss flora of stones in a stream or the rock surface of a mountain stream, there is still the considerable problem of the size of the "stand" to be sampled. Little consideration has been given to this problem; much sampling has been carried out at single fixed stations, e.g., the stations from which the data for graphs (Figs. 13 and 14) have been obtained. In the case of Fig. 13 the stations were all along a very short length of stream and the floristic data are very similar for each.

One method of procedure would be to select a sampling site in the center of an apparently uniform region of sediment and to analyze samples radiating out from this, until there was an obvious change in the flora, i.e., construct an association table for a number of sites in this relatively small area. From this the area covered by a relatively uniform flora could be determined and also the number of samples, i.e., the "stand" necessary to characterize the association would be obtained. This could then be repeated in comparable regions to give the basis of an association table. The results may reveal a "continuous network pf variation" down a stream or along the shore of a lake or down a depth profile in a lake, but more frequently there are clear breaks due either to change in the environment and the concomitant change in the dominants, or to the occurrence of less tolerant indicator species, e.g., in streams the epipelic flora in one zone may be dominated by *Surirella ovata,* in another by *Gyrosigma acuminatum,* and in yet another by *Oscillatoria limosa,* etc. On a small scale, mosaics do indeed occur, as is obvious in soil floras where patches of Cyanophyta can be distinguished on light-colored soil, but in an aquatic environment these are probably less well developed and would be unrecognizable if the "stand" is large.

The quantitative estimation of the flora is a much more difficult problem since the question of cell numbers, cell volumes, growth rates, etc., ought all to be considered. Assuming that unit area can be sampled effectively (and this in itself is often a problem), estimation of cell (or filament or colony) numbers is usually possible and essential if the association is to be described accurately. It is satisfying to find that many deductions based on cell numbers have now been verified by other techniques. The epipelic flora can be estimated by placing coverglasses on the surface of sediment in a petri dish (see Lund, 1942; Round, 1953) and a fairly accurate estimate of cell numbers made. Objections to this method are that the few nonmotile components of the flora may be underestimated, since the method depends on the phototactic response of the motile forms which rise onto the undersurface of the coverglass; also some species are lost when the coverglass is removed and some fast-moving forms may move out from beneath it before counting is done—usually 24 hr after setting up the samples. Nevertheless the method does give fairly reproducible results when only the gross flora is required.

The epiphytic and epilithic floras are much more problematical. Loosely attached species can be removed by agitation, e.g., Knudson (1957) found that epiphytic *Tabellaria* could be estimated in this way. Margalef (1949) used a method of applying a collodion film to the surface and stripping this off. Heavy growths and encrusted growths such as occur in calcareous habitats are a problem, however, and no entirely satisfactory methods have been evolved.

The concept of dominance as used for land floras is difficult to apply to these benthic communities, since seasonal changes are often rapid with a succession of dominants. Also cell numbers may show that, as is often the case, the small *Nitzschia* species are dominant on sediments (see Fig. 12), but in terms of biomass, these may not be as important as larger *Navicula* or *Pinnularia* species. A distinction must also be made between species present in large or small numbers throughout the years and a brief occurrence of large numbers of a single species; average cell counts sometimes obscure this and give the impression of equal importance of the two components in the flora, while percentage occurrence, i.e., the percentage of dates on which the species is encountered, does reveal this. Ideally then, the floristic data should involve cell numbers $\times$ cell volume coupled with some estimate of seasonal constancy. Until this is achieved comparisons between floras are not reliable and result in species lists all too frequently suggesting a similar flora for many quite different habitats.

In the comparative study of floras in the benthos the concept of fidelity to the community used by land plant ecologists is a necessity, e.g., the epipelic flora of peat pools is characterized by the occurrence and fidelity to the habitat of *Frustulia rhomboides* and *Chroococcus turgidus,* although few other benthic floras can be so confidently described. If fidelity is a necessity, then equally it is not possible to define an association until many examples have been analyzed.

The third requisite, i.e., separation of casual species from the association tables, is a real problem in dealing with the aquatic environment. However, careful consideration of the life form of each species should remove the majority of casuals, e.g., the obvious planktonic forms caught up in epiphytic material. The decision is more problematical, however, when considering components of the metaphyton and here only lengthy detailed sampling of well-defined habitats will resolve the issue.

FACTORS AFFECTING THE BENTHIC FLORAS

These factors can be conveniently divided into four groups: spatial, time, internal, and biotic. The first is concerned with the position of the flora in relation to latitude and longitude, altitude, depth, rate of flow, chemistry of water and substrata. The second deals with the flora at various geological times, with more recent changes and long-term cyclical phenomena, with annual and diurnal cycles; all these are affected in varying degree by the spatial factors. The third group, internal factors, are concerned with the mode of nutrition of the algae, the growth rates, reproductive cycles, movement and phototaxis, etc. The fourth group involves competition, the production of extracellular products, parasitism, and grazing of the community. To consider the effect of these factors on the whole benthic flora would be a large task and in some areas the data are scanty or lacking, so a limited number of aspects will be discussed.

Spatial Factors

Lakes and rivers lie in regions of differing geology and soil formations and hence are characterized by the chemical and physical features of the region. Basically a consideration of these factors can be used to characterize and classify natural waters or, more narrowly, the benthic habitats. However, an almost continuous range of chemical and physical features can be recorded and although the extremes are separable and many of the factors are linked, some other method of classifying the habitats is required and here the flora can be utilized.

In a region of apparently similar geology, soil type, and vegetation, the benthic flora of the lakes and streams tends to conform to a type but is always found to differ in detail. Of the spatial factors operating on the flora, chemical composition of the water must be one of the most basic since it modifies the planktonic flora which is not even in contact with any interfaces; secondarily the effect of the component species on the chemistry of the environment and on each other must be taken into account (see below). From the apparent ease of dispersal of algal species, a great similarity of floras would be expected and although the same species may be found over a wide range of benthic habitats the balance of the flora is always different in different sites. Also a smaller number of species are peculiarly restricted in their distribution and hence must require very special conditions which as yet are not measured in the usual analytical methods applied to the habitat.

The benthic habitats, unlike the planktonic, are nonisotropic and hence a physical selective factor is involved. This factor is responsible for the diverse flora and the greater number of species in the benthic than in the planktonic habitat. It also means that a greater number of niches are available for colonization, each with its own advantages for particular species. Since in the isotropic environment of the plankton there are no two plankton floras alike and indeed often adjacent lakes differ considerably, the possibility of variation in the benthic floras is greater owing to the interplay of water chemistry—physical nature of the substratum—niche structure—interaction of species and of species with the environment.

Nevertheless, perhaps the basic factor which varies spatially is the chemistry of the environment. Here a distinction must be drawn between the chemical control of the population structure and the chemical control of seasonal variation of the population. The former is in some way the outcome of the complex balance of ions in the environment with minor elements probably playing an important but as yet unknown role, while the latter is related to the variation of major elements involved in synthetic processes, e.g., silicon for diatoms; consideration of this latter aspect belongs with the time factors (see p. 163). In a simple environment with no other organisms present, the cﬁemical control would be dominant and an equilibrium population would be established, possibly leading to a population of a single species since the principle of *competitive exclusion* might apply, although there are arguments against this even for the simpler case of the plankton (Hutchinson, 1961). However, the products of other species, acting in a chemical manner, and more important still the presence of parasites (e.g., chytrids) and of predators, selective or otherwise (e.g., protozoa on sediments and tadpoles in the epiphytic subcommunity), disturb the equilibrium. These latter factors and those operating toward *competitive exclusion,* e.g., growth rates, interaction of species, reproductive rates, mating types, etc., have as yet hardly been considered, while stress has been laid on chemical factors of a somewhat crude nature such as alkalinity/acidity, calcium/magnesium content, halide concentration, and organic matter content. The latter factor is now known to be important not only in indicating a generally higher nutrient status in eutrophic* water or a lower status in dystrophic waters, but also as a source of metabolites for many flagellates and some genera of diatoms, Chlorophyta, etc., which are not completely photoautotrophic. Correlation between any ion and elements of the flora is most obvious at extreme concentrations, but in the mesotrophic range it is more difficult and here the careful analysis of the flora is the only means of separating the habitats.

In Table III a brief résumé of the epipelic algal flora of lakes is given and clearly there is a segregation of species in the alkaline and acidic lakes. Some common species extend over both types but rarely into the extreme conditions, and these have been omitted from the table, while certain indicators of the extremes have been included. Summarizing the data is difficult because both abundance and occurrence need to be taken into account; *Amphora ovalis* is often abundant in alkaline but still present in acid waters, whereas *Cymatopleura* species are rarely abundant in alkaline waters but frequently present and almost certainly absent from acid waters, and the converse is true of *Stenopterobia intermedia.*

In addition a distinction can be made between the indicators of alkalinity or acidity and extremes of these, and tentatively the following scheme may be suggested incorporating only the most obvious species (Table IV). Similar segregation of epipelic species occurs in running waters, where for example *Amphora ovalis, Caloneis amphisbaena, Navicula cryptocephala, N. gregaria, N. radiosa, Gyrosigma acuminatum, Nitzschia sigmoidea, Cymatopleura solea,*

* The use of the terms, *eutrophic, mesotrophic, oligotrophic,* and *dystrophic* is hotly disputed and they are used here only in a general sense to refer to waters of high, moderate, and low nutrient status and highly humic waters, respectively.

TABLE III

Occurrence of Epipelic Algae in Lakes of Alkaline and Acid Type

ALKALINE			ACID		
MALHAM TARN	MASSOMS SLACK	IRISH LAKES (ALK.)	ENGLISH LAKE DIS.	IRISH LAKES (ACID)	MALHAM TARN
Amphora ovalis	*Amphora ovalis*	*Opephora martyi*	*Frustulia rhomboides*	*Fragilaria undata*	*Amphipleura pellucida*
Neidium dubium	*Navicula pupula*	*Fragilaria intermedia*	*F. rhomboides* v. *saxonica*	*F. virescens* v. *elliptica*	*Cymbella prostata*
Navicula pupula	*N. placentula*	*F. harrissonii*	*Anomoeoneis exilis*	*Eunotia robusta* v. *tetraodon*	*C. lanceolata*
N. placentula	*N. dicephala*	*F. construens*	*A. serians*	*E. robusta* v. *diadema*	*Desmids*
N. cryptocephala	*N. oblonga*	*Gyrosigma acuminatum*	*Neidium bisulcatum*	*E. pectinalis* v. *undulata*	
Cymbella ehrenbergii	*N. cryptocephala*	*G. attenuatum*	*N. affine*	*E. pectinalis* v. *ventralis*	WELSH RESERVOIRS
Cymatopleura solea	*Cymbella aspera*	*Neidium dubium*	*N. iridis*	*Frustulia rhomboides*	
C. elliptica	*Diploneis ovalis*	*Navicula cari*	*Stauroneis phoenicenteron*	*F. rhomboides* v. *saxonica*	*Neidium iridis*
	Nitzschia sigmoidea	*N. scutelloides*	*S. anceps*	*Anomoeoneis serians*	*N. affine*
Begiatoa	*N. angusta* v. *acuta*	*N. rostellata*	*Navicula pupula*	*A. exilis*	*Anomoeoneis exilis*
Achromatium	*Cymatopleura elliptica*	*N. pupula*	*N. cryptocephala*	*Neidium iridis*	*Frustulia rhomboides*
		N. cryptocephala	*N. radiosa*	*N. productum*	*Stauroneis phoenicenteron*
		N. lanceolata	*Pinnularia mesolepta*	*N. hitchcockii*	*Navicula pupula*
		N. tuscula	*P. interrupta*	*Stauroneis phoenicenteron*	*Pinnularia gibba*
WINTERBOURNE PONDS	FINNISH LAKES (ALK.)	*N. reinhardtii*	*P. divergens*	*S. anceps*	*P. polyonca*
		N. bacillum	*P. maior*	*Pinnularia undulata*	*P. mesolepta*
		N. dicephala	*P. viridis*	*P. appendiculata*	*Nitzschia dissipata*
Fragilaria construens	*Fragilaria construens*	*N. menisculus*	*P. gibba*	*P. subcapitata* v. *hilseana*	*Surirella robusta*
Caloneis amphisbaena	*Navicula cuspidata*	*N. placentula*	*Cymbella naviculiformis*	*P. gibba*	*S. biseriata*
Amphora ovalis	*N. graciloides*	*N. oblonga*	*Nitzschia dissipata*	*P. interrupta*	*S. linearis*
Navicula pupula	*N. reinhardtii*	*Mastogloia smithii*	*N. palea*	*P. microstauron*	*Stenopterobia intermedia*
N. oblonga	*N. placentula*	*Cymbella sinuata*	*Surirella robusta*	*P. globiceps*	
N. cryptocephala	*N. gastrum*	*C. cuspidata*	*S. linearis*	*P. divergentissima*	FINNISH LAKES (ACID)
N. rhynocephala	*N. lanceolata*	*C. helvetica*	*Stenopterobia intermedia*	*P. divergens*	
N. hungarica v. *capitata*	*N. tuscula*	*C. leptoceros*		*P. acrospheria*	
Gyrosigma acuminatum	*Cymbella cesatii*	*C. ehrenbergii*		*P. nodosa*	*Melosira italica*
Nitzschia sigmoidea	*C. lanceolata*	*Nitzschia sigmoidea*		*P. dactylis*	*M. distans*
N. palea	*Nitzschia angustata*	*N. dissipata*		*P. viridis*	*Fragilaria virescens*
Cymatopleura solea	v. *acuta*	*N. denticula*		*Cymbella cesatii*	*F. undata*
C. elliptica	*N. denticula*	*Cymatopleura solea*		*C. gracilis*	*F. constricta*
	Cymatopleura solea	*C. elliptica*		*C. aequalis*	*F. polygonata*
				Surirella birostrata	*Eunotia polyglyphis*
				S. gracilis	*E. bactriana*
				S. delicatissima	*E.* spp.
				S. robusta	*Frustulia rhomboides*
				S. elegans	*Anomoeoneis serians*
				Stenopterobia intermedia	*Navicula radiosa*
					Pinnularia spp.
					Cymbella rupicola
					C. gracilis
					Stenopterobia intermedia
					Surirella linearis

TABLE IV

Characteristic Diatoms of the Epipelic Associations of Alkaline and Acid Waters

Extreme alkalinity	Alkaline	Acid	Extreme acidity
	Abundant	**Abundant**	
Mastogloia spp.	*Fragilaria construens*	*Frustulia rhomboides*	*Melosira distans*
Navicula costulata	*Amphora ovalis*	*F. rhomboides* v. *saxonica*	*Fragilaria virescens*
N. reinhardtii	*Gyrosigma acuminatum*	*Anomoeoneis exilis*	*F. undata*
Cymbella leptoceros	*Neidium dubium*	*Neidium* spp.	*F. constricta*
Nitzschia denticula	*Navicula dicephala*	*Pinnularia gibba*	*F. polygonata*
N. sinuata	*N. placentula*	*P. mesolepta*	*Eunotia polyglyphis*
	N. oblonga	*P. maior*	*E. bactriana*
	Nitzschia sigmoidia	*P. viridis*	*E. pectinalis* v. *ventralis*
		Nitzschia dissipata	*Pinnularia undulata*
	Present	**Present**	*P. episcopalis*
	Gyrosigma attenuatum	*Eunotia robusta*	*P. alpina*
	Navicula gastrum	*E. pectinalis*	*P. stomatophora*
	N. tuscula	*Anomoeoneis serians*	
	N. menisculus	*Pinnularia acrospheria*	
	Cymatopleura solea	*P. nodosa*	
	C. elliptica	*Cymbella gracilis*	
		Surirella spp.	
		Stenopterobia intermedia	

etc., are common in alkaline situations and *Eunotia* spp., *Actinella punctata, Frustulia rhomboides, Pinnularia* spp., and *Surirella* species occur in acid waters.

In addition to the general pattern of distribution based on chemistry of the waters, there is a pattern established in some waters by the nature of the actual sediments. Usually these are of the same general chemical nature as the waters, but in Malham Tarn (Yorkshire), for example, at least four different sediments exist; these are the peat sediment and the coarse calcareous sediment listed in Table III above, plus a calcareous silt with a flora similar to the coarse calcareous sediment and also a black sediment beneath beds of *Chara*, which is almost devoid of an epipelic flora owing to the hydrogen sulfide released by the decaying organic matter. In Windermere the sediments on the two sides of Pull Wyke Bay differ in structure, one being a soft silt and the other a firm sand. On the sand *Gyrosigma acuminatum, Neidium dubium, Diploneis fennica, Stauroneis smithii, Navicula radiosa, Pinnularia mesolepta, Amphora ovalis* var. *pediculus, Cymbella naviculiformis, C. prostrata, Nitzschia ignorata, Surirella robusta, S. linearis,* and *Stenopterobia intermedia* are all more abundant, as also are the nondiatoms *Hemidinium nasutum* and *Cryptomonas ovata*; on the silt *Stauroneis phoenicenteron, Pinnularia polyonca, P. hemiptera, P. acrospheria, P. cardinaliculis, Nitzschia acicularis, Euglena* spp., *Phacus pleuronectes, P. pyrum, Trachelomonas volvocina, T. hispida,* and *Cryptomonas erosa* are more abundant (Round, 1957c). This relationship between the physical nature of the substrate and the epipelic algal flora is also found on salt marshes where, for example, *Nitzschia obtusa* var. *scalpelliformis,* and *Navicula cincta* var. *heufleri* are only common on sticky silt at various levels on the marsh; while *Caloneis amphisbaena* var. *subsalina, Amphora coffeaeformis* var. *acutiuscula,* and *Navicula viridula* are associated with coarse sand (Round, 1960b).

These data suggest that chemical and physical structure of the sediments need to be considered in more detail in the future and this will result in more detailed division of the epipelic association. The habitats could be divided on a chemical/physical basis, but this would of necessity be an arbitrary subdivision, if indeed a subdivision is possible in a system without natural breaks, e.g., sediments with extractable calcium values ranging from 310 mg/liter to 1070 mg/liter are found among the larger lakes of the English Lake District and although no two sediments are alike in flora it is easier to subdivide on a floristic basis rather than chemical (Round, 1957b). Thus, the percentage occurrence of *Amphora ovalis* on the very low calcium sediments varies from 0-13, while that on the medium to high is from 61-100 with one exception (which may have been due to rate of flow at this lake station situated near an outflow). Other examples are the restriction of *Pinnularia cardinaliculis* to the lakes of the Windermere drainage system; of *P. acrospheria* to two stations in Windermere and abundant at one station in Loeswater; of *P. polyonca* at isolated stations only in Windermere, Blelham Tarn, Esthwiate Water, and Grasmere; and of *Navicula oblonga* only in Coniston.

These examples and many others indicate the subtlety of the

problems of distribution. The presence of sodium chloride in inland waters results in the occurrence of halophilic species, e.g., in saline streams around Droitwich, Worcestershire, the normal flora is modified by the growth of species such as *Gyrosigma distortum, Navicula pygmaea, N. gregaria, N. cincta, N. salinarum, Amphiprora alata, A. paludosa, A. ornata, Amphora coffeaeformis, A. lineolata, Cylindrotheca gracilis, Nitzschia tryblionella, N. hybrida, N. closterium,* and *Surirella ovata* var. *salina.* Furthermore, species common in neighboring streams, such as *Gyrosigma acuminatum, Navicula cuspidata, N. pupula, Amphora ovalis, Cymatopleura solea, C. elliptica, Nitzschia sigmoidea,* and *N. dissipata,* are either prevented from growing by the presence of sodium chloride, i.e., are halophobic, or are suppressed by competition with the halophilic species (Round, unpublished).

High amounts of organic matter tend to increase the flagellate and Cyanophyta counts on sediments; the unnatural state of organic pollution of streams and rivers has the same effect and ultimately reduces the flora to a few or a single species, e.g., of *Nitzschia* or *Oscillatoria*; somewhat surprisingly species of *Closterium* are sometimes abundant in such habitats possibly indicating the requirement of an organic factor for this desmid, a feature already reported for other groups of the Chlorophyta. Unlike the other chemical factors, organic matter is a complex of chemicals and itself interacts with ionic components, so that there is no such simple correlation as there is, for instance, with salinity. Thus, high organic content of moorland waters or dystrophic lakes and streams results in an entirely different effect from that in eutrophic waters. The flora of these dystrophic waters, which contain species such as *Chroococcus turgidus, Frustulia rhomboides,* and many desmids, is generally related to the low pH of the water, but equally it may be developed as a result of the type of organic matter or of low calcium; no exhaustive studies of the reaction of the benthic flora to these environmental features has been made.

Depth distribution of species is most clearly marked by the absence or scarcity of species at shallow depths in lakes. This is comparable to the reduction in assimilation noted in planktonic communities in the upper 1 m of water. Whether or not absence of certain epipelic species is also combined with a reduction in rate of assimilation has not been determined. Figure 2 shows the depth distribution curve of the epipelic diatoms and blue-green algae

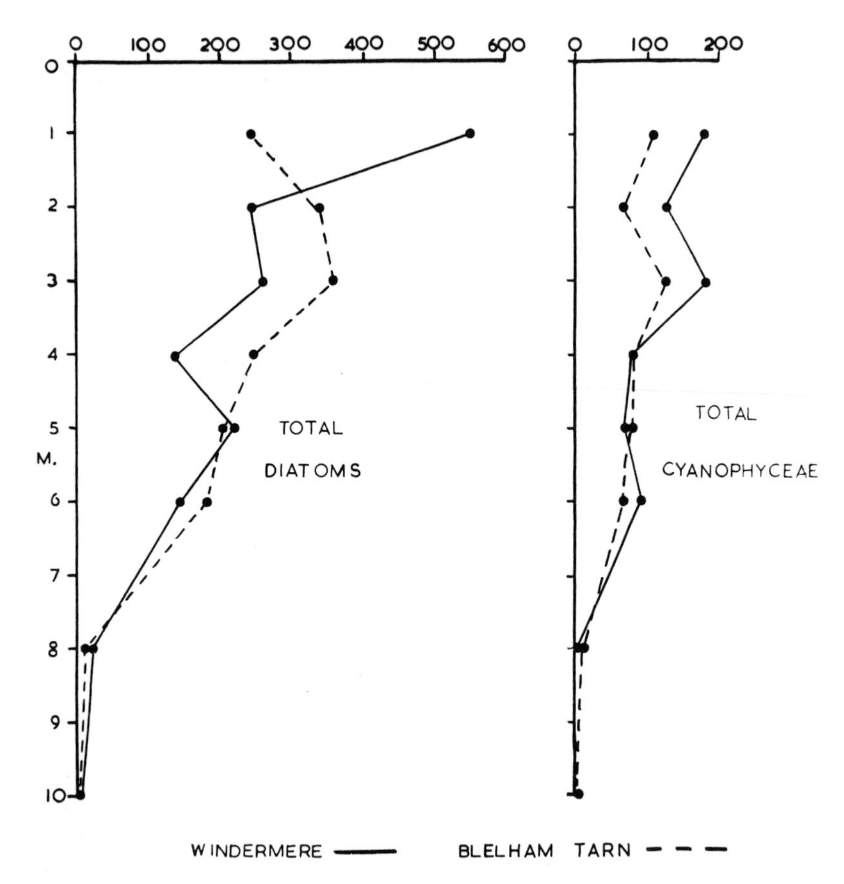

FIG. 2. Depth distribution of the total diatom and blue-green algal populations from 1 m down to 10 m along transects in Windermere and Blelham Tarn. The data for each depth are the average cell (or filament/colony for the blue-green algae) count from a twelve-month sampling period.

calculated as annual averages for each depth. Only in Blelham Tarn is there any suggestion of lower cell numbers at 1 m, and this may be due to the position of the sampling transect which commenced adjacent to a tree-lined shore. The very low cell counts at 8 m are comparable to the low rates of phytoplankton assimilation in these lakes at this depth (Talling, 1955, 1957).

The only member of the benthic flora subjected to such experimental studies is *Tabellaria flocculosa* var. *flocculosa* (Cannon, Lund, and Sieminska, 1961), which was shown to be capable of only one cell division or less per week between 6 m and 8 m from April to July and below 8 m growth was not possible, at least as measured by cell division. In fact growth was very slow even between 4 and 6 m. These studies on a planktonic and an epiphytic alga suggest that if similar factors are operating on the epipelic community then there should be a smooth decrease of cell numbers with increase in depth. The variation down the depth transect is indeed basically related to light intensity, but minor variations are possibly associated with current movements and the subsequent scouring and sorting of the sediments; this is probably more obvious in the larger lake Windermere, where the transect also involved a greater longitudinal spread than did that in Blelham. The steepness of the curves also suggests that the diatoms are more sensitive to light than are the blue-greens, and the absence of any appreciable growth below 6 m indicates that the metabolism of the organisms is strictly phototrophic.

The depth distribution of some individual species (Fig. 3) indicates a different behavior of some species at the shallowest depth in the two lakes. However, the factors involved in maintaining a large population of *Navicula pupula* and *Navicula rhynchocephala* at 1 m in Windermere but not in Blelham and the converse for *Navicula*

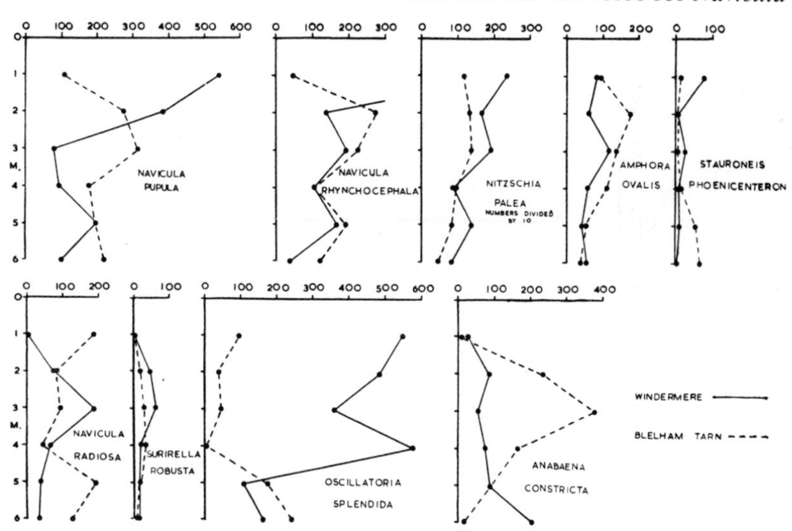

FIG. 3. Depth distribution of some common epipelic algae in Windermere and Blelham Tarn. Data compiled as in Fig. 2.

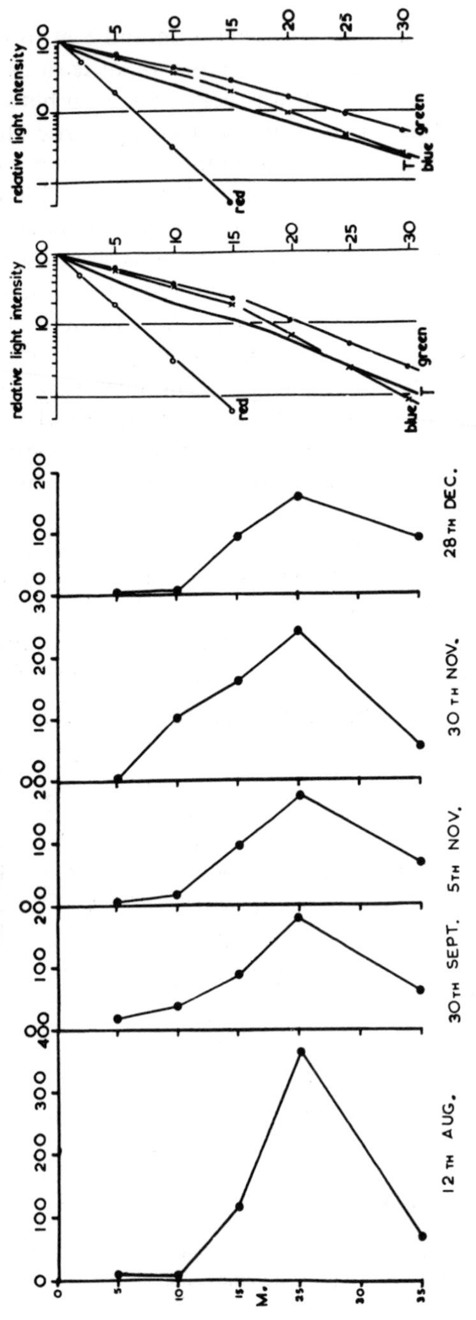

FIG. 4. Total epipelic diatom cell counts at depths from 5 m down to 35 m off the coast at La Jolla, California (Round, unpublished), together with two graphs showing light penetration in these waters. (From Talling, 1960.)

radiosa are not at all clear; this may be a feature related to sediment structure and chemistry. The slight increase in cell numbers of some species at 5 and 6 m could be due to sedimentation of disturbed species at these depths, but the increase in *Oscillatoria splendida* at 5-6 m in Blelham and of *Anabaena constricta* at these depths in Windermere may be more significant. In general the peak of cell numbers occurs either at the shallowest depth or at 3 m. This may be an interaction between light stimulation of some species and inhibition of others.

Studies on marine sediments at La Jolla (Fig. 4) show a distinct peak of cell numbers at 25 m which decreases at 35 m, although even at this depth there is a considerable population compared to that at 6-8 m on the fresh-water sediments in the English Lake District. At 25 m off La Jolla the light intensity is approximately 3% of surface illumination and at 35 m is less than 1%. The absolute amount of light at these depths may not be too low for photosynthesis since the results of Talling (1960) show that the planktonic diatom *Chaetoceros affinis* can photosynthesize at 30 m in this region. It is possible that diatoms living at those depths are partially heterotrophic (see below). The very low numbers at 5 and 10 m are due to surf and current action, which keeps the population low and at 5 m results in flora consisting of a single *Pleurosigma* species.

Time Factors

Time factors influence both the whole flora and the individual species over shorter periods but both are indirectly the result of temporal changes in chemical and physical components of the habitat. These changes have been proceeding in northern lakes since the retreat of the ice during the closing period of the last ice age. The initial effect on the epipelic flora was the exclusion of all but the base tolerant species. This was caused by the high calcium content of the water draining from the late-glacial land surface in which thawing and freezing fragmented the rock, yielding a continuous supply of bases. The epipelic flora thus contained species such as *Fragilaria construens, Navicula oblonga, Gyrosigma attenuatum, Cymatopleura elliptica*, etc., and all acidophilic species were absent. As the climate improved and the rock surface became insulated with a layer of soil, the base content decreased (Figs. 5-7) and the species complement gradually changed to an acidophilic

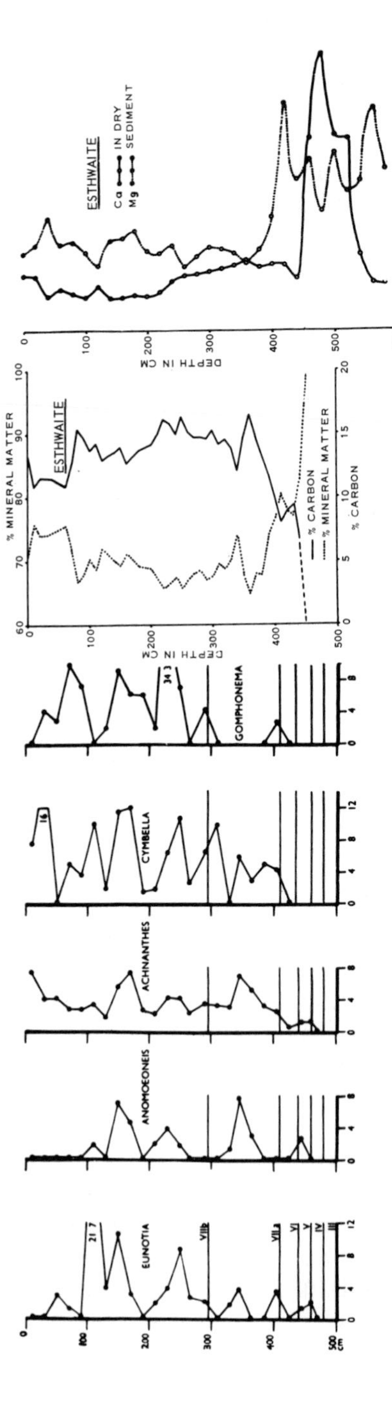

FIG. 5. Distribution of cells of the genera *Eunotia*, *Anomoeoneis*, *Achnanthes*, *Cymbella*, and *Gomphonema* in a 500-cm-long core from the center of Esthwaite Water (Round, 1961c), together with some chemical features of the core sediments (data kindly supplied from unpublished data of Mr. F. J. Mackereth).

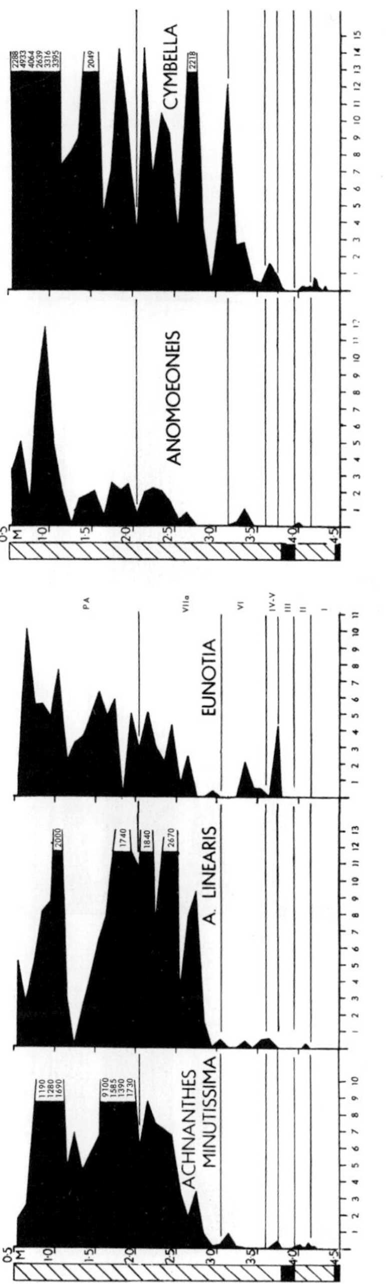

FIG. 6. Distribution of *Achnanthes minutissima*, *A. linearis*, and the genera *Eunotia*, *Anomoeoneis*, and *Cymbella* in a core from Kentmere. (From Round, 1957a.) Pollen analytical zones, PA—I.

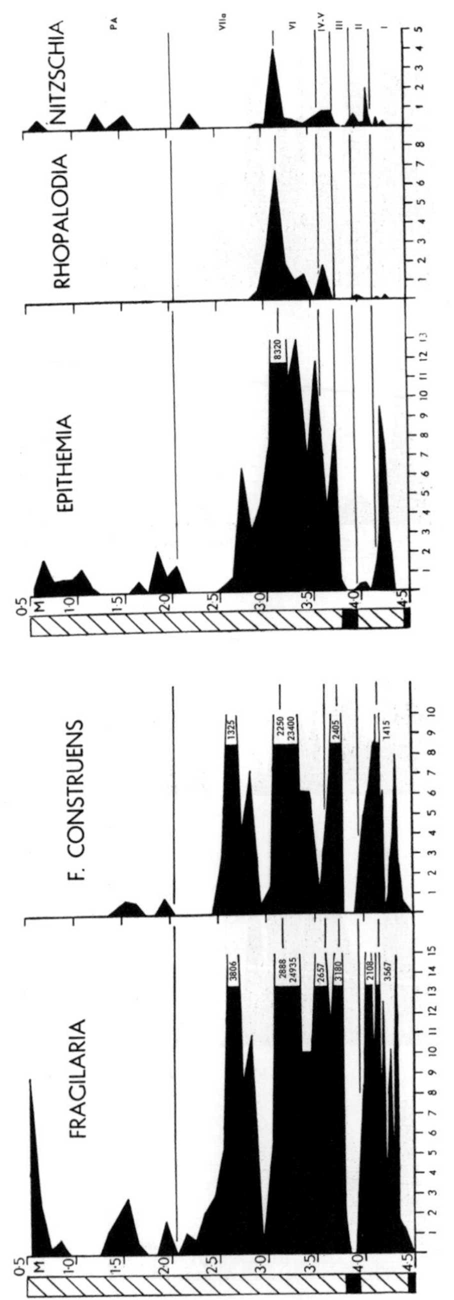

FIG. 7. Distribution of *Fragilaria construens* and the genera *Fragilaria*, *Epithemia*, and *Rhopalodia* in a core from Kentmere. (From Round, 1957a.)

stage (except in lakes in limestone regions where presumably there has always been a supply of bases). Many lakes are still in this stage, although some, e.g., Esthwaite, are beginning to undergo a further change due to man-induced eutrophication; this is reflected in the epipelic flora, which in the uppermost layers of cores from the sediments contains *Navicula rhynchocephala*, a species absent or rare in acidophilic habitats in the English Lake District.

The present-day flora of the sediments, which is not yet sufficiently represented in the top layers of the deep cores, contains many of the species of more base rich waters than those of the Lake District and some of the species more common in the late-glacial period. These studies indicate that the acidophilic species of the epipelic flora probably invaded the region from the south some time during the Boreal-early Atlantic Period. Also species with narrow distribution ranges and therefore possibly late invaders of the region can be confirmed from a study of the postglacial epipelic flora, e.g., *Pinnularia cardinaliculis* is confined to the Windermere drainage area, which includes Esthwaite, and is known to have appeared relatively recently, at least in the latter lake.

The attached benthic algal flora is also represented in the cores, but it is not possible to distinguish between the epiphytic and epilithic subcommunities. In both Kentmere and Esthwaite it is not well developed in the late-glacial period when conditions were generally unfavorable for diatom growth (Fig. 8). The smaller

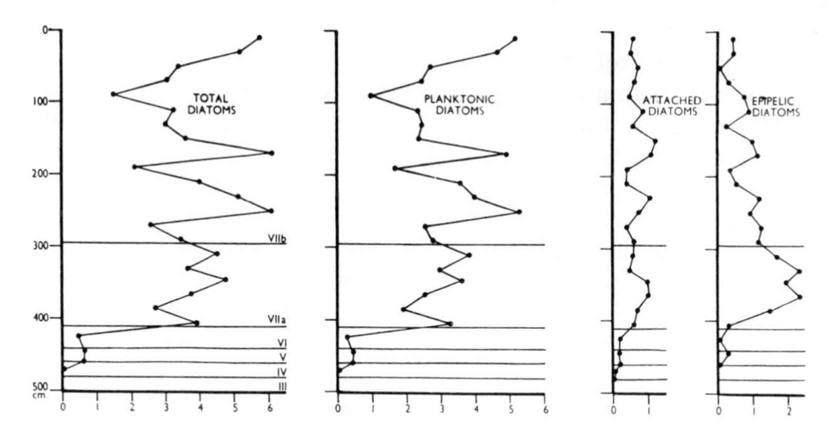

FIG. 8. Distribution of total diatoms and the planktonic, epipelic, and attached diatoms in a core from Esthwaite Water. (From Round, 1961c.)

and shallower lake basin of Kentmere probably favored the growth of this attached community such that in Atlantic/Post-Atlantic time it was dominant in the sediment and underwent a series of cyclical changes (Round, 1957a). In the larger Esthwaite Water the attached flora is subordinate to the planktonic in the remains in the sediment, either expressing a lower productivity of these subcommunities in this lake or reflecting a lower count of these essentially littoral forms in a core from the center of a large lake; probably a combination of the two factors is responsible. It would appear, however, that the lower representation in Esthwaite is unaccompanied by the drastic fluctuations which characterized the attached community in Kentmere (Round, 1961c).

Lake floras also undergo cyclical changes recognizable over a period of years, e.g., those recorded by Bethge (1952-5) for the plankton of ponds; as far as I am aware no long-term studies have been made of the epipelic flora but over periods of two or three years certain species do appear to vary in occurrence independently of seasonal cycles. In a three-year study of the epipelic algal flora of two very small ponds (Round, unpublished) there was evidence of long-term, nonseasonal distribution of certain species. For example, after being absent for two years, *Cymatopleura elliptica* appeared in the flora of one pond while *Euglena* spp., *Cryptomonas ovata,* and *Chroomonas nordstedtii* were present in the spring of one year and not in the succeeding two; in the other pond *Closterium leibleinii* and *Synura uvella* appeared in one year only and *Caloneis amphisbaena, Navicula cuspidata,* and *N. oblonga,* though sporadically present, only produced a pulse of cells in one year. Since the two ponds were within a few meters of one another, there would appear to be no physical barrier to distribution and often a species absent from one pond for at least two years was present in the adjacent one.

Annual or seasonal cycles are much more marked and occur on all lake sediments which I have studied. The complexity of these cycles is shown in Figs. 9-11. The benthic diatom flora as a whole commences growth in February-March and a peak of cell numbers is built up in April-May, after which a fall occurs in June-July prior to a smaller autumn growth which is over by November, after which numbers fluctuate until the next spring. This over-all pattern is similar to that of planktonic species and suggests that similar factors are operating on both communities. If anything, the epipelic flora is slower to react to the changing chemical regime than is the planktonic, due no doubt to the association of the epipelic with the sediments. Individual genera and species exhibit essentially the same cycle, although the species may be somewhat out of phase (Fig. 12). The Cyanophyceae behave similarly to the diatoms in Windermere, but in Blelham Tarn the main growth period tends to extend to midsummer (Round, 1961a). On some shallow sediments there is a tendency for early spring growth peaks followed by midsummer peaks, but there is considerable variation in the time taken to attain maximum numbers for a given species on different sediments. Although the general pattern of seasonal growth is probably dictated by the climatic variation coupled with the availability of nutrients and the rate at which these are utilized, there are in addition other factors or combinations of factors specific to each species which are controlling when nutrients, etc., are favorable.

In ponds there is some similarity of the seasonal cycles to those of lakes, but in general growth tends to be less concentrated in the spring and autumn periods, and the largest cell counts may be found in midsummer (e.g., Round, 1955). Also the initial growth may start earlier in the year than in lakes, e.g., in January or even December (Fig. 13), when the water temperature is below 5°C, and large populations may be built up under ice, e.g., in February of both years. However, the large pulses here, as in lakes, tend to be followed by low periods due to utilization of essential nutrients. In a stream which flowed into a pond from which the data of Fig. 13 were obtained, the seasonal cycles of diatoms on the stream sediments followed a very much simpler pattern with early spring and late spring growths and little growth during the midsummer or autumn periods. In the adjacent pond, growth started at the beginning of the year when water temperature was very low and the decrease after the spring growth period coincided in both stream and pond, suggesting that similar factors are involved.

In some nearby saline streams similar cycles were recorded (Fig. 14), suggesting that this may be a fairly regular pattern for the flora of stream sediments. It is interesting that Douglas (1958) also found that epilithic diatoms on rock and stones in a rapidly flowing stream showed a decrease in the population during periods of low flow which would correspond to the low summer population period of the epipelic flora. Douglas concluded that fluctuations in cell numbers were in fact dependent mainly on the rate of water flow, which may well be so in such a steep rocky stream, but is I think unlikely to be more than an incidental factor in slow-flowing streams, e. g., floods at the end of March caused a setback in the second spring growth of the epipelic community plotted on Fig. 13.

The epipelic subcommunity on salt marshes is exceptional in that the major growth of the diatom population occurs in early summer (Fig. 11) at a time when exposure and desiccation is at a maximum in such a habitat.

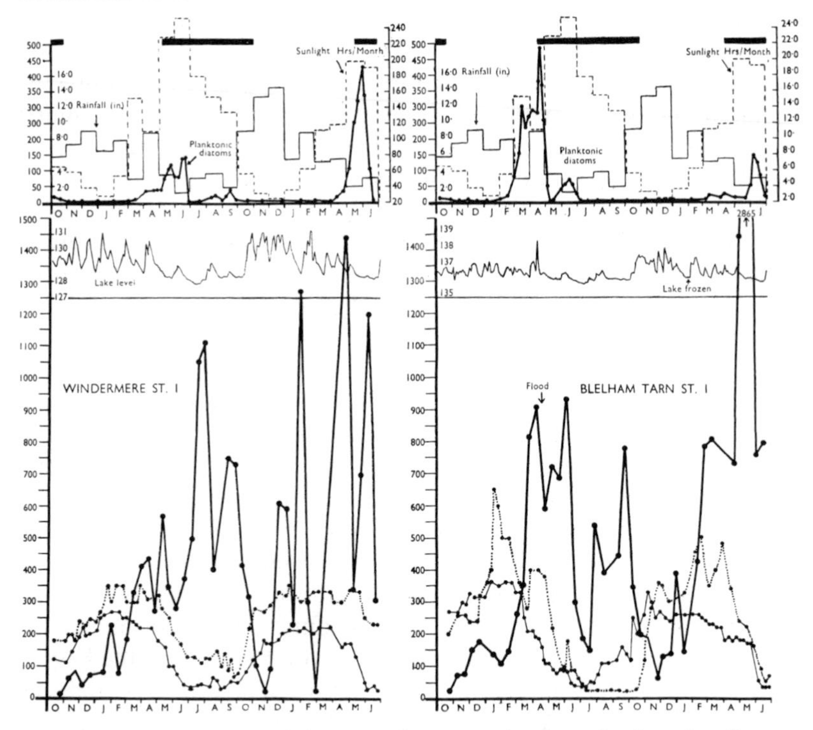

FIG. 9. Seasonal counts of diatom cells on samples from the littoral sediments of three stations in Windermere and three in Blelham Tarn. (Counting methods as in Round, 1953.) On the graphs for Windermere Station I and Blelham Tarn Station I the values for nitrate nitrogen (-- ● --) and silicate silica concentrations (—●—) in the water are also plotted. The scales are the same as those for cell numbers but divided by 100 to give silica in mg/liter and divided by 1000 to give nitrate in mg/liter. Also plotted at the top of these graphs are the water-level changes for the two lakes. Above this is a combined graph of planktonic diatom cell numbers (—●—) and histograms for monthly rainfall □ and sunlight hours ⁚⁚. The period during which the lakes were stratified is shown thus ■. On the graphs for Windermere Station III and Blelham Station III are plotted the surface water temperatures in °C.

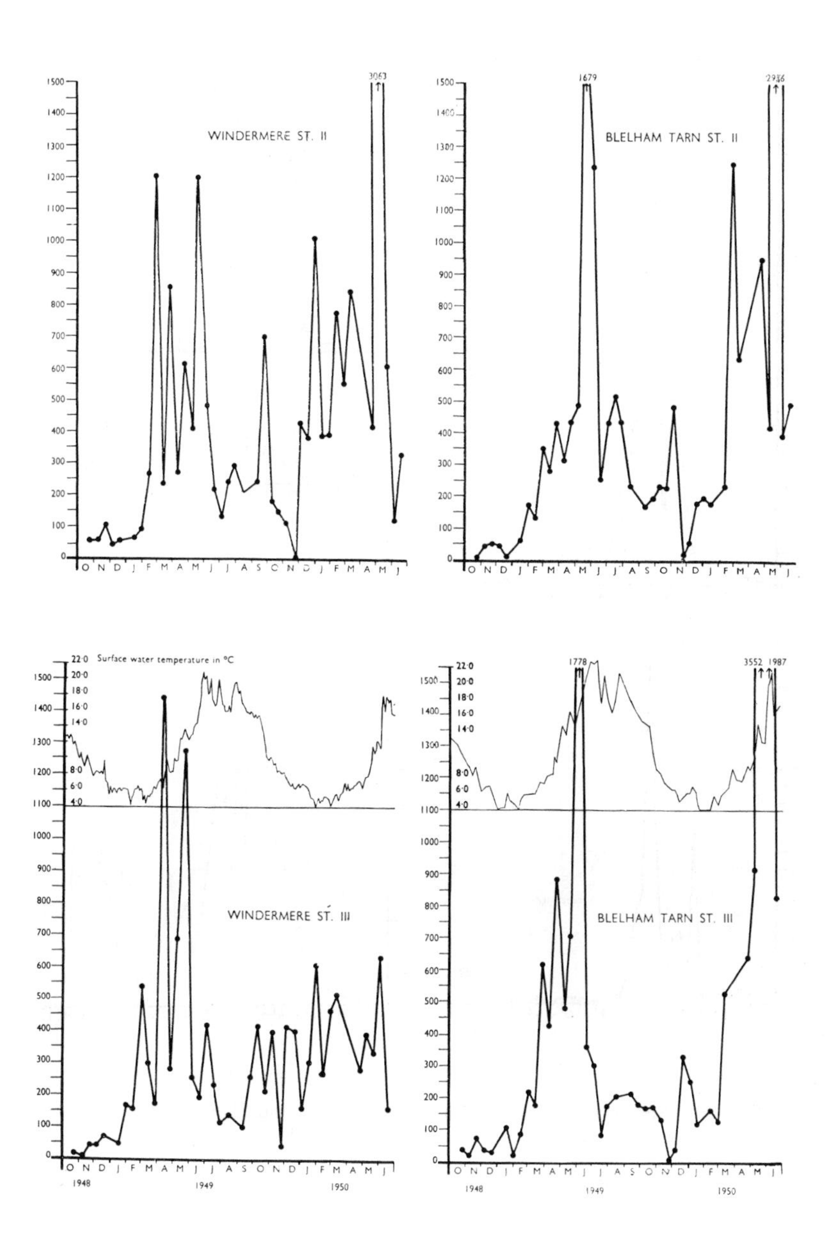
WINDERMERE ST. II
BLEHAM TARN ST. II
WINDERMERE ST. III
BLELHAM TARN ST. III
Surface water temperature in °C
3063
1679
2986
1778
3552 1987
O N D J F M A M J J A S O N D J F M A M J
1948
1949
1950

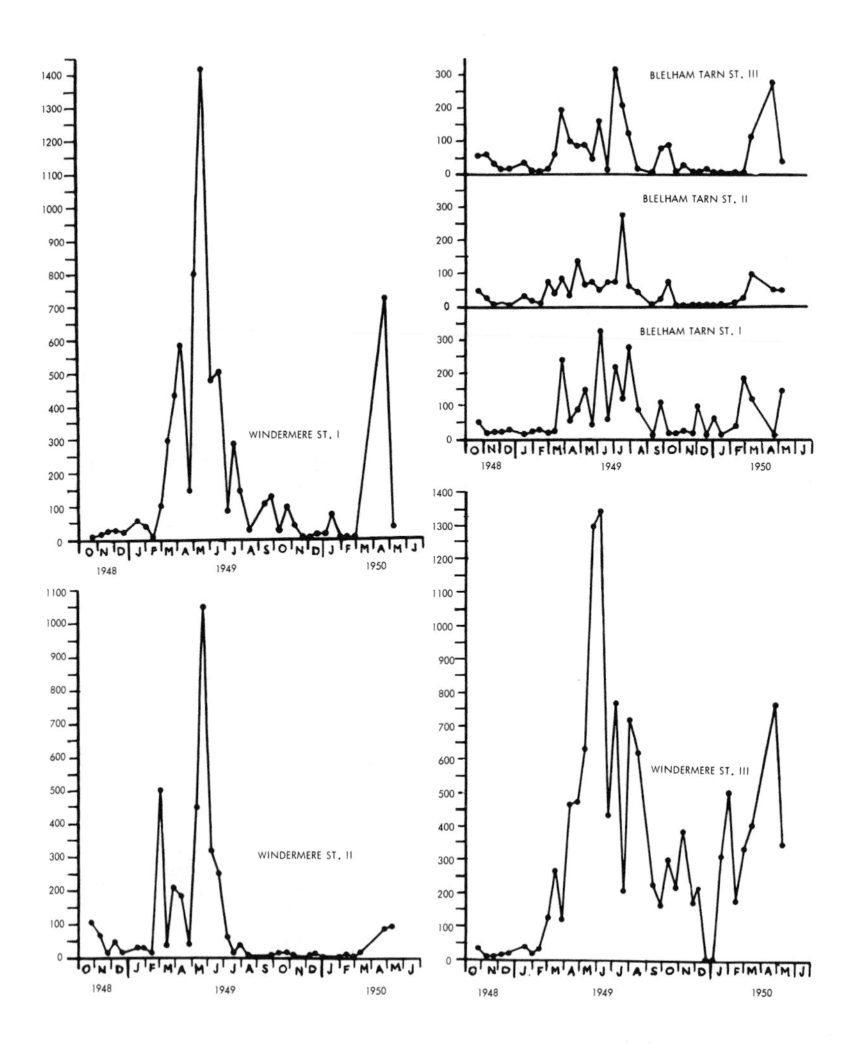

FIG. 10. Seasonal cycles of Cyanophyta colonies and filaments on three sediments in Windermere and three in Blelham Tarn. (Round, 1961a.)

The seasonal cycles of lake epiphytic algae recorded by Jørgensen (1957) showed a large spring and smaller, more extended autumn growth of diatoms on *Phragmites* stems in Furesø and a spring growth only in Lyngbysø. Chlorophyll extraction as a

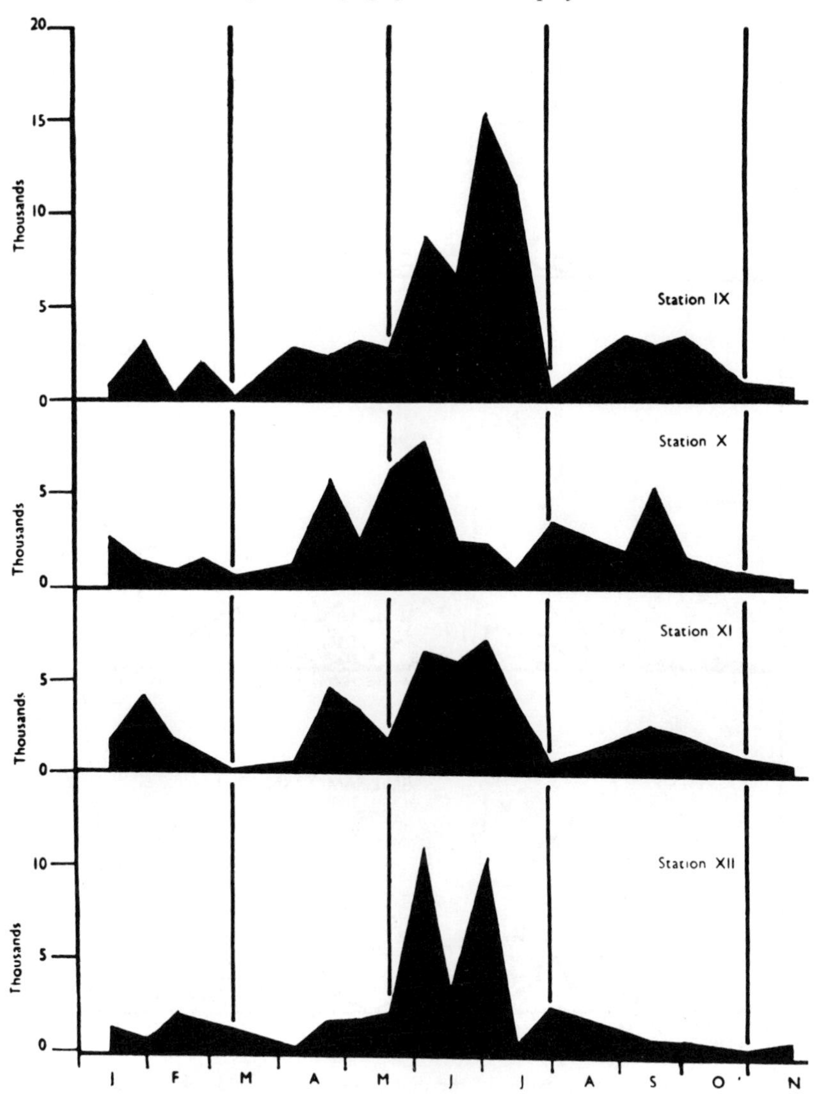

FIG. 11. Seasonal cell counts of diatoms on the estuarine sand at four stations on the lower salt marsh at Neston, Cheshire. (From Round, 1960b.)

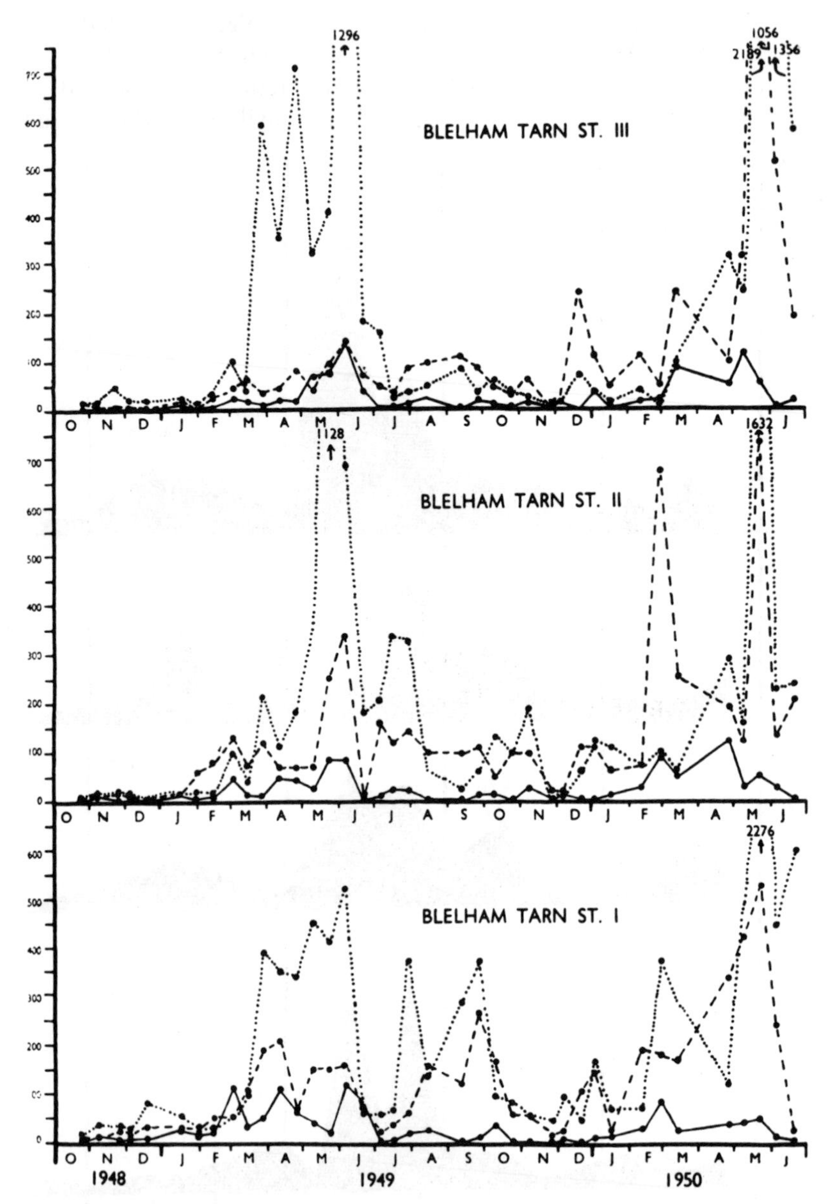

FIG. 12. Seasonal cycles of the diatom genera *Nitzschia* (● .... ●), *Navicula* (● – – – ●), and *Pinnularia* (● —— ●) at three stations in Blelham Tarn. (From Round, 1960a.)

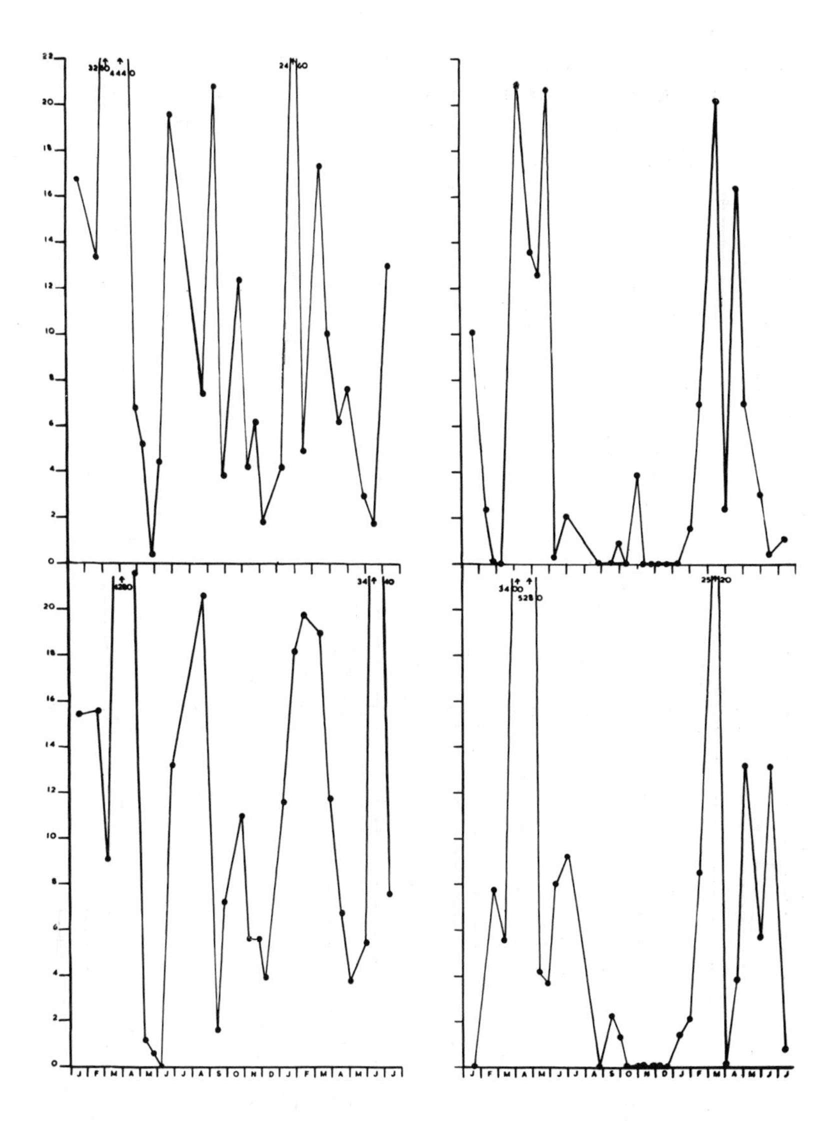

FIG. 13. Seasonal cycles of epipelic diatoms at two stations in a small pond at Hopwood, Worcestershire (left-hand graphs) and at two stations in a small stream supplying this pond (right-hand graphs). (Round, unpublished.) Cell numbers in hundreds/18-mm transect across a coverglass.

measure of the epiphytic subcommunity gave results similar to those of cell counts. In these lakes the maxima of the epiphytic subcommunity never coincided with those of the plankton and the author suggested that the growth of plankton inhibited the

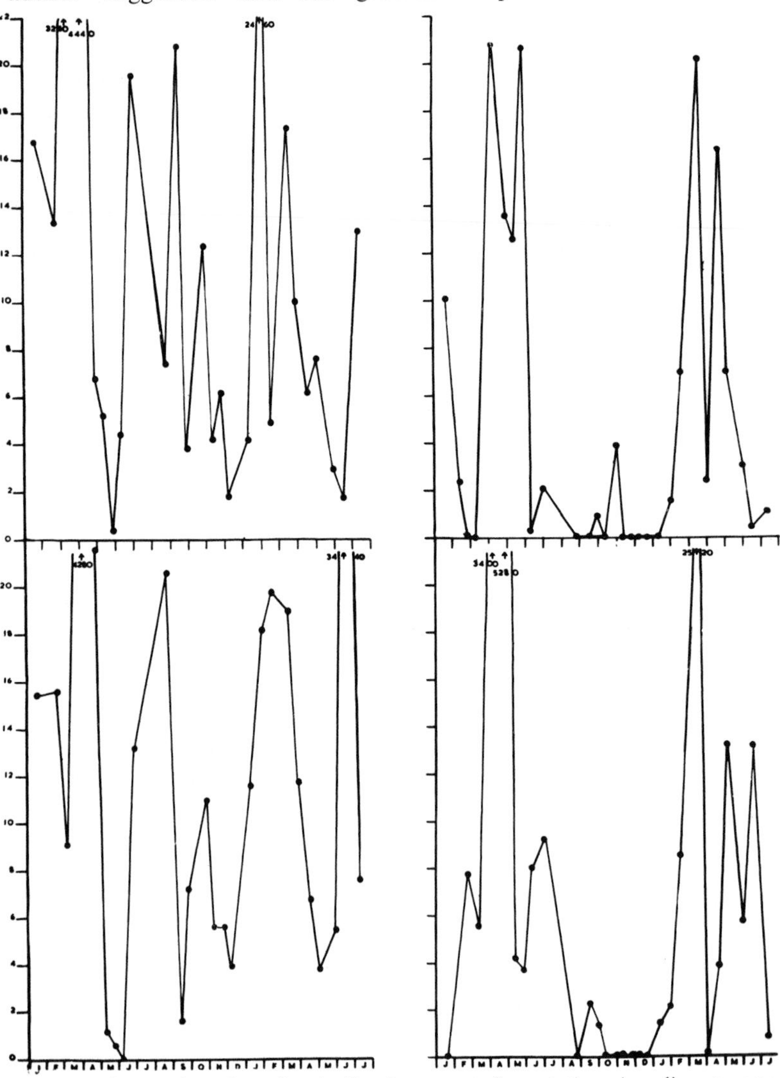

FIG. 14. Seasonal cycles of epipelic diatoms at four stations in saline streams in the Droitwich area, Worcestershire. (Round, unpublished.) Cell numbers in hundreds/18-mm transect across a coverglass.

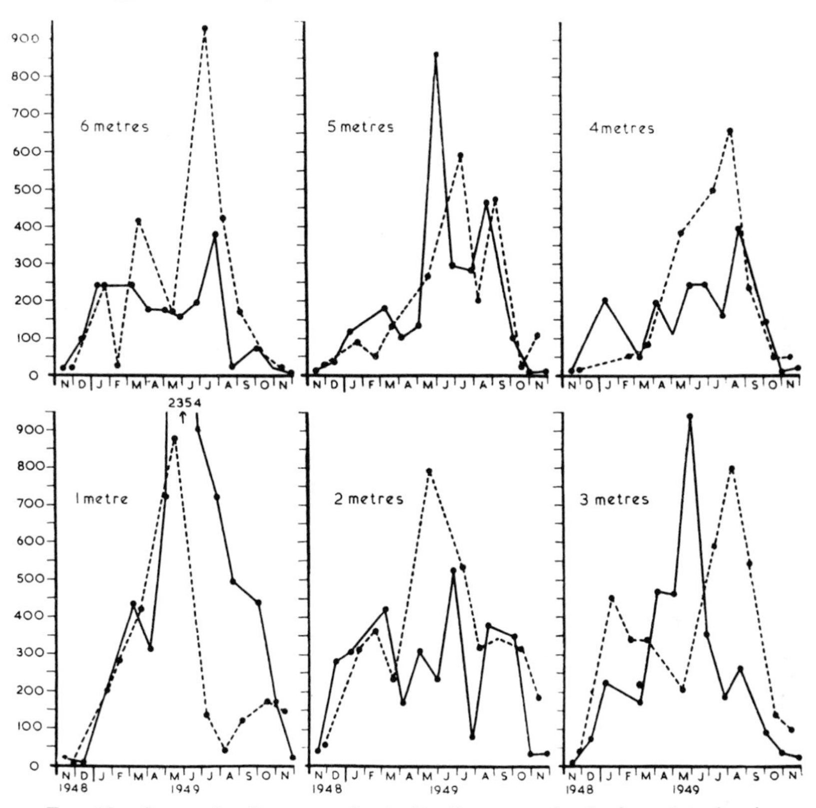

FIG. 15. Seasonal cell counts of epipelic diatoms at depths from 1 to 6 m in Windermere (●——●) and Blelham Tarn (● - - - ●). (From Round, 1961b.)

epiphytes in some way. The epiphytic diatoms were also assumed to utilize silica from the *Phragmites* stems, since this was easily dissolved and decreased from spring until summer. The epiphytic alga *Tabellaria flocculosa* var. *flocculosa* tends to have spring and autumn growths with the latter tending to be the largest—a distinct change from that of epipelic or planktonic communities (Knudson, 1957).

The study of seasonal cycles of epipelic algae is complicated by the effects of depth. In general samples have been taken in shallow water at a depth of about 1 m. In Windermere (Fig. 15) the diatom growth peaks at 1, 3, and 5 m occur in May-June, while at 2, 4, and 6 m there are much less pronounced peaks, suggesting that at these depths there is a continuous operation of some factor(s) which is depleting the population. One of the factors may be current combining with bottom topography causing a scouring at certain depths. In Blelham, which is smaller and almost certainly has less current movement, the seasonal profiles are more alike at each depth and here the tendency is for the shallow station to have a spring peak and the lower stations a midsummer peak. If, however, the total seasonal variation for the 1-6 m stations is plotted, i.e., all the counts at each station are totaled for each sampling date, then the over-all pattern is in accord with the general light and temperature cycle of the region (Fig. 16), the peak of growth of both diatoms and Cyanophyceae being slightly later in the year in Blelham than in Windermere. The only attempt to investigate the productivity of a benthic community by C^{14} uptake is that of Grøntved (1960), and the correspondence between his data for the potential seasonal gross production of the epipelic diatom flora of a Danish fiord and that of Windermere and Blelham estimated by cell counts (Fig. 17) is striking.

Short-term or diurnal fluctuations have not as yet been investigated, but there is no doubt that very frequent sampling would reveal some rapidly changing components in the population, especially among the flagellates.

Internal Factors

The third group of factors are the internal ones, which are a function of the metabolism or of the response to stimuli by the algae. The type of metabolism will be bound up with the distribution of these epipelic algae; thus, many of the flagellates are more abundant on the richer organic sediments from which they may obtain organic metabolites. The greater abundance of some species

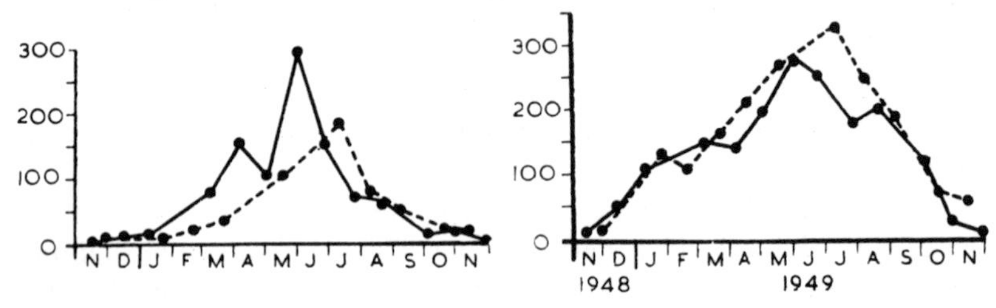

FIG. 16. Seasonal cell counts of the epipelic diatoms (right-hand graph) and Cyanophyta (left-hand graph) at all depth stations (1–6 m) totaled for each sampling date, giving a measure of the annual productivity in the photosynthetic zone of the littoral Windermere (● —— ●) and Blelham Tarn (● - - - - ●). (From Round, 1961b.)

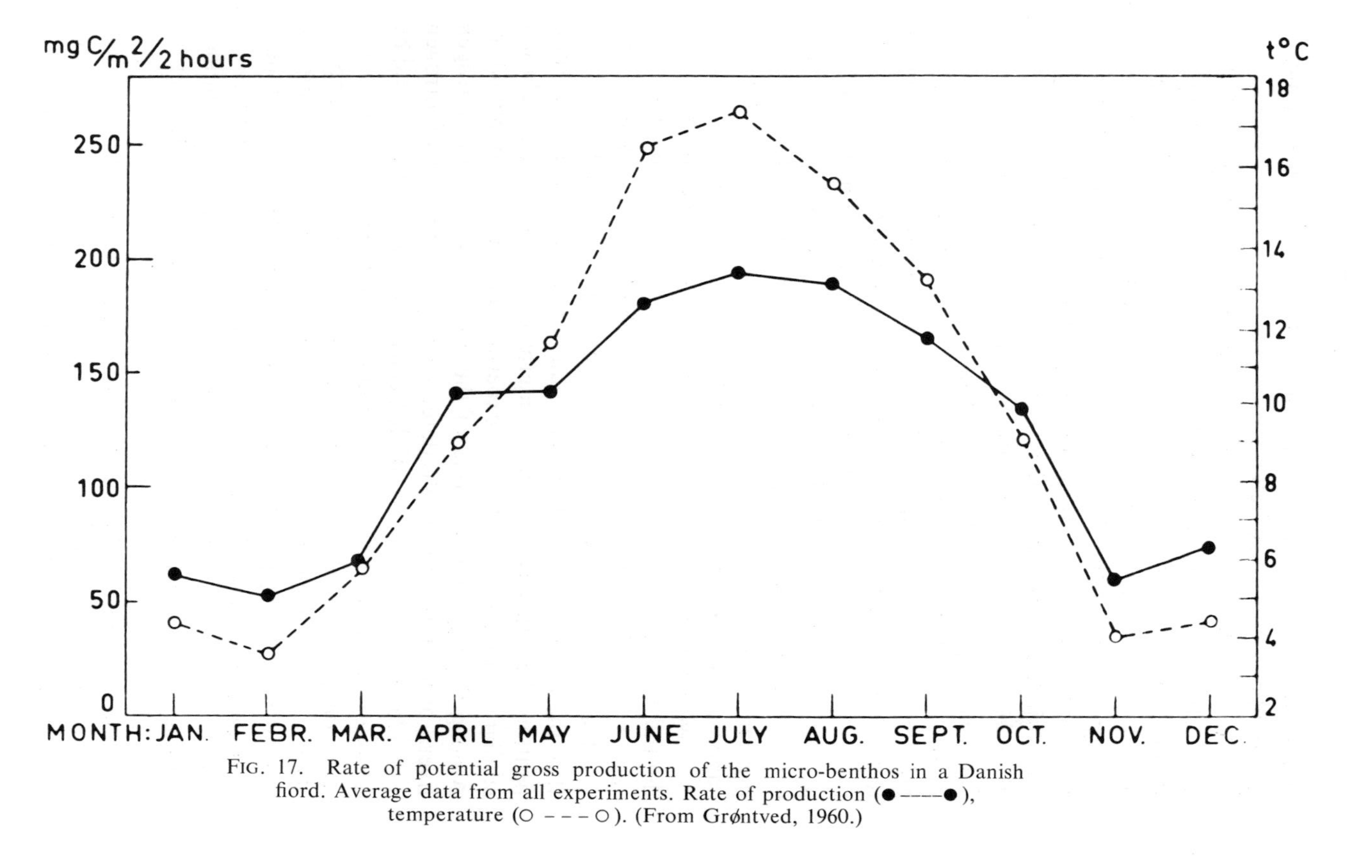

FIG. 17. Rate of potential gross production of the micro-benthos in a Danish fiord. Average data from all experiments. Rate of production (●——●), temperature (○ - - - ○). (From Grøntved, 1960.)

at the lower depths, e.g., some of the Cyanophyta, may indicate the ability to synthesize heterotrophically.

Among the diatoms, requirements for organic factors have been shown for several marine littoral species, and in some instances strains of the same species require different organic compounds (Lewin, 1953). This ability to utilize organic molecules may greatly assist the flora at 25 m and 30 m off the coast at La Jolla and almost certainly affects the growth and composition of the flora on the more polluted sediments in San Diego Bay, where *Nitzschia* and *Amphora* spp. are abundant and *Navicula, Diploneis,* and *Pleurosigma* less frequent. Suggestions have been made that there is an increase in the pigment content of epipelic algae at greater depths and certainly the chromatophores appear visually to be deeper in color and extending over a greater area of the cell; this is particularly noticeable in the marine littoral diatoms and in a marine *Holopedium* which is pink deepening to red in the deeper water. Since in planktonic algae there is a relationship between pigment content and rate of photosynthesis, it is reasonable to suppose that this holds for the epipelic species.

Whether or not this is a biological advantage at the lower depths remains to be investigated. Conversely in the shallowest regions the pigment content appears to be lower; this may be due to bleaching and thus here the rate of synthesis may be reduced. If in the epipelic associations cell numbers are a measure of carbon fixation then this is greatest at intermediate depths, a conclusion similar to that obtained from experimental studies on plankton. Among the Cyanophyta none of the epipelic species are nitrogen fixers, at least as far as this is detectable by the presence of nitrogen vacuoles in the cytoplasm. Biologically, nitrogen fixation would be a disadvantage in the epipelic habitat, since it would confer buoyancy on the cells and remove them from the habitat (see above). Reports that nitrogen-fixing blue-green algae grow in the epipelic subcommunity are found in the literature and indeed flocs of algae including blue-greens do rise off the sediments in some waters and float to the surface, but this is seen only during periods of intense photosynthesis when gas bubbles become trapped in and under the algal mat.

Other important but uninvestigated aspects of the benthic flora are concerned with the rates of growth, the light requirements and light saturation of the species, and for some species the frequency of sexual reproductive processes. If size is related to rate of cell division, and I suspect it is, then the abundance of small *Nitzschia* spp. (see Fig. 12) is not surprising. A study of the relative rates of carbon fixation of the various species would be most instructive and if this were coupled with estimates of light saturation might lead to an explanation of some of the distribution patterns. Apart from a few flagellates, the epipelic flora consists of species which never or rarely undergo sexual reproduction, or at least this has not been detected in studies of the natural populations, although it may be responsible for some of the unexplained fluctuations in cell numbers. The importance of this aspect is likely to be much greater in small bodies of water where flagellates and Chlorococcales are more abundant.

A characteristic of almost all epipelic species is the power of movement, which applies even to the apparently nonmotile colonial Cyanophyta. This movement is a prerequisite of life on the sediments, which are continually being disturbed by currents and animals. It is assumed that the movement is a phototaxis, either a phobotactic or topotactic response, and recent work on both diatoms and Cyanophyta has determined some of the features of this movement, e.g., the action spectra (Haupt, 1959; and Nultsch, 1956, 1962). This reaction is used in the estimation of cell numbers, and since phototaxis is generally, although not always, associated with pigmented cells, it is not known how much of the epipelic community consists of unpigmented species which may still be motile. Certainly many unpigmented flagellates are found in the samples, but they are not concentrated by the method as are the pigmented forms.

In many communities where movements occur, there is also a rhythm induced by the alternate light/dark phases and there may be a reversal of the direction of movement, e.g., in the epipelic community on salt marshes; however, this is also connected with tidal movements. For instance, on the Avon banks *Euglena, Pleurosigma,* and *Nitzschia* only rise to the surface of the sediment in the light, while *Surirella* does so whatever the light conditions. Whether or not there is any such movement out of and down into the sediments when they are permanently covered is not known.

A further problem concerns the movement of species up or down the depth profile. It would be expected that over a period of time cells would all move up the slope into the region of most favorable illumination, and in fact they may attempt to do this, but they are being constantly moved out of this zone by other agents. Coupled with this is the movement of flagellates, which move to the surface of the sediment after disturbance. But do these continue out into the water mass and become lost to the sediments? Alternatively they may be retained at the surface of the sediment by some chemotactic effect. On the other hand, it may be that in large bodies of water the flagellates are merely casuals among the epipelic flora and really belong to the littoral plankton or to the metaphyton. Indeed, wherever there is more current flow, e.g., in streams or in the marine littoral, the flagellate population of the sediments is extremely small.

Competition, etc.

The interrelationships between the species and parasitism and grazing in the epipelic is an important but relatively uninvestigated aspect of the flora. It is possible to detect what appear to be interreactions between species, but this may be merely a response to different factors. The interrelations are also bound up with the topics mentioned above, such as relative growth rates and size. It would be surprising if there were no such phenomena in a community growing at an interface between solid and liquid media, such as are detectable in soil floras where mosaics of blue-green algae, diatoms, and green algae can be observed on some favorably colored soils.

On the other hand, in the aquatic medium there is much greater mixing of the flora. Parasitism by fungi and animals does occur, but has only been observed regularly in small bodies of water. Grazing is obvious when sampling the epipelic community, however, and large numbers of cells may be removed by ciliates, etc. This is coupled with selective feeding, so that species of genera such as *Nitzschia*, small species of *Navicula*, and Chlorococcales may be removed. As far as I am aware no attempt has been made to evaluate this. In some small ponds the epipelic flora may be conditioned by the grazing so that only species resistant to passage through the animals survive, e.g., in small ponds on skerries off the south coast of Finland, where only species of *Scenedesmus* are abundant (Round, 1959).

LITERATURE CITED

Behre, K. 1956. Die Algenbesiedlung einiger Seen um Bremen und Bremerhaven. *Ver. Inst. f. Meeresforsch. Bremerhaven* 4:221-283.

Bethge, H. 1952-1955. Beiträge zur Kenntnis des Teichplanktons I, II, III. *Ber. d. Deut. Bot. Ges.* 65:187-196; 66:93-100; 68:319-330.

Cannon, D., Lund, J. W. G., and Sieminska, J. 1961. The growth of *Tabellaria flocculosa* (Roth) Kütz. var. *flocculosa* (Roth) Knuds. under natural conditions of light and temperature. *J. Ecol.* 49:277-287.

Cooke, W. B. 1956. Colonization of artificial bare areas by microorganisms. *Bot. Rev.* 22:613-638.

Douglas, B. 1958. The ecology of the attached diatoms and other algae in a small stony stream. *J. Ecol.* 46:295-322.

Düringer, I. 1958. Über die Verteilung epiphytischer Algen auf den Blättern wasserbewohnender Angiospermen. *Öst. Bot. Z.* 105:1-43.

Grøntved, J. 1960. On the productivity of microbenthos and phytoplankton in some Danish fiords. *Medd. Dan. Fish. og Havundersøg. N.S.* 3:55-92

Haupt, W. 1959. Die Phototaxis der Algen. *Encycl. Plant. Phys.* 17(1):318-370.

Hustedt, F. 1959. Die Diatomeenflora der Unterweser von der Lesummündung bis Bremerhaven mit Berücksichtigung des Unterlaufs der Hunte und Geeste. *Ver. Inst. Meeresforsch. Bremerhaven* 6:13-176.

Hutchinson, G. E. 1961. The paradox of the plankton. *Am. Nat.* 95:137-146.

Jørgensen, E. G. 1957. Diatom periodicity and silicon assimilation. *Dansk. Bot. Ark.*t 18:1-54.

Knudson, B. M. 1957. Ecology of the epiphytic diatom *Tabellaria flocculosa* (Roth) Kütz. var. *flocculosa* in three English lakes. *J. Ecol.* 45:93-112.

Lewin, J. C. 1953. Heterotrophy in Diatoms. *J. Gen. Microbiol.* 9:305-313.

Lund, J. W. G. 1942. The marginal algae of certain ponds with special reference to the bottom deposits. *J. Ecol.* 30:245-283.

Lund, J. W. G. 1949. Studies on *Asterionella.* I. The origin and nature of the cells producing seasonal maxima. *J. Ecol.* 37:389-419.

Lund, J. W. G. 1954. The seasonal cycle of the plankton diatom *Melosira italica* (Ehr.) Kütz. subsp. *subarctica* O. Müll. *J. Ecol.* 42:151-179.

Margalef, R. 1949. Une nouvelle méthode limnologique pour l'étude du periphyton. *Verh. Int. Ver. theoret. u. angew. Limnol.* 10:284-285.

Margalef, R. 1953. Materiales para la hidrobiología de la isla de Mallorca. *Inst. de Biol. Applicada, Barcelona* 15:5-11.

Nauman, E. 1920. Några synpunkter angående de limniska avlagvingarnas terminologi. *Svensk. Geol. Unders. Årsbok (Ser. C.)* 14.

Nauman, E. 1927. Zur Kritik des Planktonbegriffe. *Ark. f. Bot.* 21A:1-18.

Nultsch, W. 1956. Studien über die Phototaxis der Diatomeen. *Arch. Protistenk.* 101:1-68.

Nultsch, W. 1961. Der Einfluss des Lichtes auf die Bewegung der Cyanophyceen. I. Mitteilung. Phototopotaxis von *Phormidium autumnale. Planta* 56:632-647.

Round, F. E. 1953. An investigation of two benthic algal communities in Malham Tarn, Yorkshire. *J. Ecol.* 41:174-197.

Round, F. E. 1955. Some observations on the benthic algal flora of four small ponds. *Arch. f. Hydrobiol.* 50:111-135.

Round, F. E. 1956. The phytoplankton of three water supply reservoirs in central Wales. *Arch. f. Hydrobiol.* 52:457-469.

Round, F. E. 1957a. The late-glacial and the post-glacial diatom succession in the Kentmere valley deposit. *New Phytol.* 56:98-126.

Round, F. E. 1957b. Studies on bottom-living algae in some lakes of the English Lake District. Part I. Some chemical features of the sediments related to algal productivities. *J. Ecol.* 45:133-148.

Round, F. E. 1957c. Studies on bottom-living algae in some lakes of the English Lake District. Part II. The distribution of Bacillariophyceae on the sediments. *J. Ecol.* 45:342-360.

Round, F. E. 1959. The composition of some algal communities living in rock pools on skerries near the Zoological Station at Tvarminne, S. Finland. *Soc. Sci. Fenn. Comm. Biol.* 217:1-13.

Round, F. E. 1960a. Studies on bottom-living algae in some lakes of the English Lake District. Part IV. The seasonal cycles of the Bacillariophyceae. *J. Ecol.* 48:529-547.

Round, F. E. 1960b. The diatom flora of a salt marsh on the River Dee. *New Phytol.* 59:332-348.

Round, F. E. 1960c. The epipelic algal flora of some Finnish lakes. *Arch. Hydrobiol.* 57:161-178.

Round, F. E. 1961a. Studies on the bottom-living algae in some lakes of the English Lake District. Part V. The seasonal cycles of the Cyanophyceae. *J. Ecol.* 49:31-38.

Round, F. E. 1961b. Studies on the bottom-living algae in some lakes of the English Lake District. Part VI. The effect of depth on the epipelic algal community. *J. Ecol.* 49:245-254.

Round, F. E. 1961c. The diatoms of a core from Esthwaite Water. *New Phytol.* 60:45-59.

Round, F. E. Unpublished. The benthic diatom flora of the marine littoral sediments off La Jolla, California.

Round, F. E. Unpublished. A seasonal survey of the benthic algal flora of saline and nonsaline streams in the Droitwich, Worcestershire area.

Round, F. E. Unpublished. The flora, its seasonal cycles and the chemistry of a pond and stream at Hopwood, Worcestershire.

Round, F. E. Unpublished. A three-year survey of the benthic algal flora and the chemistry of the waters of two small ponds.

Schlüter, M. 1961a. Zur Bedeutung der litoralen Diatomeen in unseren Gewässern. *Zeitschr. f. Fisch. u. deren Hilfswissensch.* 10(N.F.):351-359.

Schlüter, M. 1961b. Die Diatomeen-Gesellschaften des Naturschutzgebietes Strausberg bei Berlin. *Int. Revue ges. Hydrobiol.* 46:562-609.

Sládečková, A. 1960. Limnological study of the reservoir Sedlice near Želiv. XI. Periphyton stratification during the first year-long period (June, 1957-July, 1958). *Sci. Papers Inst. Chem. Tech. Prague* 4:143-261.

Sládečková-Vinnikova, A. 1956. Der Aufwuchs und seine Bedeutung in der Hydrobiologie. *Biológia* 2:724-730.

Symoens, J. J. 1957. Les eaux douces de l'Ardenne et des regions voisines: Les milieux et leur vegetation algale. *Bull. Soc. Roy. Bot. Belg.* 84:111-314.

Talling, J. F. 1955. The relative growth rates of three plankton diatoms in relation to underwater radiation and temperature. *Ann. Bot., N.S.* 19:329-341.

Talling J. F. 1957. The phytoplankton population as a compound photosynthetic system. *New Phytol.* 56:133-149.

Talling, J. F. 1960. Comparative laboratory and field studies of photosynthesis by a marine planktonic diatom. *Limnol. and Ocean.* 5:62-77.

Wesenberg-Lund, C. 1908. Plankton investigations of the Danish lakes. General part: The Baltic fresh-water plankton, its origin and variation. 389 pp. Gyldendalske Boghandel, Copenhagen, Denmark.

A Discussion of Natural and Abnormal Diatom Communities

Ruth Patrick

Chairman, Department of Limnology
Academy of Natural Sciences
Philadelphia, Pennsylvania

Diatoms, which are unicellular plants with cell walls of silicon, have long been of interest to biologists. They belong to the Chrysophyta, which have as one of their characteristics the storage of oil rather than carbohydrate as a reserve food.

Their distribution throughout the world has long been of interest to students in this field. Ehrenberg (1854), in his *Mikrogeologie,* was among the first to present tables showing the geographical distribution of various species. As time went on not only was their geographical distribution of interest, but also the ecological conditions under which they occurred. That is, was the water warm or cold; was it marine, brackish, or fresh? In the first half of the twentieth century elaborate systems were worked out by Kolbe (1932), Petersen (1943), and Hustedt (1953) to describe the chloride content that various species of diatoms could tolerate. The tolerances of diatoms to other ecological factors have been treated by various workers, such as Krasske (1939), Hustedt (1937-38, 1957), Foged (1947-48, 1950), Jørgensen (1948, 1950), Møller (1950), and Fjerdingstad (1950, 1960). These studies led to the realization that there were associations of diatoms which were characteristic of various types of water such as cold arctic seas or temperate waters, marine or brackish water or fresh water. P. T. Cleve and Astrid Cleve-Euler were among the first to show the value of the use of diatom associations in determining the geological history of an area. They and later workers, such as Backman and Cleve-Euler (1922), Lundquist (1927), Fontell (1917), Hustedt (1924), Lindberg (1910), Halden (1929), and Hyyppa (1936), have contributed much to our knowledge of glaciation in Scandinavia by studies of the diatom associations in various fossil sediments.

More recently, diatom associations have been used to describe the history of the development of lakes. Such studies as those of Krieger (1929), Patrick (1936, 1939, 1943, 1946), Pennington (1943), and Fjerdingstad (1954) have pointed out physical and chemical changes in the lake history as indicated by changes in the association of diatoms. Working independently, Hutchinson, Patrick, and Deevey (1956) have shown that the changes in Lake Patzcuaro, Mexico, indicated by the chemical and physical analyses of the sediments, corresponded with the changes indicated by diatom communities. In a similar way Vallentyne (1953), working on the pigments in the sediments, and Patrick (1954), working with the diatoms, have reached the same conclusions as to the past history of the flora of Bethany Bog, Connecticut.

The importance of diatom associations in describing recent bodies of water has also been studied. For example, Cleve (1896), Hart (1935), and Hendey (1937) have pointed out that currents in the seas can be identified by the diatom associations found in them at a given season of the year. The characterizing of various lake conditions by diatom associations is well illustrated by work such as Jørgensen's (1948) on diatom communities in some Danish lakes and Patrick's (1945) on lakes and streams of the Pocono Plateau, Pennsylvania.

In all of these studies the emphasis has been on relating an association of diatoms rather than specific species to a general ecological condition such as fresh, brackish, or marine water; arctic or temperate floras; and oligotrophic or eutrophic water. Also, the emphasis has been to have these associations indicate general ecological conditions rather than specific chemical concentrations.

More recently the importance of considering the structure of the diatom community has also been emphasized as an important aspect of diatom associations. Patrick (1949) showed that in natural streams of the Conestoga Basin in Pennsylvania the diatom community (in fact the total algal population) was characterized by the presence of many species, most of which were represented by relatively small populations. Patrick, Hohn, and Wallace (1954) found that the truncated normal curve was a reasonably good expression of the structure of a sample of a diatom community in a eutrophic stream such as Ridley Creek in Pennsylvania, and probably the normal curve is a fairly good representation of the complete flora; but since in a river it is practically impossible ever to study all specimens composing the flora, this will be difficult to prove. Furthermore, as pointed out in 1954 and confirmed by several hundred later studies, the number of species composing this total community in any given region does not change greatly from season to season or from year to year if drastic changes in the environment do not occur. However, the kinds of species show great change (Patrick, 1961). An analysis of the diatoms from the same area in the Savannah River, which supports a natural diatom flora, showed that only 39 species or 12% of the species identified in January, May, July, and October 1958, were common to all four studies. Furthermore, of the 150 species determined in January, 34% occurred only in that study; of the 154 for May, 28% were found only at that time; of the 149 identified in July, 30% were found only then; and of the 139 identified in October, 25% were found only then. Furthermore, when we compared the change in kinds of species in the same area from year to year during October, we found that only 11% were common to all three studies, and the percents of the populations confined to one study ranged from 32% to 44%.

In an effort to determine how the structure of diatom communities varies in different types of water, a brackish water estuary of Middle River, Maryland, was studied (Patrick and Hohn, 1956). The results of these studies showed that the structure of the diatom community was quite similar to that found in Ridley Creek (Figs. 1, 2), although the number of species composing the community was a little less (Middle River, 132; Ridley Creek, 160-225). Of course, the kinds of species were very different as one would expect in a brackish water condition. Likewise the structure of the diatom population in an oligotrophic-to-mesotrophic stream (McMichaels Creek, Pocono Plateau, Pennsylvania—Fig. 3) was quite similar to those of soft eutrophic streams.

This similarity of the structure of the diatom community for brackish water estuaries and eutrophic fresh-water streams has been confirmed by many later studies. Furthermore, because the structure of these communities does not change greatly unless severe changes occur in the environment, we have been able to calculate the 95% and 99% confidence intervals for these curves (Patrick and Strawbridge, in press). The ellipses describing these

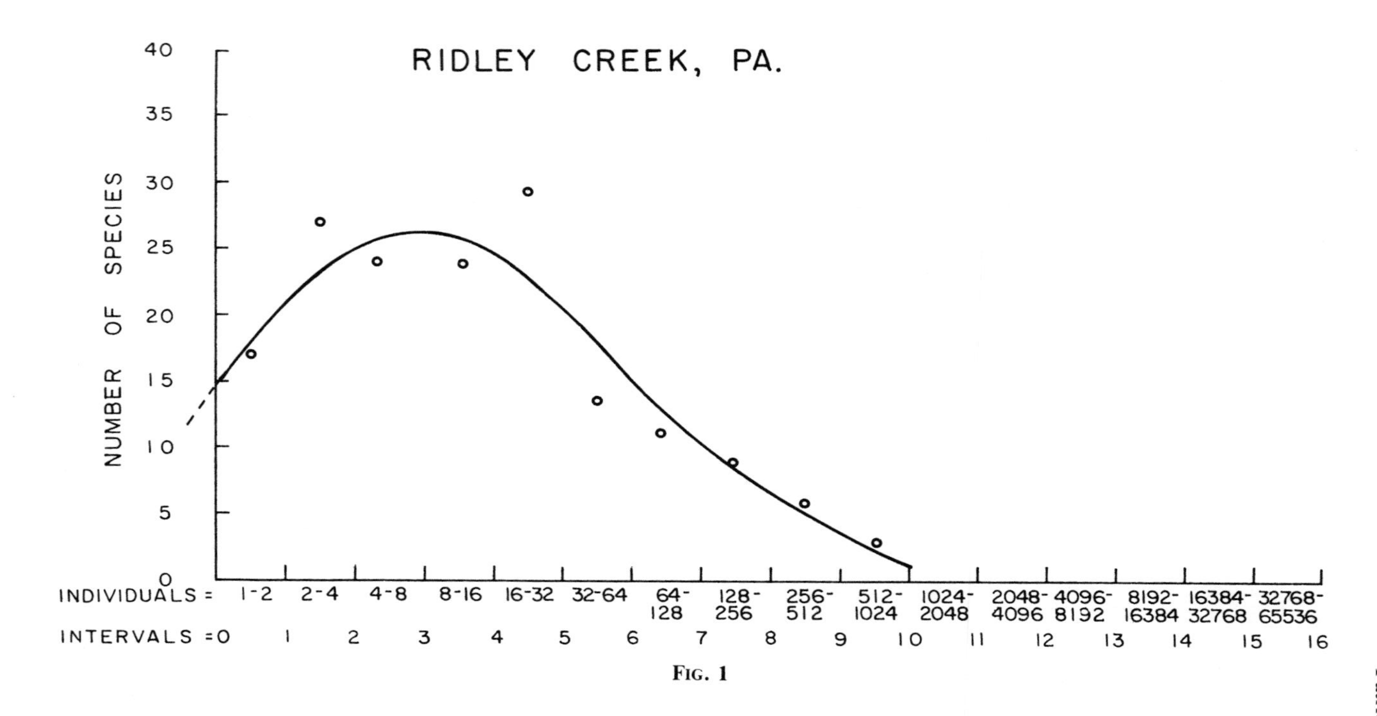
RIDLEY CREEK, PA.
NUMBER OF SPECIES
40
35
30
25
20
15
10
5
0
INDIVIDUALS =
1-2
2-4
4-8
8-16
16-32
32-64
64-128
128-256
256-512
512-1024
1024-2048
2048-4096
4096-8192
8192-16384
16384-32768
32768-65536
INTERVALS = 0
1
2
3
4
5
6
7
8
9
10
11
12
13
14
15
16

FIG. 1

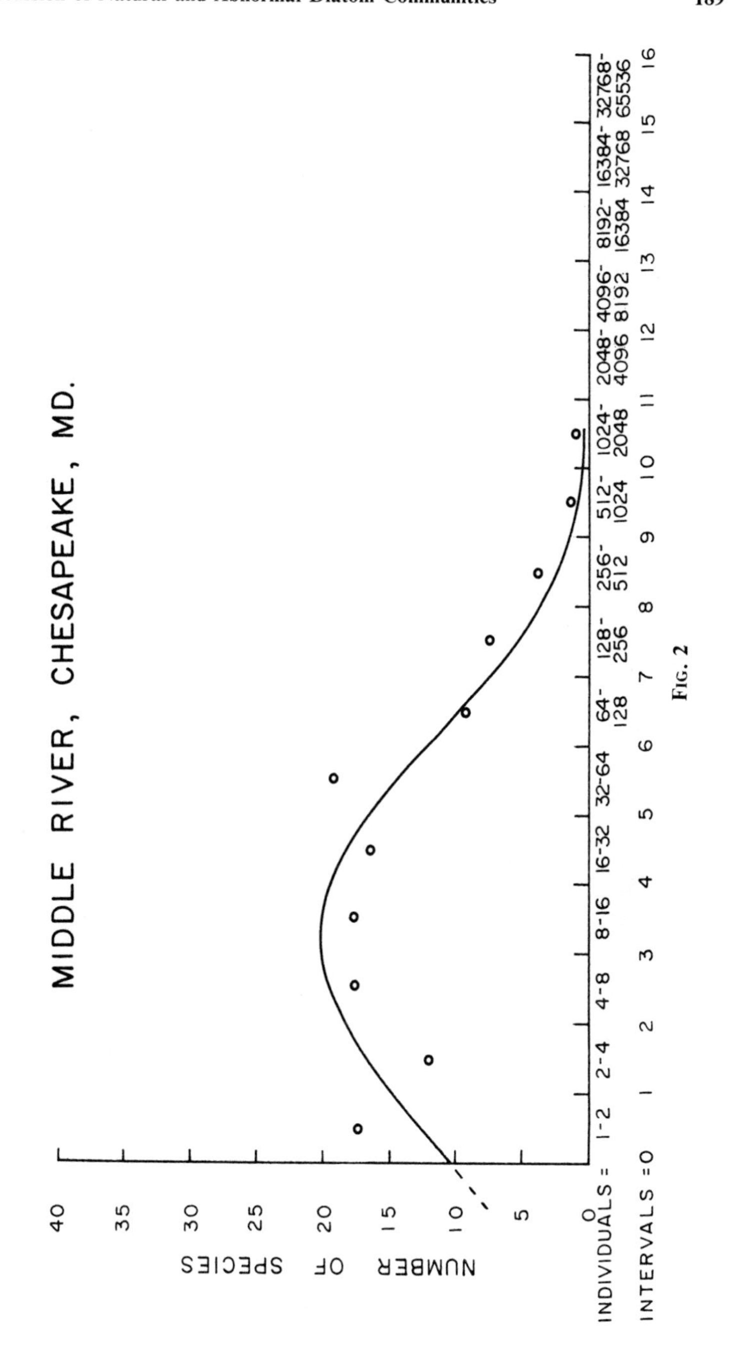
MIDDLE RIVER, CHESAPEAKE, MD.
NUMBER OF SPECIES
40
35
30
25
20
15
10
5
0
INDIVIDUALS =
1-2
2-4
4-8
8-16
16-32
32-64
64-128
128-256
256-512
512-1024
1024-2048
2048-4096
4096-8192
8192-16384
16384-32768
32768-65536
INTERVALS = 0
1
2
3
4
5
6
7
8
9
10
11
12
13
14
15
16

Fig. 2

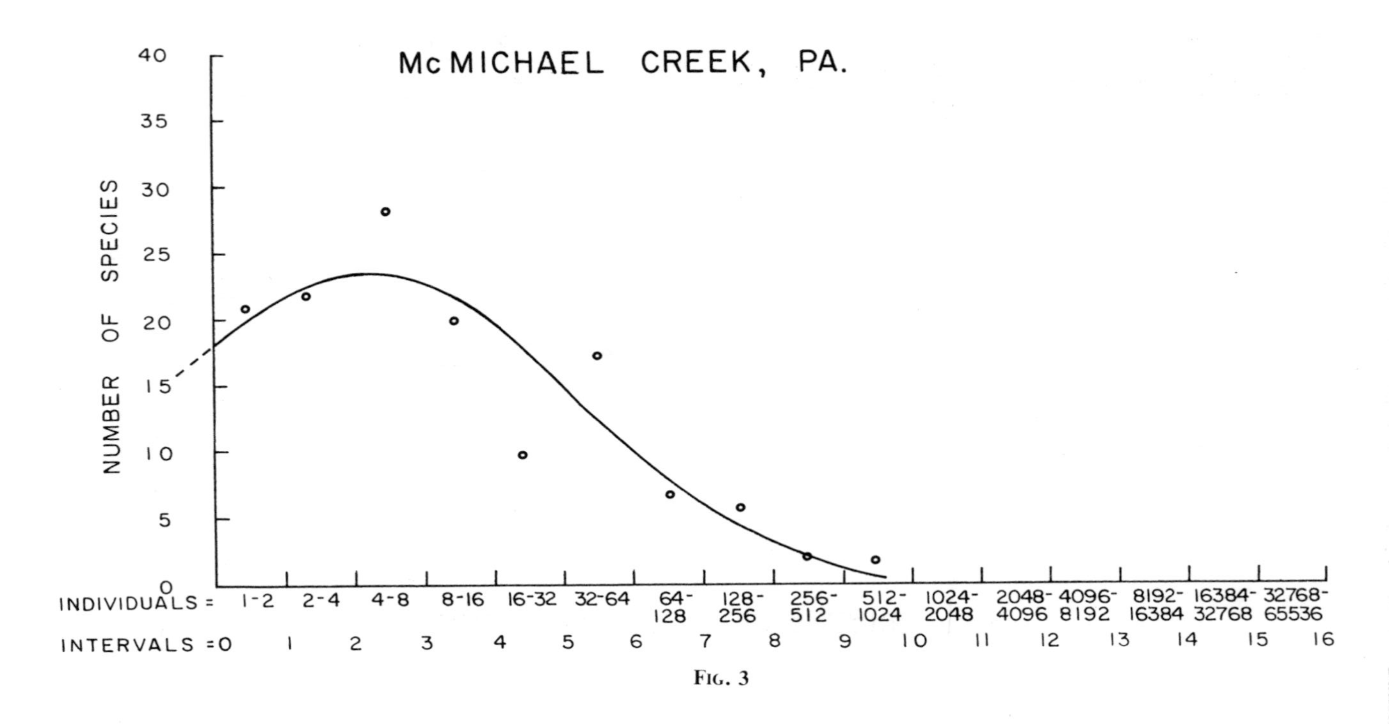
Mc MICHAEL CREEK, PA.
NUMBER OF SPECIES
40
35
30
25
20
15
10
5
0
INDIVIDUALS =
1-2
2-4
4-8
8-16
16-32
32-64
64-128
128-256
256-512
512-1024
1024-2048
2048-4096
4096-8192
8192-16384
16384-32768
32768-65536
INTERVALS =0
1
2
3
4
5
6
7
8
9
10
11
12
13
14
15
16

Fig. 3

variables, based on 50 fresh-water studies and 25 brackish water studies, are shown in Fig. 4 and described in the paper referred to above. It is interesting to note that there is an inverse correlation between the height of the mode and sigma squared. That is, the higher the mode which indicates a large number of species composing the community, the more similar the sizes of the populations of most of the species.

Studies of the structure of the diatom community in a dystrophic stream, however, show a very different picture (Fig. 5). In order to place the mode of the curve in the interval where the populations of the species are represented by two to four specimens, 17,709 specimens had to be counted. A similar count in a eutrophic stream (Patrick *et al.,* 1954) would place the mode in the interval where the species have populations of 8 to 16 specimens. The height of the mode was about 11 species instead of 24 for the eutrophic stream, and the number of observed species was 70 instead of 173 when similar counts were made of the two types of streams. Also, in the dystrophic stream a few species were represented by a great many specimens; as a result the curve covered 13 intervals rather than 10.

The kinds of species composing the communities were very different. This would be expected as the pH, when measured, was always about 5 in the dystrophic stream whereas it was circumneutral in the eutrophic stream. Likewise, the chemical constituents of the water were very different, the greatest difference being that one was a typical dark water rich in humates and the other one was a typical clear eutrophic soft water. The dominant species in the dystrophic stream were *Eunotia sudetica* O. Müll., *Eunotia tenella* (Grun.) Hust. *in* Pascher, *Eunotia pectinalis* var. *ventralis* (Ehr.) Hust., and *Eunotia pectinalis* var. *minor* (Kütz.) Rabh. Also very common was *Fragilaria virescens* var. *capitata* Østr. In natural eutrophic streams the dominant species belong to such genera as *Achnanthes, Navicula, Nitzschia,* and *Synedra.*

Recent data which we have accumulated indicate that springs of mesotrophic streams which have relatively constant chemical and physical characteristics have a similar structure to that found in dystrophic streams, That is, they are characterized by a community consisting of relatively few species, and many of these species are characterized by large to very large populations. More work is being done to try to determine the cause of this type of structure.

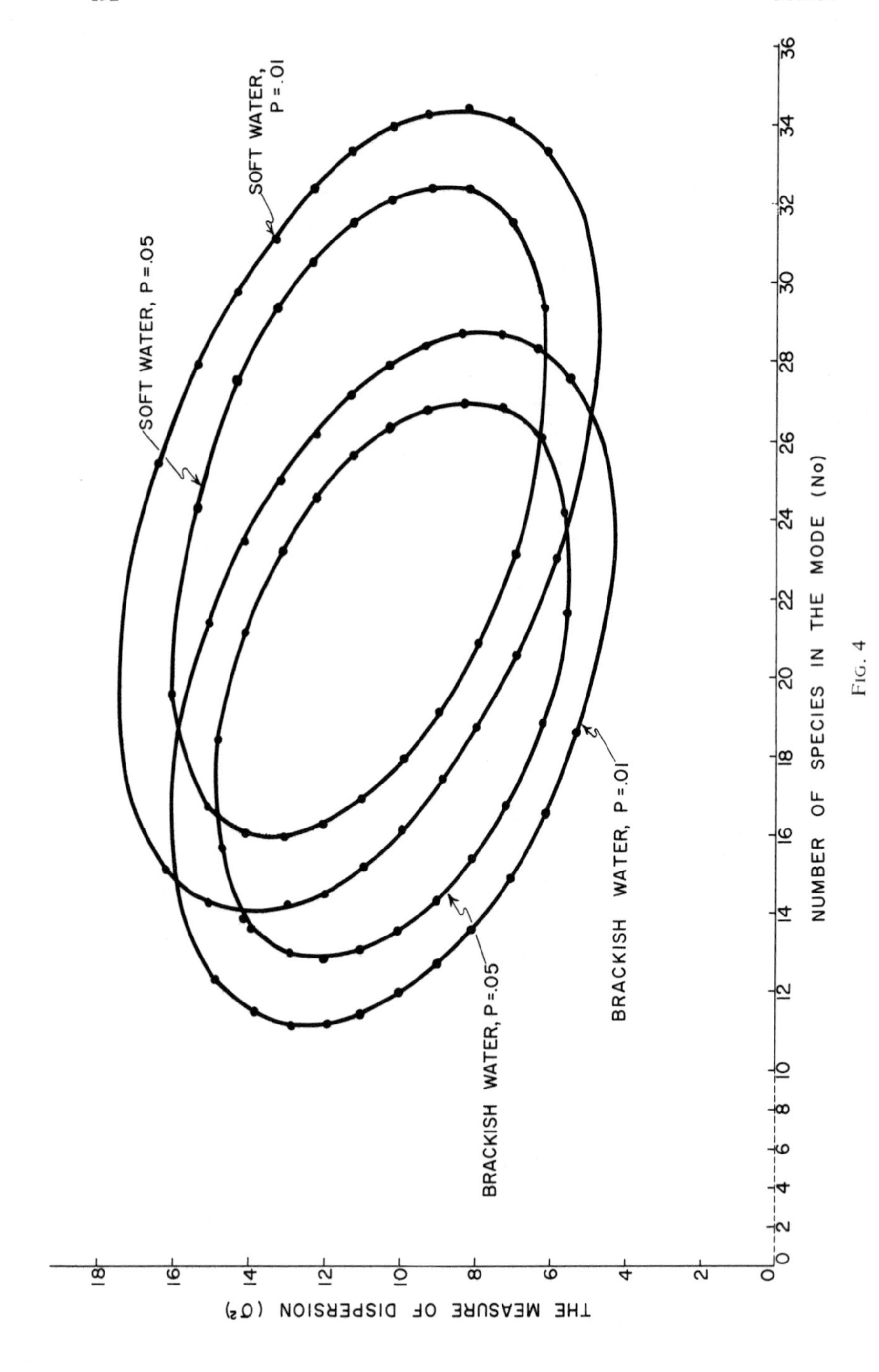
SOFT WATER, P = .05
SOFT WATER, P = .01
BRACKISH WATER, P = .05
BRACKISH WATER, P = .01
NUMBER OF SPECIES IN THE MODE (No)
THE MEASURE OF DISPERSION (σ^2)
0
2
4
6
8
10
12
14
16
18
20
22
24
26
28
30
32
34
36

FIG. 4

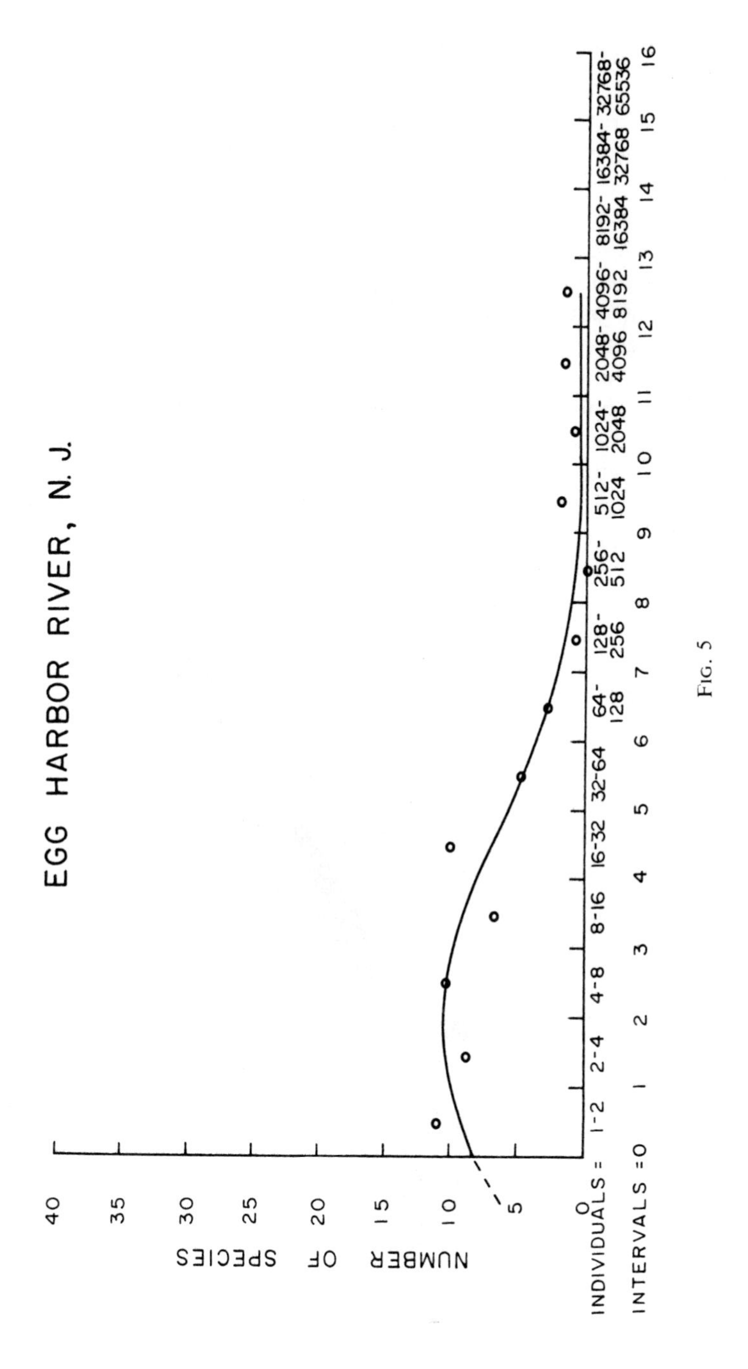
EGG HARBOR RIVER, N. J.
NUMBER OF SPECIES
0
5
10
15
20
25
30
35
40
INDIVIDUALS =
1-2
2-4
4-8
8-16
16-32
32-64
64-128
128-256
256-512
512-1024
1024-2048
2048-4096
4096-8192
8192-16384
16384-32768
32768-65536
INTERVALS = 0
1
2
3
4
5
6
7
8
9
10
11
12
13
14
15
16

FIG. 5

DIATOMS AS RELATED TO POLLUTION

So far we have dealt with the structure of diatom populations under natural conditions. However, since the beginning of the twentieth century a great many workers have concerned themselves with the occurrence of diatoms in the presence of pollution. During the course of these studies, just as our approach has changed, so has the meaning of the word "pollution." Originally pollution was organic in nature and referred mainly to sanitary wastes. Today it is a collective noun referring to many different concentrations of very different chemical and physical conditions.

It is undoubtedly this change in the nature of "pollution" which has caused many of the conflicting results of various workers.

Kolkwitz and Marsson (1908) were among the first to set forth the premise that the occurrence of certain diatom species indicated the presence of pollution. They gave a list of species characteristic of various stages in the assimilation of organic pollution

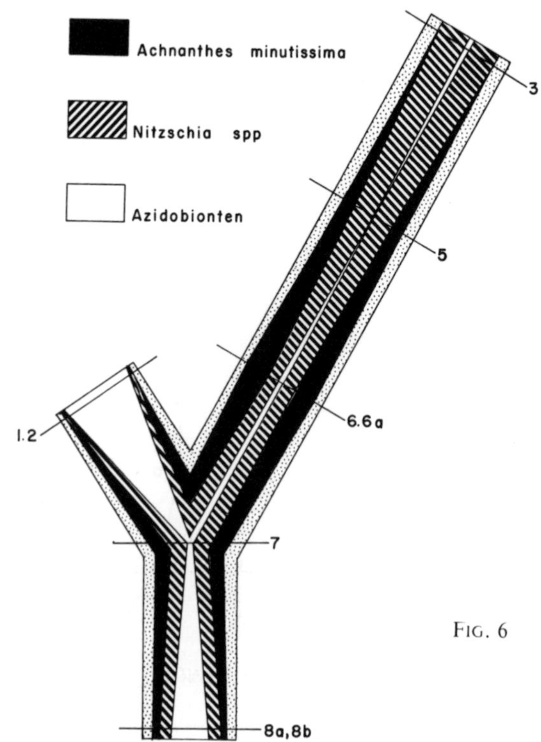

FIG. 6

by a body of water. These stages were known as polysaprobic, alpha-mesosaprobic, beta-mesosaprobic, and oligosaprobic. Later workers, such as Hentschel (1925), Naumann (1925), Butcher (1947), and Liebmann (1951), used and revised this system. Considerable disagreement arose as to whether or not the mere presence of a given group of species indicated the stage of pollution. This disagreement led to the necessity of recognizing associations of species and particularly the abundance of certain species if one was to judge the stage of pollution.

Thienemann (1939) recognized three types of organisms composing an association or community: "coenobiont" species, which occurred in great numbers only in one biotype or one type of ecological conditions; "coenophile" species, which were those species which had their best development in one type of biotype but might be found in other associations in which they were represented by smaller populations; and "coenoxene" species, which were those species which did not seem to represent any particular biotype but were found in small numbers in many different biotypes. This concept of dominant and associated species has been followed by many workers, such as Fjerdingstad (1950, 1960).

More recently Cholnoky (1958) has developed a system, based upon the relative dominance of certain species, to judge quantitatively the acid or organic pollution present in the Olifantsvlei Swamp and the Klipspruit Creek near Johannesburg in South Africa (Fig. 6). He charts the sum of the frequencies of *Eunotia* spp., *Frustulia magaliesmontana* Choln., *Frustulia rhomboides* var. *saxonica* (Rabh.) DeT., and *Pinnularia acoricola* Hust. and from these determines changes in acidity. Likewise, by the frequency of *Nitzschia* spp. he plots the amount of nitrogen in the water. The increase in the abundance of *Achnanthes minutissima* Kütz. he uses to indicate the increase in oxygen present in the water.

Whereas this system may be useful in a stream in which the pollution is due to acidity and sanitary wastes and in which the general ecological characteristics of the area do not vary greatly, which Dr. Cholnoky says is the case in Africa, we have not found this system useful in the rivers of the United States. For example, in a study of an area of the Wateree River in South Carolina just below the outfall of an industry which by industrial standards had a similar type and quantity of waste discharge during the period of study, we find considerable variation in the kinds of species found. This waste was composed of a mixture of chemicals such as is usually the case in waste discharges of industries in the United States. A study of the dominant species* (Table I) shows that no species was consistently dominant throughout the year; those species most consistently dominant were *Achnanthes minutissima* (66% of the time), *Gomphonema parvulum* Kütz. (75% of the time), and *Melosira distans* var. *alpigna* Grun. *in* Van Heurck (58% of the time). Furthermore, the abundance of these species varied greatly (Table I); *Achnanthes minutissima* from not being dominant to 52% of the specimens studied, and *Gomphonema parvulum* from not being dominant to 53% of the specimens studied. Certain species, although not dominant through most of the year, were very common during certain months; for example, *Cymbella tumida* (Breb.) V.H. was represented by over 15,000 specimens or 36% of the specimens counted in February; *Nitzschia filiformis* (W. Sm.) Hust. was represented by approximately 15,800 specimens or 38% of the specimens counted in April; and *Synedra pulchella* (Ralfs) Kütz. was represented by approximately 27,400 specimens or 51% of the specimens counted in December 1958, approximately 142,000 specimens or 97% of the community in January 1959, and about 17,813 specimens or 37% in February 1959. In March 1959, no species was dominant.

It is true that the flow of the river was variable during this period, but since this river is subject to severe weekly variations in flow due to lack of flow from a power dam over weekends, this factor of variation in flow is not such an irregularly occurring ecological factor as would normally be the case. If we had depended on variation in the abundance of certain dominant species to indicate degrees of pollution, our results would have been very erroneous.

It would appear from these results that unknown factors in the natural environment or unknown variation in the chemical or physical characteristics of the wastes are having a greater influence on the dominant species than the known quality and amount of the effluent. The importance of considering the effect of other environmental effects when determining the amount of pollution present was first pointed out by Hentschel (1925).

However, if we consider the total percent of the specimens

* By dominant species is meant a species represented by more than 1000 specimens when similar segments of a community are studied. When no species are excessively dominant, this usually means counting 8000 specimens.

TABLE I

Dominant* Species Found in a River in South Carolina during Monthly Studies of One Year's Duration

(To show how dominant species over 1000 vary during the year on Station 2 of the Wateree River, 1958-1959, based on the semidetailed readings.)

	Percent of Community Composed of Each Species for Date of Study											
	1958			1959								
	Oct. 25	Nov. 21	Dec. 19	Jan. 30	Feb. 27	Mar. 25	Apr. 24	May 22	July 4	July 31	Aug. 28	Sept. 25
Achnanthes minutissima	16.5	51.6	18.3	.75	.	No species with over 1000 specimens	8.8	.	41.4	31.0	.	10.7
Cocconeis placentula	.	.	.	.	.		.	.	.	.	.	13.5
Cymbella tumida	.	.	8.7	.81	35.8		.	.	.	.	.	.
Eunotia pectinalis var. *minor*	.	.	.	.	.		.	.	.	2.2	.	10.9
Synedra miniscula	.	.	.	.	.		4.7	16.3	.	.	.	.
Synedra rumpens	.	.	.	.	.		.	.	6.5	.	.	.
Synedra rumpens var. *meneghiniana*	1.5	.	.	.	.		.	.	.	4.2	.	.
Gomphonema affine var. *angustatum*	.	.	.	.	3.1		.	.	.	.	.	.
Gomphonema gracile	4.0	.	.	.	.		.	.	.	.	8.2	.
Gomphonema intricatum var. *pumile*	6.7	.	.	.	.		.	.	.	.	.	.
Gomphonema parvulum	52.5	6.9	1.9	.	3.5		14.1	.	35.2	45.2	40.1	6.3
Gomphonema LL 8	.	.	2.7	.	.		.	.	.	.	.	.
Melosira distans var. *alpinga*	.	4.9	4.7	.	2.8		8.9	20.6	5.0	.	16.7	.
Melosira varians	.	.	2.3	.	11.7		6.4	.	.	.	.	.
Navicula spicula	.	.	.	.	.		.	.	.	.	.	7.5
Navicula AP1	.	.	.	.	.		.	.	.	.	.	13.5
Nitzschia confinis	2.4	.	.	.	.		.	.	.	.	.	.
Nitzschia filiformis	.	.	3.1	.	.		38.0	14.4	.	.	.	.
Nitzschia fonticola	.	.	.	.	.		.	.	.	.	.	5.5
Synedra affine	.	10.4	.	.	.		.	.	.	.	.	.
Synedra pulchella	.	.	51.0	97.2	37.3		.	.	.	.	.	.
Synedra rumpens	.	.	.	.	.		.	.	.	.	.	7.4
Synedra ulna var. *acus*	.	.	.	.	.		10.2	.	.	.	.	.
Synedra vaucheriae	4.7	6.3	.	.	.		.	12.7	2.7	.	6.7	.
Synedra JW1	.	.	.	.	.		.	.	.	3.2	.	.
Percent of Community Composed of Dominant Species	88.3	80.1	92.7	98.8	94.2		91.1	64.0	90.8	85.8	71.7	75.3

*A dominant species is one represented by 1000 specimens when 8000 or more are counted.

counted which compose the dominant species we have a very different result. The total percent of the population due to dominant species (except in March, when no species was dominant) was usually between 71 and 92%, with one occurrence of 64% and one of 98%. In other words, the kinds of species composing the dominant segment of a population may change, but the percent of the community composed of dominant specimens does not usually vary greatly if the pollution load does not change greatly in quality or quantity. The lack of any dominant species in March was due to very poor diatom growth resulting from cold weather and very high flow.

In another study of the Wateree River we considered all of the species identified in three samples taken from this same area on March 28, May 9, and August 15, 1958, and we had some interesting results. First we considered the variation in species numbers and kinds. The total number of species identified were 86, 78, and 116, respectively. Of the 86 species identified in March, 36 or 42% were found only in that study; in May of the 78 species, 27 or 35% were only in that study; and in August of the 116 species, 65 or 56% were only in that study. Of the 190 species in all three studies 67% were found only in one study. Thus we see that variation in kinds is much greater than variation in numbers of species. Only 27 species or 13% were common to all three studies. Of these 27, two—*Gomphonema parvulum* and *Synedra vaucheriae* Kütz.—were dominant in all three studies. When we consider the number of species the similarity is much greater, 116 to 78 or 62% (38% difference).

Furthermore, if we compare the diatom floras for the same time of the year, that is, when flow and temperature conditions are about the same, i.e., August 1957, August 1958, and early September 1959, we see similar kinds of variations. The numbers of species were roughly the same, being 125, 116, and 117, respectively. The number of species which occurred only in August 1957 were 65 or 52% of the total number of species identified; in August 1958, 45 or 39% of the total identified; and in September 1959, 51 or 44% of the total identified. Of the 242 species identified in all three studies 66.5% were found only in one study. Of the 34 species common to all studies none was dominant in all three studies.

A similar study made each month for one year of the diatom flora in a brackish estuary of the Neches River in Texas gave the following results as to the abundance of dominant species (Table II). One species, *Nitzschia filiformis,* was a dominant throughout

TABLE II

Dominant* Species Found in a River in Texas during Monthly Studies of One Year's Duration

(To show how dominant species over 1000 vary during the year on Station 3 of the Neches River, 1959, based on the semidetailed readings.)

	Percent of Community Composed of Each Species for Date of Study											
	Jan. 19	Feb. 16	Mar. 31	Apr. 27	May 26	June 22	July 20	Aug. 31	Sept. 28	Oct. 29	Nov. 24	Dec. 21
Bacillaria paradoxa	.	.	.	.	.	11.7	9.9	6.3	22.1	13.9	3.7	4.7
Cyclotella meneghiniana	.	.	.	.	.	.	.	.	3.8	.	.	.
Gomphonema parvulum	.	4.0	.	4.2	2.6	.	.	.	.	.	.	.
Navicula cryptocephala var. *veneta*	.	.	.	.	.	.	7.9	.	.	.	.	.
Navicula tripunctata	.	.	.	.	.	1.9	12.9	4.3	12.9	22.6	28.7	21.9
Nitzschia clausii	.	5.8	11.8	3.8	1.3	.	.	.	.	.	.	.
Nitzschia filiformis	3.7	34.9	29.2	33.4	54.5	21.2	18.5	54.0	8.5	6.8	44.8	6.3
Nitzschia sp. 7	.	.	.	.	.	.	.	.	17.7	26.5	.	.
Synedra sp. 1	.	49.6	54.5	54.0	40.0	.	35.7	.	.	.	.	.
Synedra sp. 2	91.5	.	.	.	.	62.3	.	31.2	28.9	.	17.1	58.8
Percent of Community Composed of Dominant Species	95.2	94.3	95.5	95.4	98.4	97.1	84.9	95.8	93.9	69.8	94.3	91.7

*A dominant species is one represented by 1000 specimens when 8000 or more are counted.

the year. Two species, *Bacillaria paradoxa* Gmel. and *Navicula tripunctata* var. *schizonemoides* (V.H.) Patr., were dominant 58% of the time. Although *Nitzschia filiformis* was always a dominant, the percent of the population it composed varied from 3.7% to 54.5%. Other species, although very abundant during a part of the year, were not dominant during other months. When we consider the percent of the population composed of dominant species throughout the year we find much more similarity. The variation ranged only from 70% to 98%.

If, as with the Wateree River, we examine the structure of the samples of the diatom flora collected in the same area of the Neches River during February, May, August, and November 1957, we find the numbers of species were 86, 89, 109, and 142, respectively. The kinds of species occurring only during a particular study were 26 (30%) in February, 24 (27%) in May, 34 (31%) in August, and 50 (35%) in November. Of the 230 species identified in the three studies, 134 (58%) were found only in one study; 24 species were common to the four studies and only one of these, *Nitzschia filiformis*, was a dominant in all four studies.

When we compare the studies of the diatom floras for the same area made in August 1957, 1958, 1959, 1960, we find the number of species identified in similar segments of the flora to be 109, 139, 110, and 103, respectively. Of the species found only in one study we have 31 (28%) August 1957, 46 (33%) August 1958, 35 (32%) August 1959, and 39 (38%) August 1960. Of the 253 species identified, 151 (60%) were found only in one study. 26 species were common to all four studies and only one, *Nitzschia filiformis*, was consistently a dominant.

Thus, it would appear that until we know more about the physiological requirements and tolerance to pollution of these diatom species, it is difficult to know whether it is some factor in the environment or some unknown variation in the type of pollution which is causing the variation in species abundance and composition. Until that time, when we are studying the effects of pollution characterized by a wide variety of chemical and physical properties, changes in the percent of the population composed of all the dominant species and changes in the number of species seem to be more reliable criteria for judging the general pollution load of a river than changes in the percent dominance of one or a few species.

CONCLUSIONS

From the many studies by various workers it has become evident that if we wish to compare results we must study similar segments of the population, and the results from any one study must be reproducible on re-examination. If this is done we find that the numbers of species composing a diatom community and the relative sizes of the populations of the species do not change greatly over time so long as the environment does not change greatly—i.e., is stable. However, from season to season and from year to year the kinds of species found in any one area vary greatly. This is probably due to the fact that we have such a large species pool available, each species with slightly different preferences for this variable yet stable environment. Also the rapid rate of reproduction makes great changes in population size possible. No doubt if we had a much smaller species pool available for the ecological conditions, as in the case of some of the lower invertebrates such as Bryozoa, we would not see so much change in kinds of species.

It is evident that when we are considering the natural environment there are many associations of species which are reliable indicators of general ecological conditions such as temperature, salinity, and nutrient level of the water. The variations in these conditions have existed over geological time, and there has been time for taxa to evolve which are characteristic of these general conditions. Even in such cases the diatom associations have value mainly as qualitative indicators.

When we are dealing with pollution in its present complex form we have quite a different problem. The complex pollution which we encounter today is of recent origin, and time has not been sufficiently long for specific taxa to evolve which are characteristic of its many components. Rather, we have certain species which are tolerant to one or more of a wide variety of chemical and physical conditions commonly associated with pollution. Thus, any one species is not equally tolerant to all the conditions we class as pollution. For these reasons we must be more concerned with a more general type of change to indicate the many aspects of pollution, such as the number of species present and the relative sizes of the populations of the species rather than to the specific species present and the sizes of the populations of specific species. For example, as previously shown, the structure of the community from a dystrophic stream and one which is badly polluted by man are quite similar. In each case some factor of the environment has greatly decreased the number of species which can tolerate the environment, and some of those that can have become very abundant. The causes of the development of large populations of certain species are many—e.g., less competition with other species, the removal of predators, and the increase in amount of nutrients available for the species. Once one determines that this condition exists, one can then examine the kinds of species to see if one can determine the cause of this type of community structure. Where the cause is similar to conditions which have occurred in nature over geological time, we will probably find species which will indicate it. For example, there are certain species of *Eunotia* found only in acid waters rich in humates. Much more needs to be learned about the physiology of individual species before we can determine accurately what their presence—or, perhaps more important, their absence—indicates.

LITERATURE CITED

Backman, A. L., and Cleve-Euler, Astrid. 1922. Die fossile Diatomeenflora in Österbotten. *Acta Forestalia Fennica* 22. Kuopio.

Butcher, R. W. 1947. Studies in the ecology of rivers. VII. The algae of organically enriched waters. *J. Ecol.* 35:186-191.

Cholnoky, B. J. 1958. Beitrag zu den Diatomeenassoziationen des Sumpfes Olifantsvlei südwestlich Johannesburg. *Ber. Deut. Bot. Gesell.* 71(4):177-187.

Cleve, P. T. 1896. Diatoms from Baffin Bay and Davis Strait; Collected by M. E. Nilsson. *Bih. K. Svenska Vet.-Akad. Handl.* 22, Afd. 3(2):1-22.

Ehrenberg, C. G. 1854. Mikrogeologie. Leipzig, Germany.

Fjerdingstad, E. 1950. The microfauna of the river Mølleaa; With special reference to the relation of the benthal algae to pollution. *Folia Limnol. Scandinavica* No. 5. 123 pp.

Fjerdingstad, E. 1954. The subfossil algal flora of the lake Bølling Sø and its limnological interpretation. *Det Kgl. Danske Videnskab. Selsk. Biol. Skr.* 7(6). 56 pp.

Fjerdingstad, E. 1960. Forurening af vandløb biologisk bedømt. Reprint from *Nordisk Hygienisk Tidskrift* 41:149-196.

Foged, N. 1947-1948. Diatoms in water-courses in Funen. I-VI. *Dansk Bot. Ark.* 12(5):1-40; 12(6):1-69; 12(9):1-53; 12(12):1-112.

Foged, N. 1950. Diatomevegetationen i Sorte Sø. En dystrof skovsø i Sydfyn. *Af Fyns Flora og Fauna* 3.

Fontell, C. W. 1917. Süsswasserdiatomeen aus Ober-Jämtland in Schweden. *Arkiv f. Bot.* 14(21). 68 pp.

Halden, B. E. 1929. Kvartärgeologiska diatomacéstudier belysande den postglaciala transgressionen a svenska Västkusten. I. Höganästrakten. *Geol. Förening. Stockholm Förhandl.* 51(3):311-366.
Hart, T. J. 1935. On the diatoms of the skin film of whales, and their possible bearing on problems of whale movements. *Discovery Repts.* 10:247-282.
Hendey, N. I. 1937. The plankton diatoms of the Southern Seas. *Discovery Repts.* 16:151-364.
Hentschel, E. 1925. Abwasserbiologie. Abderhalden's Handb. d. Biol. Arbeitsmethoden, Abt. 9, Teil 2, 1 Häfte, pp. 233-280.
Hustedt, F. 1924. Die Bacillariaceen-Vegetation des Sarekgebirges. *In* Hamberg. Naturwissenschaftliche Untersuchungen des Sarekgebirges in Schwedisch-Lappland, Vol. 3, Bot., pp. 525-626.
Hustedt, F. 1937-1938. Systematische und ökologische Untersuchungen über die Diatomeen-Flora von Java, Bali und Sumatra. *Arch. f. Hydrobiol. Suppl.* 15(1): 131-177; 15(2):187-295; 15(3):393-506; 15(4):638-790.
Hustedt, F. 1953. Die Systematik der Diatomeen in ihren Beziehungen zur Geologie und Ökologie nebst einer Revision des Halobien-Systems. *Svensk Bot. Tidskr.* 47(4):509-519.
Hustedt, F. 1957. Die Diatomeenflora des Flusssystems der Weser im Gebiet der Hansestadt Bremen. *Abh. Naturw. Ver. Bremen* 34(3):181-440.
Hutchinson, G. E., Patrick, Ruth, and Deevey, E. S. 1956. Sediments of Lake Patzcuaro, Michoacan, Mexico. *Bull. Geol. Soc. Am.* 67(11):1491-1504.
Hyyppa, E. 1936. Über die spätquartäre Entwicklung Nordfinnlands mit Ergänzungen zur Kenntnis des spätglazialen Klimas. *Bull. Comm. Geol. Finlands.* No. 115, pp. 401-465.
Jørgensen, E. G. 1948. Diatom communities in some Danish lakes and ponds. *Det Kgl. Danske Videnskab. Selsk.* 5(2):1-140.
Jørgensen, E. G. 1950. Diatom communities in some Danish lakes and ponds II. *Dansk Bot. Ark.* 14(2):1-19.
Kolbe, R. W. 1932. Grundlinien einer allgemeinen Ökologie der Diatomeen. *Ergeb. d. Biol.,* 8:221-348.
Kolkwitz, R. and Marsson, M. 1908. Ökologie der pflanzlichen Saprobien. *Ber. Deut. Bot. Gesell.* 26a:505-519.
Krasske, G. 1939. Zur Kieselalgenflora Südchiles. *Arch. f. Hydrobiol.* 35:349-468.
Krieger, W. 1929. Algolisch-monographische. Untersuchungen über das Hochmoor am Diebelsee. *Beitr. z. Naturdenkmalpflege* 13(2):231-300.
Liebmann, H. 1951. Handbuch der Frischwasser- und Abwasser-biologie. Verlag R. Oldenbourg, Munich. Vol. 1, 539 pp.
Lindberg, H. 1910. Phytopaläontologische Beobachtungen als Belege für postglaziale Klimaschwankungen in Finnland. Compte rendu de la XIe session du Congrès géologique international. Stockholm.
Lundquist, G. 1927. Bodenablagerungen und Entwicklungstypen der Seen. Binnengewässer 2.
Møller, M. 1950. The diatoms of Praestø Fiord. *Folia Geographica Danica* 3(7): 187-237.
Naumann, E. 1925. Die Arbeitsmethoden der regionalen Limnologie. Abderhalden's Handb. d. Biol. Arbeitsmethoden. Süsswasserbiologie. 1.
Patrick, Ruth. 1936. Some diatoms of Great Salt Lake. *Bull. Torrey Bot. Club* 63(3):157-166.
Patrick, Ruth. 1939. The occurrence of flints and extinct animals in Pluvial deposits near Clovis, New Mexico. V. Diatom evidence from the Mammoth Pit. *Proc. Acad. Nat. Sci. Philadelphia* 90:15-24.
Patrick, Ruth. 1943. The diatoms of Linsley Pond, Connecticut. *Proc. Acad. Nat. Sci. Philadelphia* 95:53-110.

Patrick, Ruth. 1945. A taxonomic and ecological study of some diatoms from the Pocono Plateau and adjacent regions. *Farlowia* 2(2):143-221.

Patrick, Ruth. 1946. Diatoms from Patschke Bog, Texas. *Not. Nat. Acad. Nat. Sci. Philadelphia* No. 170. 7 pp.

Patrick, Ruth. 1949. A proposed biological measure of stream conditions, based on a survey of the Conestoga Basin, Lancaster County, Pennsylvania. *Proc. Acad. Nat. Sci. Philadelphia* 101:277-341.

Patrick, Ruth. 1954. The diatom flora of Bethany Bog. *J. Protozool.* 1(1):34-37.

Patrick, Ruth. 1961. A study of the numbers and kinds of species found in rivers in eastern United States. *Proc. Acad. Nat. Sci. Philadelphia* 113(10):215-258.

Patrick, Ruth, and Hohn, M. H. 1956. The diatometer—a method for indicating the conditions of aquatic life. *Proc. American Petroleum Inst., Sect.* 3, 36:332-338.

Patrick, Ruth, Hohn, M. H., and Wallace, J. H. 1954. A new method for determining the pattern of the diatom flora. *Not. Nat. Acad. Nat. Sci. Philadelphia,* No. 259. 12 pp.

Pennington, Winifred. 1943. Lake sediments: The bottom deposits of the North Basin of Windermere with special reference to the diatom succession. *New Phytol.* 42(1):1-27.

Petersen, J. B. 1943. Some halobion spectra (diatoms). *Det Kgl. Danske Videnskab. Selsk. Biol. Meddel.* 17(9):1-95.

Thienemann, A. 1939. Grundzüge einer allgemeinen Ökologie. *Arch. f. Hydrobiol.* 35:267-285.

Vallentyne, J. R. W. 1953. Organic pigments in lake sediments (doctoral dissertation). Yale University.

The Ecology of Plankton Algae

James B. Lackey

Phelps Laboratory
University of Florida
Gainesville, Florida

INTRODUCTION

Not too long ago the use of the word "algae" in a conversation usually met with a blank look, and a need for definition. Nowadays the layman has a fair understanding of the term, because he has heard it so often in discussions of good water, of space travel, and of toxic or nuisance effects. The biologist recognizes the much wider coverage of the word than the "microscopic green plants" of the layman; he too is cognizant of the above connotations, but is also interested in algae as the primary producers. That interest must of necessity include those factors which influence primary production, which factors are the environment.

THE TERM "PLANKTON ALGAE"

We regard plankton algae as microscopic suspended plants, generally unicellular, containing chlorophyll and usually other pigments, whose habitat is the zone of water penetrated by light. There are many exceptions to this statement. Some plankton algae (*Valonia, Lyngbya majuscula*) are not microscopic. Others (*Volvox*) form colonies of large size. These are easily seen by the unaided eye. Some are colorless but may simply have learned to live without chlorophyll. Species of *Euglena* have their chlorophyll masked by red pigments so they may form a red blanket on ponds, while some dinoflagellates may color the surface waters of Gardiners Bay, Long Island, New York, a vivid purple at times.

For definite purposes the following groups are included in this discussion: Chlorophyceae; Volvocales; Euglenophyceae; Cryptophyceae; Chrysophyceae; Dinoflagellata; Chloromonadida; Silicoflagellata; Coccolithophora; Bacillarieae; and Zooflagellata. These include many nonplankton forms, but embrace most of the Thallophyta.

It is not always easy to say categorically that a particular species is a plankter. Many organisms, diatoms for example, occur in nature principally on the bottom, i.e., on a solid substrate. They are therefore termed benthic. If 100 ml of Lake Santa Fe (Florida) water is centrifuged and the catch examined, certain organisms, especially species of *Navicula,* are not found. But if a clean slide is suspended to a depth of one foot, where the water is nine feet deep, within 24 hr the slide is colonized by various benthic organisms, not seen in the centrifuged catch, or very rarely so. Evidently a clean separation into benthic or planktonic algae is difficult.

Spirogyra is usually regarded as an attaching green alga, yet it often forms huge colonies extending well toward the surface or even floating just beneath the surface. *Lyngbya majuscula* and some other blue-green algae tend to blanket the bottom. But gases either evolved in the mud beneath them, or within the algal cells, float them to the surface in masses, and they bring other species of their association with them. The reverse may also occur. In winter *Cryptomonas,* which swims freely, tends to accumulate near the bottom.

No longer are bacteria and other colorless plants looked upon as the sole organisms decomposing and mineralizing organic matter. There are many algae which are either colorless or whose chlorophyll content is so low they appear virtually colorless. There are other chlorophyll-containing algae which lose their color in the presence of abundant organic matter in solution, and still others which despite a vivid color of chlorophyllaceous and carotenoid pigments capture and ingest solid food. Examples of the first group are *Polytoma uvella* and *Astasia klebsii;* of the second, *Euglena acus, E. gracilis*, and *Eutreptia lanowii;* and of the third, *Ochromonas ludibunda.* Interest in algae as primary producers turns up some very intricate problems as to what and how varied are the substances they utilize.

Since most plankton algae are microscopic it may be a matter of wonder as to whether they are really competitive to the huge attached algae such as kelp, or rooted plants such as *Potamogton* or *Zostera.* The answer lies in the much greater mass of water plankton algae occupy and in their rapid cell division. The attached seaweeds grow only in a narrow zone of shallow water, and grow slowly. Plankton algae, on the contrary, have the vast acreage of the seas and other waters, and can grow at the same depths as rooted aquatic plants, because light penetration is the greatest limiting factor. In addition, plankton algae are food for the microscopic protozoa, rotifers, worms, filter feeding crustaceans, and shellfish. Not the sole food, because many of these appear to be omnivorous, hence eat bacteria and small animals. But attached plants may be too large and inaccessible for the animals.

The biomass of plankton algae is difficult to assay. Measurement of chlorophyll is subject to the limitation that algae vary in chlorophyll content, not only from species to species, but also under varying nutritional levels. There is also patchy occurrence. On a windy day a bay into which the wind blows may have a dense aggregate of plankton algae, whereas the opposite side has few. This is only one instance of the cause and occurrence of patchiness. Measurement of O_2 production is subject to the same limitation, plus the decreased light as depth increases and the overlying aggregate becomes more and more dense.

In attempting to arrive at volumetric relations in the Scioto (Ohio) River, the different species of organisms were counted. A sufficient number of each species was measured to get an average species size. They were then classified as to whether they were spheres, cones, cylinders, or whatever geometric shape, and their cubic volume was calculated. This was a laborious method, but was accurate enough to give the biomass, in a given river sample. Some organisms such as *Trachelomonas cylindrica* were almost perfect cylinders; others were cylinders with a hemisphere at each end or pyramids or cones, etc. Where a very large number of species is encountered any use of this method is obviously time-consuming. About 400 species were found in the Scioto, and the engineer in charge of the work, Mr. Kehr (1), established volumetric relationships for many of the organisms, the amount of tabular and descriptive material was so great, however, that in the end only a brief report was published.

DIRECT ALGAL EFFECTS ON MAN

The most evident relationships tend to be those which directly affect us. Pathogenic bacteria we accept as a matter of course in many contagious diseases. There are very few plankton algae which directly affect man, however. Most of these apparent direct effects are such as those of the so-called red tides. For many years now, it has been known that when the dinoflagellate *Gonyaulax catenella* "blooms" in Pacific coastal waters, the edible mussel *Mytilus* becomes toxic. The poison is known to be accumulated from the dinoflagellate, to which it is apparently harmless, but is lethal to man. When the dinoflagellate occurs in large numbers, the mussel beds are quarantined. Despite such precautions illness and death still occur. This poisoning is not confined to the Pacific coast but is commonest there.

There are also other aspects of algal poisoning. One of these is direct toxicity to other organisms. It has long been known that when certain algae bloomed—occurred suddenly in enormous numbers—fish were killed.

An early record is one by Hornell (2) off the Malabar coast of India, apparently due to *Eutreptia.* Kills in fish ponds in Israel due to the small flagellate *Prymnesium* have been studied by Shilo (3) and others. The most exhaustive studies have probably been those on the Gulf of Mexico Red Tide, the dinoflagellate *Gymnodinium breve,* which produces massive kills of any and all animals. This has involved the University of Miami (4), the University of Florida (5), the U.S. Fish and Wildlife Service (6), the Florida State Board of Conservation (7), and others. Its effects on man have been primarily on his pocket-book; no fishing is done in a Red Tide area for the duration of the bloom, and miles-long windrows of decaying fish on the beaches drive tourists away. There are also some indications, not well authenticated, of skin and respiratory irritations in Red Tide areas.

In the summer of 1960 *Gymnodinium flavum* appeared in bloom proportions around San Diego, California. Lackey and Clendenning (8) studied a small fish kill in Mission Bay, San Diego, during this time. This occurred at a public bathing beach, and at once caused alarm. There were no drastic temperature changes at the time, dissolved oxygen was normal, and there was no evidence of any "spill" of chemicals or sewage into the Bay. The water contained

22 or more species of plankton algae and protozoa (not all the small green cells and minute colorless flagellates were identified) and the number per ml was 3504. Of these 1776 were *Gymnodinium flavum.* It is a medium-sized, distinctive dinoflagellate, and this was its first recorded appearance since 1921. While the evidence is circumstantial it is believed to be valid, especially since *G. breve* is known to be a killer.

PROBLEMS OF TOXICITY TO DOMESTIC ANIMALS

Not only do plankton algae kill fish and other marine animals, but there is abundant evidence that *Microcystis aeruginosa* liberates a toxin in fresh water lakes and ponds which readily kills or sickens cattle and other domestic animals. Olson (9) has summarized current knowledge of this matter. The phenomenon is world-wide and quite serious in certain areas of Australia and Africa. According to Professor Sundaresan (10) it does not occur in the "tanks" of India, where *Microcystis* is a common bloom component.

INDIRECTLY INDUCED TOXICITY

Microcystis poisoning is still not well understood. Investigations by Wheeler, Lackey, and Schott (11) failed to pinpoint the toxic agent. The gelatinous envelope of this alga harbors an interesting collection of bacteria (and other organisms) which might be causative agents. When the bloom is decaying the oxygen level is low and nutritive substances are abundant. These might well be the conditions under which botulism develops, causing duck sickness. Some years ago, a plankton specimen was collected from a western waterfowl refuge where ducks were dying from presumed botulism. The sample contained a heavy concentration of dinoflagellates and *Microcystis*. Both gave evidence of decomposition, and the bacterial population was very heavy. As early as 1934, Kalmbach and Gunderson (12) described this duck disease as a botulinus-caused illness.

Other examples could be cited, but these show that algae may have both direct and indirect harmful effects on man and animals through the production of toxins.

OTHER RELATIONSHIPS TO DISTRIBUTION

The relationships cited above are dramatic and obvious. But there are others which are less apparent, and which are not generally thought of as algae-human relationships but as part of the normal environmental relationships. This of course does not include the possible use of algae such as *Chlorella* in spaceships, which merits a whole environmental and physiological study in itself.

Plankton algae are found in streams, ponds, lakes, and oceans. They tolerate extremes of temperatures, altitude, hypo- and hypersalinity, variety of composition in the surrounding water, little light or much. In short, there is hardly an environmental situation except those of intense heat and extreme chemical toxicity that they do not inhabit. It seems hardly necessary to document cases of *Haematococcus* growing on snow, or of *Gomphosphaera aponina* growing in abundance in the small pool termed Bitter Waters in Death Valley, California, where the concentration of magnesium salts is so intense that the mud is crusted with it. The colors of the rocks around the hot springs in Yellowstone National Park are due principally to blue-green algae. They turn salinas pink and quarry pools become brown with *Ceratium hirundinella* or yellow-green with *Uroglena* when the water there contains so little dissolved solids it would seem impossible for more than a very few to grow. *Chlorella* sometimes grows well in a carboy of distilled water on a laboratory shelf, and *Tribonema* and *Ulothrix* (which aren't normally planktonic) thrive in the presence of extreme amounts of radioactivity at Oak Ridge. Colorless *Euglena gracilis* and species of *Phacus* have been taken from waters of caves in Florida where light never penetrates.

These examples are all personal experiences, except the one for Yellowstone National Park. They do not apply to species which are peculiar to the conditions cited. Rather, the examples are those of algae common to a great many habitats from many parts of the world. There are species which have been found in only a few or even one situation. Years ago, Stokes (13) described *Trentonia flagellata* from near Trenton, New Jersey. Apparently this was the sole record for it. In fact it was regarded as a dubious entity by Dangeard (14), probably because it had not been seen since. But in 1956 routine examinations of the biota of Warm Mineral Springs in South Florida (15) showed the organism to be a common plankter there. Such "rare" species are seldom encountered. Usually an examination of a new situation reveals that the organisms there are those found in many other locations. There are truly pelagic species in the high seas; neritic species, many of them also common to estuaries; fresh-water species common to waters whose pH is under 7.0; and fresh-water species prefering hard water. These four major groups are not hard and fast groups since many organisms tend to cross the boundaries for each group.

FOOD RELATIONSHIPS AND ALGAE

The presence of an organism and its abundance or scarcity is a result of the action of many factors. First, its food requirements must be met. We generally consider chemical analysis of the organism, or of the water in which it lives, as a partial answer to this. Axenic cultures and defined chemical media are few in number, however; nor do they always provide an answer. They may not inform us of substances synthesized by the organism or of the pathway of metabolism. Some progress is being made, however; we have used C^{14} in acetate to determine which of the two carbons in the acetate molecule is used. We have also used other radioisotopes to determine which ones are concentrated by algae. Those which accumulate large amounts of radioactive entities tend to scavenge it from the environment and if we devise ways of harvesting the crop and disposing of it the process becomes a decontamination method. Man may be also adversely affected if, on the other hand, there is a food chain from a "hot" alga to man via fish.

Man provides algal foods in the disposal of wastes—human, animal, or industrial. Our sewage is a rich food, so rich in fact that it must be diluted if it is to be mineralized in a few hours. It is becoming a common practice, when land is cheap, populations sparse, and time is not too important, to build shallow basins (oxidation ponds) into which sewage flows and is decomposed by bacteria, probably on or near the bottom, and blooms of algae on or near the top. Very little nuisance or health menace is caused in this case, but where the same process occurs in a stream, both odors (mostly H_2S) and a health menace result unless the sewage is greatly diluted. In addition, the heavy bacterial growths below the entrance of the sewage may cause an oxygen depletion until re-aeration occurs. Such depletions kill animal life and prevent regrowth. Re-aeration is accomplished in several ways, one being algal growth.

Depletions of O_2 may occur from many commercial wastes, such as those from distilleries or pulpwood processing. But they may occur also from too heavy algal growths or blooms. Occasionally such blooms form a surface layer so dense that light does not penetrate, and the water below the surface is anaerobic. A recent case of an O_2 demand in the Forge River where green plankton was abundant has been reported by Barlow and Myren (16). In these cases too much food was present. Another example of too much food has been cited in the Great South Bay studies (17-19). Here great numbers of ducks were produced along Carman's River, and their wastes were carried into the shallow Bay, where the famous Blue Point oysters were grown. The wastes resulted in a tremendous and long-lasting bloom of the small *Nannochloris bacillaris,* which choked off the growth and succession of a whole series of algal species which served as food for oysters. Or it may have produced antimetabolites. At least in some manner the oysters were affected and failed to fatten or died.

Many efforts have been made to eliminate algal food from wastes. The major effort is directly aimed at reducing PO_4 and NO_3. Sewage treatment plants generally decompose the organic matter in about 8 hr with no use of algae in the process. But the effluent is still so rich in PO_4 and NO_3 that algal blooms may occur, with resulting nuisances such as algal scums over the surface of receiving waters, and often a secondary BOD.

It should be pointed out that in nature, i.e., with no treatment of wastes, the first algae to appear are usually blue-greens, Volvocales, and Euglenophyceae. Presumably these organisms compete with bacteria for soluble organic food. In some cases at least chlorophyll is lost and the organisms live and become abundant as in darkness. Personal experience with such cases includes *Eutreptia viridis, Euglena gracilis, E. acus,* species of *Phacus,* and *Trachelomonas reticulata.* There are still other algae (*Ochromonas ludibunda,* several dinoflagellates) which ingest solid food, despite being chlorophyll-bearers. Perhaps there are as many preferences for specific foods or specific combinations in plankton algae as we recognize when we purchase a specific fertilizer for strawberries or for camellias. The difficulties caused by most man-produced wastes stem from their general nature—they contain everything.

In natural waters, largely unaffected by man (and they are increasingly difficult to find), a biological balance is usually achieved. Such waters tend to be low in nutrients, without extreme conditions or concentrations of limiting substances, and to have many species, but rarely large numbers of individuals. Table I shows the numbers of plankton (not all algae) in the Washougl River, a tributary of the Columbia in Oregon, and a similar analysis for the Wabash River of Indiana. The Washougl is low in nutrients, its flow being largely from granitic rocks. The Wabash is in a limestone area, and besides flowing through an area rather high in human population, is subject to washed-in agricultural fertilizer. The very large numbers of species and individuals in the Wabash are direct evidence not only of the influence of man but also of the warmer temperature and more soluble substrate than in the Washougl.

The effects of man may be indirect insofar as affecting algae. Table II shows the plankton of the Santa Fe River near Mikeville, Florida, at a point where the current is greatly slowed and where cattle come for water, so that their droppings are added; and a station immediately above Mikeville. The effects are obvious, and largely due to the presence or absence of enrichment by the cattle.

NONFOOD FACTORS AND GROWTH

The predominance of many euglenids in ponds with barnyard leachings, or streams below the entrance of sewage, seems clearly due to organic food in such situations. One such *Euglena* is the species *sanguinea,* one of the few red ones. It often forms dense red blooms on ponds in late summer, and these were usually attributed to cattle. However, three ponds in East Tennessee were examined in 1957 and none showed any indication of cattle or other fertilizing effects. Evidently this *Euglena* shows its greatest blooming response to other factors than mere food concentration.

Oxygen and CO_2 are rarely problems although dissolved O_2 is a necessity for most green plankton, while the colorless plankton includes a few anaerobic forms. We have not found a report of the use of glycogen as an oxygen source for any colorless alga. CO_2 is just as much a necessity for algae as O_2, and cultures often

TABLE I

Number of Plankton Organisms (per ml) in the Washougl River (Washington) and the Wabash River (Indiana), on Comparable Dates

Organism	Washougl (near mouth), August 25, 1941	Wabash (near mouth), Sept. 13, 1940
1. *Gomphosphaeria aponina*	–	140
2. *Lyngbya contorta*	–	16
3. *Merismopedia glauca*	–	4
4. *Merismopedia tenuissima*	1	24
5. *Microcystis incerta*	–	48
6. *Oscillatoria* sp.	–	6
7. *Chlamydomonas* spp.	2	20
8. *Collodictyon triciliatum*	–	12
9. *Thoracomonas phacotoides*	1	4
10. *Euglena* sp.	–	4
11. *Chroomonas* sp.	3	–
12. *Cryptomonas erosa*	–	8
13. *Rhodomonas lacustris*	1	–
14. *Chrysococcus rufescens*	–	–
15. *Gymnodinium sp.*	–	1
16. *Actinastrum hantschii*	1	8
17. *Ankistrodesmus falcatus* var. *mirabilis*	5	8
18. *Coelastrum microporum*	2	24
19. *Coelastrum reticulatum*	–	12
20. *Crucigenia apiculata*	–	8
21. *Crucigenia tetrapedia*	–	12
22. *Dictyosphaerium Ehrenbergianum*	–	84
23. *Dictyosphaerium pulchellum*	1	32
24. *Franceia tuberculata*	–	4
25. *Golenkinia paucispina*	–	25
26. *Kirchneriella lunatus*	–	32
27. *Oocystis* sp.	2	80
28. *Pediastrum boryanum*	1	–
29. *Pediastrum duplex*	–	4
30. *Perioniella planctonica*	–	12
31. *Scenedesmus* sp.	1	216
32. *Tetradesmus wisconiensis*	–	8
33. *Tetraedron caudatum*	–	8
34. *Tetraedron minimum*	–	8
35. *Tetrallantos lagerheimii*	–	4
36. *Tetrastrum heteracanthum*	–	16
37. *Tetrastrum staurogeniforme*	–	4
38. *Treubaria appendiculata*	–	12
39. *Trochischia reticulata*	–	4
40. *Westella botryoides*	–	8
41. *Unid.* green cells	6	600
42. *Cyclotella meneghiniana*	6	884
43. *Fragilaria* sp.	–	2

TABLE I *(continued)*

Organism	Washougl (near mouth), August 25, 1941	Wabash (near mouth), Sept. 13, 1940
44. *Gomphonema olivaceum*	–	4
45. *Gyrosigma* sp.	–	1
46. *Melosira granulata*	–	7
47. *Melosira monilata*	–	8
48. *Navicula* sp. 48-80	3	44
49. *Navicula* sp. 15	2	–
50. *Nitzschia* sp.	–	8
51. *Synedra acutissimus*	–	20
52. *Synedra capitata*	–	2
53. *Cyclidium glaucoma*	1	–
54. *Tintinnidium fluviatile*	1	–
55. *Strobilidium humile*	–	1
56. *Strombidium* sp.	–	1
57. *Urotricha farcta*	–	1
58. *Vorticella campanula*	–	1
Total species	18	51
Total no. per ml	40	2504

TABLE II

Number of Species of Microorganisms Found at Two Stations in the Santa Fe River (Florida), Scarcely More Than 10 Miles Apart

	Brooker* (Unpolluted)	Mikeville* (Cattle droppings)
Organism groups		
Blue-green algae	4	13
Chlorophyceae, Chloromonadida	13	66
Chrysophyceae	5	10
Cryptophyceae	3	5
Dinoflagellata	5	8
Euglenophyceae	12	43
Bacillarieae	22	16
Zooflagellata	1	13
Rhizopoda	7	17
Ciliata	7	17
Total number of species at each station	79	208

* Brooker was sampled 10 times, Mikeville 9.

make better growth if it is added. The green plants release O_2 during times of illumination and since some species at least can function in the presence of high BOD values, re-aeration from this source is invaluable. Wheatland (20) found ammonia as high as 7+ ppm in the worst polluted part of the Thames, with no dissolved oxygen, yet Lund found up to 1000 or more cells/ml there at the same time. Commenting on this, Westlake stated that 1 ppm of H_2S stimulated algal activity, but that at 4 ppm (common in the Thames below London Bridge), there may have been some retardation of metabolism. Aside from outright toxicity, it is not surprising to find algae, but no dissolved oxygen. There are too many demands for the oxygen. Oxidation of ammonia proceeds as follows:

$$NH_3 + 2O_2 \longrightarrow HNO_3 + H_2O$$

and requires 4.57 g of oxygen to each gram of nitrogen. Oxidizing sulfide to sulfur also uses oxygen, although in brackish water sulfate may be reduced to sulfide, so that there is a gain of oxygen.

These oxygen relationships normally occur in water subject to organic pollution and have a bearing on our ability to dispose of wastes in marine waters. There are evidences that sulfur bacteria (*Beggiatoa*) will not grow unless a trace of oxygen is present. In Warm Mineral Springs (Florida) no *Beggiatoa* grow below the zone in the escape vent at which blue-green algae appear, which is about the 30-foot depth. No dissolved oxygen is found in the upwelling water, however.

CO_2 may drive the pH into the acid range at times, but many algae tolerate low pH values. Good examples are the blooms of *Gonyostomum semen* and *Euglena velata* seen on the acid waters of cedar and cypress swamps. The latter seems to tolerate a much wider range of pH than *Gonyostomum semen. Euglena mutabilis,* not a plankter, and *Chromulina spp.,* which are, occur in abundance in acid coal mine waters, which might be termed a human industrial waste. The sulfuric acid in these streams may drive the pH to 0.9, yet these organisms and some others thrive there. They have little or no effect in raising the pH, however.

Physical factors in the environment include many not caused by man. Moore (21) lists nine of these for the oceans, some of them operative for lakes, ponds, and streams. Temperature seems to affect growth more than the mere presence or absence of a plankton alga. Low temperatures are expected to slow down, but unless extreme, not to stop metabolism, and long experience shows the same organisms, usually in the trophic state, but less in numbers, in winter as in summer, in a given locality. Nevertheless cold water species are known. The densest bloom of *Platymonas elliptica* ever observed came from beneath several inches of ice in Lake Monona, Wisconsin.

Cooling water used in manufacturing plants may so raise the temperature of a stream into which it discharges as to kill the biota in that stretch, and also to act as a plug on organisms moving through it. It is noteworthy that most alpine and arctic or subarctic lakes have small plankton populations, although blooms do occur.

Wide ranges of salinity may occur in nature, and for them there is a wide tolerance. But when man pumps brines into a natural water, unless the dilution factor is large, many plankton organisms perish. Pressure factors are not generally applicable in ponds and streams. But some of the world's great lakes are so deep that they might have a pressure factor. This point needs investigation.

Light penetration is easily influenced by man, even under oceanic conditions. It was suggested that sediment and particulate matters from the sewage of San Diego might so diminish the illumination in the depth of water at which the planktonic gametes of the kelp *Macrocystis pyrifera* settle and attach as to be at least partially responsible for the disappearance of certain beds around San Diego. The point is still under investigation. But it has been pointed out above that some plankters are independent of illumination.

Waves and currents hardly affect plankton organisms. But abrasion by sand and silt, as the mud in a river following a rainstorm, smothers plankton organisms and may kill them by grinding. In the surf at La Jolla, California, it was felt the abrasion factor might well operate. But comparative counts in the surf and in the deep water at the end of the Scripps Institution pier revealed the same species, and more of them in the surf. Tides have no physical effect other than to flush estuaries and bays but they may completely change the water therein.

PLANKTON HABITATS

In a broad way, the differences in habitats are recognized. There is no way of analyzing all the characteristics of a mass of water, whether still or moving. There is evidence for the occurrence of certain unique plankton algae in the Dead Sea, the Salton Sea, sulfuretums, and such locales, but it is largely circumstantial. There is not the slightest information on why *Trentonia flagellata* occurs in some abundance in Warm Mineral Springs or why the sulfur bacterium *Thiodendron mucosum* (neither a plankter nor an alga) should be known only from Warm Mineral Springs (22) and the Inland Waterway at Titusville, Florida. Perhaps both will eventually be found elsewhere but in greater abundance.

Strombomonas is apparently a potamophile alga, but does occur in still waters. There are other plankton algae which could just as easily be assigned to certain particular types of physical habitats, and they have been noted at various times. But even so, just what they indicate about the particular water in which they live is difficult or impossible to state.

We agree that certain species occur predominantly in the high seas, others in bays, others in hard fresh water, and others in acid fresh water. But some are so ubiquitous as to occur in three or even all four of these habitats. Examination of a large number of reasonably similar habitats reveals substantially the same organisms. A good example of this is a recent publication by Thomasson (23) on some alpine lakes in the Pacific Northwest of North America. Streams, if the water is sufficiently aged, tend to have perhaps a greater variety of plankton algae than other water masses of similar size. But they too vary—in the lower reaches there is perhaps a paucity of food and the list and numbers are reduced. In the middle reaches production tends to be highest. The headwaters have not sufficient time for great populations to develop; at least this has been the case for the Mississippi, the Ohio, and the Scioto Rivers. At once exceptions creep in. The Missouri is heavily silted for long stretches. It is "Big Muddy." The same sort of generalizations can be applied to other habitats.

There are always problems. We do not know why certain plankton algae suddenly bloom—it is not enough to say food has reached peak quantities. We do not know why the blue-green alga *Richelia intercellularis* almost invariably lives in the diatom *Rhizosolenia.* There have been a great many papers written on the plankton ecology of streams, ponds, lakes, and the oceans, but always they leave open as many questions as they answer. Blum (24) has given an excellent review of the ecology of both planktonic and benthic river algae, and recognizes many of the unanswered questions. He brings the literature quite up to date—1956—with 243 references, but many others have either appeared since then, or were not mentioned, the numbers being so large. Various other workers have considered special cases. Chandler (25) considered the fate of lake plankton in streams, and Reif (26) the fate of river plankton in lakes.

As long as man exerts such profound changes on environments as he has since the Industrial Revolution began, there will be a real need to study man-made effects, direct or indirect, on the microbiota of all waters. Such ecological studies of salt waters are just beginning to be numerous.

This paper is an attempt to call attention to ecological aspects of man's presence as influencing plankton. It is hoped it is pertinent to the topic of "Algae and Man." If such is the case the reason for its appearance will be valid.

LITERATURE CITED

1. Kehr, R. W., et al. 1941. A study of the pollution and natural purification of the Scioto River. Pub. Health Bull. No. 276. U.S. Public Health Service, Stream Pollution Investigations, Div. Public Health Methods, Nat. Inst. Health, pp. XII-153. Govt. Print. Office, Washington, D. C.
2. Hornell, J. 1917. A new protozoan cause of widespread mortality among marine fishes. Madras Fisheries Bull. 11.
3. Shilo, M., and Shilo, M. 1955. Control of the Phytoflagellate *Prymnesium parvum*. Proc. Internat. Assoc. Theoretical and Appl. Limnol, Stuttgart 12, p. 233.
4. Gunter, G., et al. 1948. Catastrophic mass mortality of marine animals and coincident phytoplankton bloom on the west coast of Florida, November 1946 to August 1947. *Ecologic Monographs* 18:309-324.
5. Lackey, J. B., and Hynes, Jacqueline A. 1955. The Florida Gulf Coast Red Tide. *Eng. Progr. at the University of Florida* 9 (2). Bull. Series No. 70. Florida Eng. and Ind. Expt. Station. Gainesville, Fla.
6. Wilson, W. B., et al. 1960. Red Tide investigations. *In* Fishery Research, Galveston Biol. Lab. Fiscal Year 1960. U.S.D. Interior, Fish and Wildlife Service, Bureau of Commercial Fisheries. Circular 92. pp. 39-54.
7. Hutton, R. F. 1956. An annotated bibliography of Red Tides occurring in the marine waters of Florida. *J. Fla. Acad. Sci.* 19, Nos. 2 and 3.
8. Lackey, J. B., and Clendenning, K. A. A possible fish killing yellow tide in California waters (in press).
9. Olson, T. A. 1960. Water poisoning. A study of poisonous algae blooms in Minnesota. Mimeographed. pp. 1-30. *Abs. Am. Jour. Pub. Health.* Vol 92, pp. 883-885.
10. Sundaresan, B. B. Personal Communication.

11. Wheeler, R. E., Lackey, J. B., and Schott, S. 1942. A contribution on the toxicity of algae. Pub. Health Reports 57.
12. Kalmbach, E. R., and Gunderson, M. F. 1934. Western duck sickness. A form of botulism. U.S.D.A. Tech. Bull. 411. Washington, D.C.
13. Stokes, A. C. 1886. Notes of new fresh water infusioria. *Proc. Am. Philosophical Soc.* 23:562.
14. Hollande, A. 1952. *In* Traite de Zool. Vol. 1, Part 1. Ed. P.P. Grasse et Cie, Paris.
15. Lackey, J. B., and Morgan, G. B. 1960. Chemical microbiotic relationships in certain Florida surface water supplies of flowing waters in Florida. *Quart. J. Fla. Acad. Sci.* 23(4):289-301.
16. Barlow, J. P., and Myren, R. T. 1961. Oxygen resources of tidal waters. *Water and Sewage Works* 109(2):68-72.
17. Lackey, J. B., Vanderborgh, G., Jr., and Glancy, J. E. 1949. Plankton of waters overlying shellfish grounds. Proceedings Nat. Shellfisheries Assoc.
18. Redfield, A. C. 1952. Report to the Towns of Brookhaven and Islip, N.Y., on the hydrography of Great South Bay and Moriches Bay (unpublished manuscript). Reference No. 52-26. Woods Hole Oceanographic Institution, Woods Hole, Massachusetts.
19. Ryther, J. H., et al. 1957. Report on a survey of the chemistry and hydrography of Great South Bay and Moriches Bay made in June 1957, for the Town of Islip, New York (unpublished manuscript). Reference No. 57-59. Woods Hole Oceanographic Institution, Woods Hole, Massachusetts.
20. Wheatland, A. B. Some aspects of the carbon, nitrogen and sulphur cycles in the Thames Estuary. 1. Photosynthesis, denitrification and sulphate reduction. Reprinted from Institute of Biology, Symposia Report No. 8. Effects of Pollution on Living Material.
21. Moore, H. B. 1958. Marine ecology. X1 + 493. John Wiley & Sons, Inc., London.
22. Lackey, J. B., and Lackey, Elsie W. 1961. The habitat and description of a new genus of sulphur bacterium. *J. Gen. Microbiol.* 26:29-39.
23. Thomasson, K. 1962. Planktological notes from western North America. *Arkiv för Botanik.* 4(14).
24. Blum, J. L. 1956. The ecology of river algae. *Bot. Review.* 22(5):291-341.
25. Chandler, D. C. 1937. Fate of typical lake plankton in streams. *Ecologic Monographs.* 7:445-479.
26. Reif, C. B. 1939. The effect of stream conditions on lake plankton. *Trans. Am. Mic. Soc.* 58:398-403.

Principles of Primary Productivity: Photosynthesis Under Completely Natural Conditions

Jacob Verduin

Chairman, Department of Biology
Bowling Green State University
Bowling Green, Ohio

The photosynthetic process under completely natural conditions is highly variable from time to time and place to place. The same can be said for practically all other processes occurring in nature. The variations are such that samples collected a few moments or a few meters apart will routinely differ by 20% and not at all unroutinely by 50% or more. The most important consequence of this variation is that an individual sample is poorly representative of the larger population from which it is drawn. To obtain valid averages representing that population one must draw numerous samples. Moreover, because the individual sample is so poorly representative there is usually no virtue in determining the characteristics of the individual sample with a certainty of 95-98%. An investigator who selects methods which measure the characteristics of the individual sample with a certainty of 80% and which permit him to study five times more samples than methods achieving a certainty of 98% would permit, has made a wise choice. This is perhaps the most valuable idea to be presented in this paper. Consequently it has been placed at the beginning. It is also appropriate here because the methods and results of investigation of photosynthesis under completely natural conditions exemplify this principle.

Any sample of water which is enclosed in a bottle is obviously not under natural conditions. It is subject to much reduced turbulence and a large glass surface surrounds it, providing a favorable attachment substrate for bacterial development. These two factors are probably of prime importance in reducing the photosynthetic rate below that which is likely to occur under natural conditions, and increasing the respiration rate above that which is representative of the natural community. Consequently, the major emphasis in the data presented here is on photosynthetic rates measured under completely natural conditions.

Such measurements can be made conveniently by determining CO_2 change or O_2 change over 4- to 6-hr intervals by day, in the natural habitat, and by determining the nocturnal changes over 8- to 10-hr intervals to provide an estimate of respiration by the total community. This respiration rate is added to the diurnal CO_2 or O_2 changes in computing total photosynthesis. Oxygen determinations can be made using the Winkler method. CO_2 changes can be most conveniently determined by measuring pH changes during the time intervals, and relating these to a differential titration curve for the water as shown in Table I.

In Table I a 0.020 N solution of NaOH was added, 0.5 ml at a time, to a 100-ml sample of water from a small pond. The pH was measured after each addition and is shown in the second column. When pH is below 8.4 the major chemical reaction involved is:

$$NaOH + H_2CO_3 \text{ - - - - - } NaHCO_3$$

At higher pH levels the major reaction is:

$$NaOH + NaHCO_3 \text{ - - - - - } Na_2CO_3 + H_2O$$

TABLE I

An Example of Differential Titration (Small Pond)*

Milliliters of 0.020 N NaOH added to 100 ml of water	pH	Micromoles CO_2 change per liter (cumulative)
0.0	7.89	
0.5	8.21	100
1.0	8.47	200
1.5	8.60	300
2.0	8.71	400
2.5	8.82	500
3.0	8.90	600
3.5	8.98	700

*Each ml of NaOH added reacts with 20 micromoles of carbonic acid, or bicarbonate.

In each case a molecule of NaOH reacts with one molecule of CO_2, raising the pH in much the same way that removal of CO_2 by the photosynthetic process raises it. However, Na ions are being added to the water, thus increasing its titratable base. To avoid a serious error from this influence a differential titration should be confined to a relatively small interval over the pH range, say 0.5 to 1.0 pH unit, so the total increase in titratable base does not amount to more than a few per cent of the original titratable base present in the water.

Table I represents data obtained during the summer of 1962 while the author was serving as guest lecturer in an Institute for High School Teachers subsidized by the National Science Foundation. The Institute was held at Thiel College, Greenville, Pennsylvania. The pond was a small duck pond near the campus, having a titratable base of 2.84 meq/liter (total alkalinity of 142 ppm). This pond and a similar one on a farm a few miles away were used to demonstrate methods of measuring primary productivity to the Institute participants. A sample computation of CO_2 change is provided below: at 0900 the pH value was 8.54; at 1400 it had risen to 8.94. Reference to Table I reveals that a pH change of 0.4 unit in this pH range represented a CO_2 change of about 400 μmole per liter, or about 80 per liter per hour.

Because most high schools are not equipped with pH meters it seemed desirable to provide a method of measuring productivity using tools more readily accessible to the participants. Such a method is available in the titrations for carbonate and for free CO_2 routinely used by limnologists. The indicator used is phenolphthalein and when it is added to a water sample the appearance of red color is regarded as evidence of carbonate. A 100-ml sample is then titrated with 0.020 N H_2SO_4until the red color disappears. Each milliliter of H_2SO_4 added will generate 20 μmole of H_2CO_3 from the bicarbonate, which is usually the predominant chemical species present. This will, of course, react with carbonate to re-form the bicarbonate. Conversely, the absence of red color after adding phenolphthalein is regarded as evidence of the presence of carbonic acid. A 100-ml sample is then titrated with 0.020 N NaOH until a faint, but permanent, pink color appears. Each milliliter of NaOH added reacts with 20 μmole of H_2CO_3 in the sample. The phenolphthalein end-point (pH 8.6, approximately) is close to the value at which neither H_2CO_3 nor carbonate are present in significant amounts.

A sample experiment applying this method was performed using *Ulva lactuca* in ocean water. The *Ulva* sample had a plant volume of 0.7 ml and was immersed in 100 ml of water. It was placed in a sunny window for 2 hr, beside a reference sample of 100 ml of water which did not contain an *Ulva* sample. After 2 hr the reference sample required 0.8 ml of 0.020 N NaOH to reach the phenolphthalein end-point, and the sample containing the *Ulva* fragment required 5.3 ml of 0.020 N H_2SO_4 for the titration. Thus an apparent photosynthetic rate (i.e. uncorrected for plant respiration) of about

$$\frac{(5.3 + 0.8) \times 20\ \mu\text{mole}}{0.7\ \text{ml} \times 2\ \text{hr}} = 87\ \mu\text{mole}$$

of CO_2 absorbed per ml of *Ulva* per hour was measured in this experiment.

This method of measuring photosynthesis, by titration to the phenolphthalein end-point of samples collected at short time intervals from the natural habitat, was then applied to the two small ponds near the Thiel College Campus. The data gathered during two nights and one day are presented in Table II.

At 5:00 pm on July 10 a 100-ml water sample from the duck pond (Young's) required 3.62 ml of 0.020 N H_2SO_4 to reach the phenolphthalein end-point. A sample collected at 9:00 pm required 3.73 ml. Apparently the autotrophic component of the community was carrying on photosynthesis at a rate which just utilized the CO_2 produced by respiration of the entire community. Consequently, neither the photosynthetic rate nor the respiratory rate could be deduced from these measurements. Between 9:00 pm and 5:15 am the next day the titration value dropped from 3.73 ml to 0.83 ml of H_2SO_4. This drop represented a respiration rate of

$$\frac{(3.73 - 0.83) \times 20\ \mu\text{mole}}{0.1\ \text{liter} \times 8.25\ \text{hr}} = 70\ \mu\text{mole}$$

of CO_2 evolved per liter per hour.

On July 11 a balance between photosynthesis and respiration was again observed between 5:15 and 8:15 am. But distinct CO_2 absorption by photosynthetic activity of the autotrophs was observed from 8:15 am to 1:45 pm, and again from 1:45 to 5:10 pm.

TABLE II

Photosynthetic Yields Obtained during the Summer of 1962 from Two Small Ponds in Pennsylvania

	Young's Pond (titratable base 2.8 meq/liter)		Brown's Pond (titratable base 1.4 meq/liter)	
Time	Amt. of 0.020 N H_2SO_4 required for titration (ml)	Change per liter per hr (μmole CO_2)	Amt. of 0.020 N H_2SO_4 required for titration (ml)	Change per liter per hr (μmole CO_2)
July 10				
5:00 pm	3.62		0.21	
9:00	3.73		0.41	
		70 (resp.)		2.9 (resp.)
July 11				
5:15 am	0.83		0.29	
8:15	0.95	67 (photosyn.)	0.32	6.7 (photosyn.)
1:45 pm	2.67		0.47	
5:10	4.26	59 (photosyn.)	0.71	14.0 (photosyn.)
9:15	4.20		0.73	
July 12				
		64 (resp.)		4.1 (resp.)
5:00 am	1.70		0.55	
Algal population				
At 9:00 am	*Glenodinium* *Trachelomonas* *Euglena*	250 μl/liter	*Ceratium*	200 μl/liter
At 11:45 am	*Glenodinium* *Trachelomonas* *Euglena*	162 μl/liter	*Ceratium*	50 μl/liter

Total photosynthetic yield, (micromoles per microliter of algae per hr):

$$\frac{63 + 68}{206} = 0.64 \qquad \frac{10.3 + 3.5}{125} = 0.11$$

These rates amounted to 67 and 59 μmole of CO_2 absorbed per liter per hour. Between 5:10 and 9:15 pm the balance between respiration and photosynthesis was observed once more and during the 9:15 pm to 5:00 am period a respiration rate of 64 μmole per liter per hour was observed. If we assume that the diurnal respiration rate of the total community is similar to its nocturnal respiration rate (a conservative assumption), then a gross photosynthetic rate of $63 + 67 = 130$ μmole/(liter $\times$ hour) is obtained. The close balance observed here between nocturnal CO_2 evolution and diurnal CO_2 removal is a typical feature of pond and lake habitats. The CO_2 budget and the O_2 budget of such an aquatic habitat is largely internal. That is, the CO_2 taken up by photosynthesis by day is replaced by respiration at night and the O_2 produced by photosynthesis by day is consumed by respiration at night. Exchange of these gases between the water and atmosphere is small, representing only a few per cent of the internal budget.

The phytoplankton community responsible for the photosynthetic activity was sampled and the plant volume of the major components was determined. This was accomplished by pouring a water sample into a Stender dish to a depth of 1 cm. A few drops of Dr. Rodhe's fixative (10 g I_2, 20 g KI, 200 ml H_2O, 20 ml glacial acetic acid) were added to kill the organisms and the samples were allowed to settle for a few hours, then 10 fields were counted, at 100$\times$ magnification, by simply immersing the 10$\times$ objective until the organisms on the bottom came into focus. The samples were also examined at 430$\times$ magnification, and 10 individuals of the major components were measured to compute their average volume. These measurements were made only roughly approximate. For example, when a cell of *Ceratium hirundinella* was measured, the volume of its "horns" was neglected and its biomass was considered concentrated near its middle. The dimensions of this region were measured in three directions (width, length, depth) and volume computed as the product of these dimensions, as though the central region (which is roughly spherical in shape) were a rectangular box. This approximation makes the volume estimate of the central region about 20% too high, but this overestimate serves to correct somewhat for the neglect of the volume of horns. Similar approximations were made with other species found among the major components. Plant volumes so obtained are presented in Table II. Although there were dozens of different species present in the community the major volume of plant matter (95% or more) was contributed by three species: one a member of the genus *Glenodinium,* one of *Trachelomonas,* and one of *Euglena.* A notable decrease in population density was observed between 9:00 and 11:45 am, suggesting a migration away from the increased light intensity near midday. The two values of population density were averaged to provide the computation of photosynthetic yield; 0.64 μmole of CO_2 absorbed per microliter of plant matter per hour.

The data from Brown's pond (Table II, columns 4 and 5) present a sharp contrast to those obtained from Young's pond. The observed changes in CO_2 were about an order of magnitude lower, amounting to photosynthetic yields averaging 10.3 and respiration rates averaging 3.5 μmole/(liter × hour). The community was dominated completely by *Ceratium* and the decrease in population density near the surface between 9:00 and 11:45 am was also observed in this pond. The mean plant volume of 125 μl/liter yielded a photosynthetic rate of 0.11 μmole/(μl × hour). In these computations of photosynthetic yield per unit of plant volume the nocturnal respiration rate of the community is added to the observed diurnal CO_2 absorption rate (in excess of respiration) to provide a measure of total photosynthesis. Such an estimate probably errs on the conservative side, because diurnal respiration rates are likely to be higher than the nocturnal rates.

The photosynthetic rates of 0.64 and 0.11 μmole per microliter of plant volume per hour are significantly lower than the average yields observed in western Lake Erie (Verduin, 1960), where rates of 2-3 μmole/(μl × hour) are commonly observed. However, these higher rates are always associated with phytoplankton densities an order of magnitude lower than those in Table II. When phytoplankton densities approach 100 μl/liter in western Lake Erie, as they often do during the peak of the August Cyanophyta pulse, the hourly yields per microliter are similar to those in Table II.

This small set of observations supports a relationship observed in numerous studies in northwestern Ohio (Verduin, 1959, 1960) as well as by other investigators elsewhere, namely, that an inverse relationship exists between phytoplankton density and the hourly photosynthetic yield per unit of plant biomass. Only in relatively sparse populations does one observe high yields per microliter of plant volume (3-6 μmole/hour), and in densely populated communities (100 μl/liter) the yields per microliter of plant volume are an order of magnitude lower. Consequently, in any attempt to estimate the primary productivity of an aquatic environment accurately the photosynthetic capacity of the autotrophic community must be measured. Any computations based on the assumption that photosynthetic yield per unit of plant biomass (chlorophyll, plant volume, organic seston, etc.) is a constant should be regarded as first approximations, and elegant refinements of the determination of daily light supply, and of the penetration of different portions of the spectrum, will not improve the estimate of productivity if the variation in photosynthetic capacity of the phytoplankton communities from different depths and locations is not determined. These facts have a special significance to oceanography. No one has yet devised a method of measuring photosynthesis under completely natural conditions in the deep regions of the ocean. The C_{14} technique, now in wide use, utilizes samples that have not been subjected to artificial concentration of the plankton community, but the influence of enclosing the community in a small bottle for six hours or more, and exposing it to constant, relatively bright light, is difficult to evaluate. It seems certain that the flagellate component of the community, swimming in response to the light stimulus, will strike the walls of the bottle and adhere to it, or even be broken up, with a consequent abnormal photosynthetic performance on the part of this group.

A LIMITING-FACTOR EQUATION APPLIED TO PRIMARY PRODUCTIVITY

The Baule-Mitscherlich (Baule, 1918) limiting-factor equation has been widely ignored by ecologists, and also by plant physiologists of the English-speaking countries. It is reproduced below:

$$E = E_{max}(1 - e^{-0.7x/h})(1 - e^{-0.7y/h})(1 - e^{-0.7z/h})$$

where E is the rate of a process, E_{max} is its rate if all factors (x, y, z, etc.) are present in abundance, and the $0.7/h$ factor is introduced to facilitate computation; e, of course, is the base of natural logarithms.

The following modification of this equation is submitted for consideration as a model of the limiting-factor relationships as applied to primary production:

$$Y = Y_{opt}(1 - 2^{-x})(1 - 2^{-y})(1 - 2^{-z}) \text{ (etc.)}$$

where Y is the photosynthetic yield per m^2 of earth surface, Y_{opt} is the yield obtainable if all factors are present at optimal intensity, and 2 is substituted for e (it is no less natural and facilitates computation considerably); x, y, z, of course, are the factors influencing the photosynthetic process.

In Table III a series of evaluations is provided for the term $(1 - 2^{-x})$ when various intensities of the factor x are considered. It is assumed that when x equals unity the value of Y is equal to $\frac{1}{2}Y_{opt}$, if all other factors are present at optimal intensity. The values in Table III are shown graphically in Fig. 1. It is evident that increasing values of x result in approximately linear yield increases at low levels of x, but at higher values the effect on yield becomes progressively smaller, while the yield curve approaches a value of $1(Y_{opt})$ asymptotically.

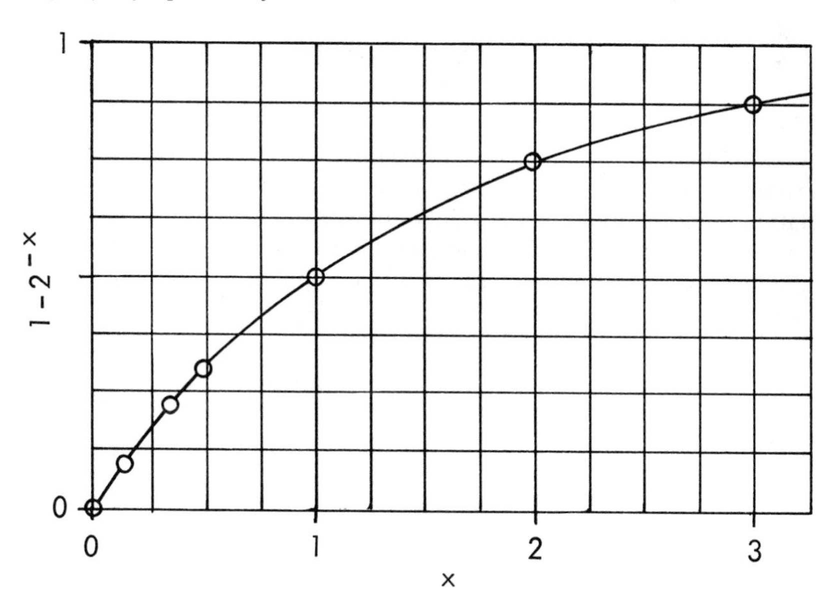

FIG. 1. A graph of the relationship between factor intensity (x) and its influence on yield $(1-2^{-x})$.

A most significant feature of the limiting-factor equation is that the overall yield is a product of numerous factors taking the form $(1 - 2^{-x})$. Consequently, if three factors simultaneously have a relative intensity of 0.33, although all others are abundant, the yield will be:

$$(0.20 \times 0.20 \times 0.20) = 0.008 Y_{\text{opt}}$$

But if all three factors are increased to 0.50, the yield will be:

$$(0.33 \times 0.33 \times 0.33) = 0.11 Y_{\text{opt}}$$

Thus, increasing these 3 factors each by 1.6 causes a yield increase of 14-fold. However, if 3 factors have an intensity of 2 the yield will be:

$$(0.75 \times 0.75 \times 0.75) = 0.42 Y_{\text{opt}}$$

Increasing these factors by 1.5 would change the yield to:

$$(0.87 \times 0.87 \times 0.87) = 0.66 Y_{\text{opt}}$$

an increase of about 1.6-fold. If an equation of this type is approximately valid, then it becomes possible to attach quantitative values to some of the factors which obviously influence photosynthesis. For example, many studies of the influence of light indicate that the 1/2 saturating value is about 1/8 of full sunlight or 1000 ft-c (10,000 lux). Thus, such a unit would be close to 1 in Table III. Similarly, in western Lake Erie, an increase in phytoplankton population from 2 μl/liter to 20 μl/liter results in only a 2- to 3-fold increase in photosynthetic yield per m^2 per day. Hence it may be postulated that 2 μl/liter represents a population intensity somewhere in the vicinity of 0.5 unit in Table III.

TABLE III

Factor Intensity Related to Its Contribution to Yield

Factor intensity	$(1 - 2^{-x})$
0.17	$(1 - 2^{-1/6}) = 0.11$
0.33	$(1 - 2^{-1/3}) = 0.20$
0.50	$(1 - 2^{-1/2}) = 0.29$
1.0	$(1 - 2^{-1}) = 0.50$
2.0	$(1 - 2^{-2}) = 0.75$
3.0	$(1 - 2^{-3}) = 0.87$
4.0	$(1 - 2^{-4}) = 0.94$

The equation also suggests that research involving addition of substances such as nitrogen, phosphorus, iron, etc., in small amounts similar to the quantities present in natural water, singly and in combination, then measuring their influence on photosynthetic rate, would provide further information concerning the relative position on the scale of Table III. The importance of adding small quantities similar to the concentrations present in nature when carrying on experiments on the influence of limiting factors must be emphasized. Much work in the laboratory culturing of algae has been done with cultures having nitrogen, phosphorus, etc., concentrations 1000-fold higher than those of natural waters. Such cultures are useful for promoting algal growth and for maintaining a population, but growth rates and rates of other metabolic processes determined in such cultures have a very limited application to the natural condition.

Many biologists still speak of limiting factors in the singular (*the* limiting factor), as though only one factor can be limiting at a time. The Baule-Mitscherlich equation indicates that several factors can be limiting simultaneously. Indeed, it seems likely that this is the rule in the process of photosynthesis in aquatic habitats. Nitrogen, phosphorus, carbon dioxide, light, and the density of the phytoplankton population are usually present in suboptimal concentrations, and the actual photosynthetic yield is probably only a few per cent of the rate that could be realized if all factors were supplied at optimal intensities. On our planet the intensity of sunlight, and the evolutionary adaptations of algae to the range of factor intensities found in nature, set the limit for Y_{opt}, at a few thousands of millimoles of CO_2 absorbed per square meter per day. The usual values encountered in ponds and lakes are of the order of a few hundreds of millimoles per square meter per day.

APPLICATION OF LAMBERT-BEER'S LAW TO AQUATIC ENVIRONMENTS

Because light is such an important factor in primary productivity, and because its attenuation in aquatic habitats is so rapid, a consideration of the Lambert-Beer's law as applied to light attenuation in aquatic media seems appropriate here.

The law is expressed as follows:

$$I = I_o e^{-kdc} \tag{1}$$

where I is light intensity at depth d, I_o is the surface light intensity, c is the concentration of light-absorbing particles, and k is a proportionality constant. Usually limnologists and oceanographers have determined light penetration without relating it to the concentration of suspensoids. They use an equation of the form

$$I = I_o e^{-kd} \tag{2}$$

In such an equation k is highly variable because it contains the variable c, and its computation is of no practical or theoretical value. One can more profitably report the per cent light attenuation per meter, or the depth at which 10% or 1% of surface light intensity is observed. But when both d and c are measured and the k in eq. (1) is computed we obtain an extinction coefficient which relates light penetration to the concentration of suspensoids, hence has both theoretical and practical significance.

For practical application of eq. (1), it is convenient to measure the depth associated with 1% of surface light (d') in meters and to measure the concentration of suspensoids (centrifuged or millipore filtered) in milligrams per liter, dry weight. Then the relationship

$$\frac{I_o}{I} = e^{kd'c} \tag{3}$$

follows, and by taking the base e log of both sides and noting that

$$\frac{I_o}{I} = 100$$

because d' is the depth associated with 1% of surface light, we obtain:

$$\log_e 100 = kd'c \tag{4}$$

thus

$$4.6 = kd'c \tag{5}$$

and

$$k = \frac{4.6}{d'c} \tag{6}$$

A correction for the light-quenching effect of water (Whitney, 1938) can be introduced as follows:

$$k = \frac{4.6/d' - 0.03}{c} \tag{7}$$

The values of k computed from eq. (7) using data from western Lake Erie, West Lost Lake, Michigan, Pymatuning Reservoir, Maumee River, and the Ohio River at Louisville lie between 0.07 and 0.21, with an average value of 0.12. The light penetration values (d') varied between 0.4 and 40 m, and the concentration of suspensoids varied between 0.8 and 87 mg/liter. Thus, variations in these parameters over two orders of magnitude yielded values of k averaging 0.12 with a standard deviation of ± 0.034. These data suggest that the light-quenching properties of suspensoids in aquatic habitats are surprisingly uniform. However, the variations encountered in this k value merit further study. Investigation of k values in habitats where the major portion of suspensoids are algal as compared to habitats containing large quantities of silt will establish extinction coefficients peculiar to each type of suspensoid, and under favorable circumstances communities dominated by algae of small volume per individual might be contrasted with communities dominated by larger forms, or studies of communities dominated by Chlorophyceae, Myxophyceae, or Chrysophyceae, may reveal k values peculiar to each.

The work reported here was based on light measurements obtained with a Weston photronic cell, which has its peak sensitivity in the blue and green portions of the spectrum and is relatively insensitive to nonvisible light. Such an instrument measures the light most effective in photosynthesis and is not influenced by the portions of the spectrum which do not promote photosynthesis. The k values quoted here are therefore valid for the visible spectral region. If light-measuring instruments employing filters are used, the extinction coefficients observed for different portions of the spectrum vary considerably, with highest extinction coefficients observed in the ultraviolet and dark red. Figure 2 presents graphically some data of this kind, gathered by graduate students at the Potamological Institute at Louisville, using an instrument recommended by Dr. Wolfgang Schmitz. The data were obtained in the Ohio River, and the contrast between the extinction coefficients for different light colors is evident in the different slopes of the lines

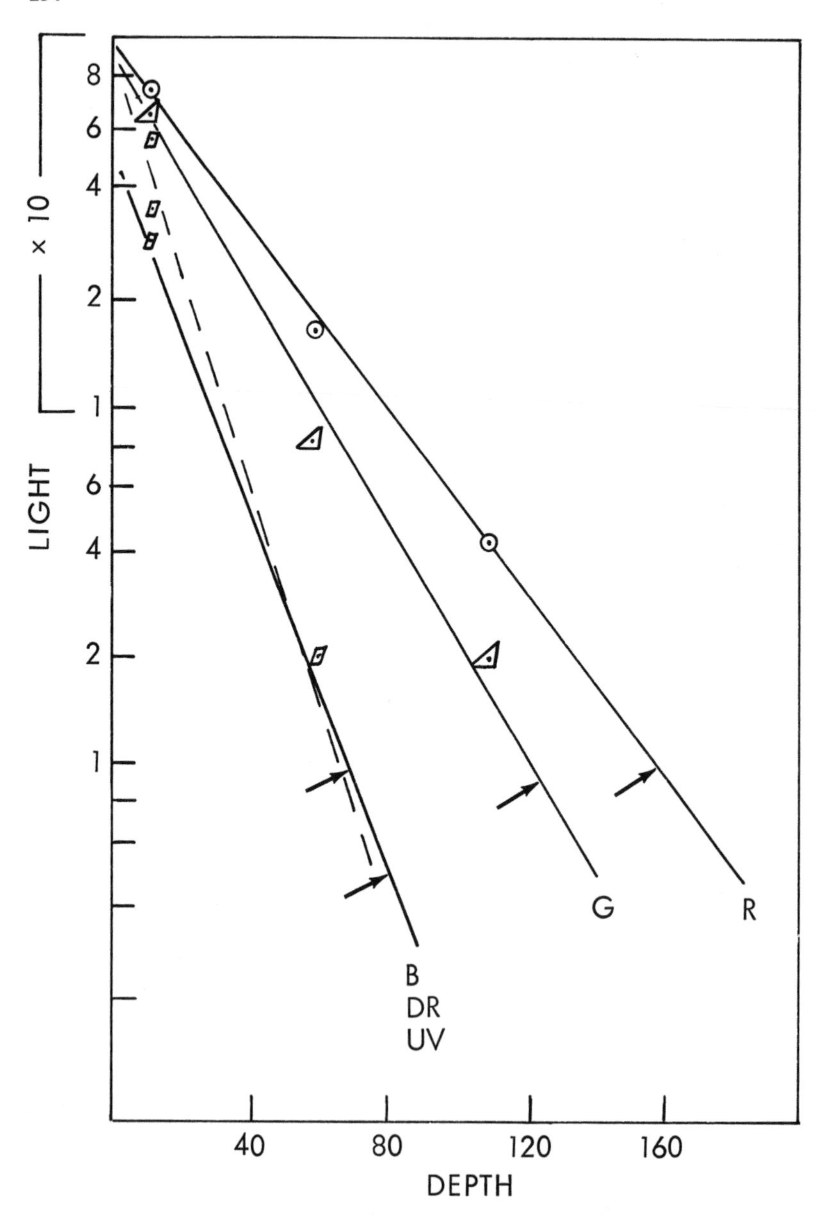

FIG. 2. A semilog graph of light vs. depth (cm) showing different attenuation rates for different light colors (R = red, G = green, B = blue, DR = dark red, UV = ultraviolet). Blue, dark red, and ultraviolet are quenched about twice as rapidly (d' = 70-80 cm) as green and red light (d' = 120 and 160, indicated by arrows).

drawn through these data. The fact that blue (B), dark red (DR), and ultraviolet (UV) light are attenuated rapidly in water causes the measurements with an instrument that measures a wide portion of the spectrum to show a different attenuation rate in the first few meters than is observed at greater depths after these components have been screened out. Consequently, I recommend that, in computing k values from data obtained with an instrument such as the Weston photronic cell, one determine the depth associated with 1% of surface light by drawing a line through the major portion of the date and extrapolating back to zero depth. In clear water the surface light intensity obtained by such extrapolation will usually lie somewhat below the actual value observed at the surface, and will represent the surface intensity of that portion of the spectrum which penetrates deeply. In this way one uses the surface layers as a light filter and ignores the unfiltered data obtained near the surface. An estimate of d', based on such extrapolation, is presented graphically in Fig. 3 (see arrow). The d' value is the depth associated with a light intensity of 0.55 because the line extrapolated to zero depth intercepts the ordinate at 55, although the light intensity observed at the surface was considerably higher.

The average k value of 0.12 quoted above can be used to obtain approximate estimates of suspensoid concentration from submarine photometer readings alone. For example, in Fig. 3 a d' value of 20 m is indicated, consequently the water must have a suspensoid concentration of

$$c = \frac{4.6/d' - 0.03}{0.12} = 1.7 \text{ mg/liter}$$

The clearest ocean waters show d' values of about 150 m. If this value is used to compute their suspensoid concentration in eq. (7) we obtain:

$$c = \frac{4.6/150 - 0.03}{0.12} = 0.005 \text{ mg/liter}$$

indicating that practically all of the light-quenching effect in these waters is attributable to the water itself and not to the suspensoids. In most ponds and lakes computations of suspensoid load from eq. (7) should yield an estimate having an average accuracy of about 80%. In waters containing dissolved pigments larger errors would be present. Determinations of the k value in eq. (7) in such waters

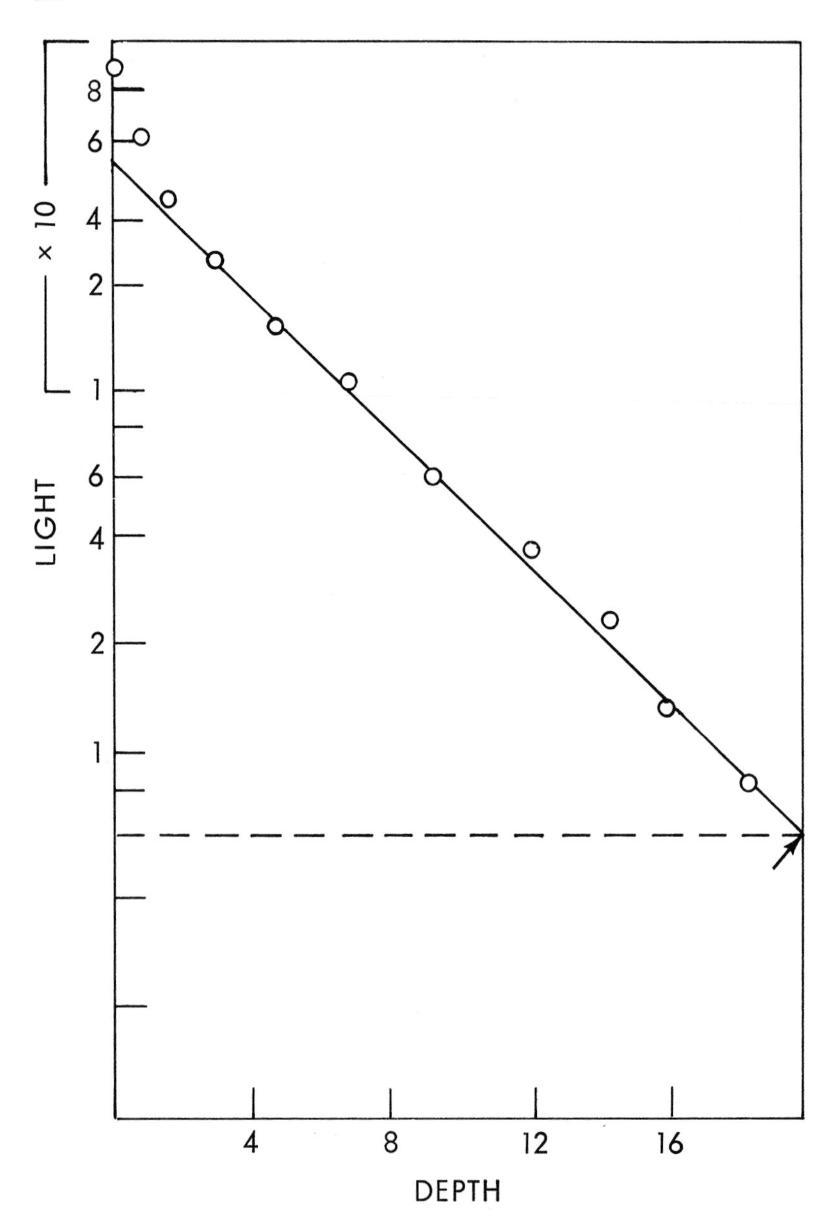

FIG. 3. A semilog graph of light vs. depth (m) demonstrating the extrapolation of the line drawn through the major portion of the data to obtain the surface light value (55) used to determine the depth (d') associated with 1% of surface light (see arrow).

would, when compared with the value reported above, yield valuable information concerning the light-quenching influence of such dissolved pigments.

SUMMARY

Primary productivity determinations are highly variable in individual samples, and approximate methods permitting study of numerous samples are superior to more precise methods which reduce the number of samples that can be analyzed. Studies under completely natural conditions are desirable, wherever possible, because samples enclosed in bottles are likely to exhibit abnormally low photosynthesis and abnormally high respiration.

A comparison of photosynthesis and respiration in two small ponds revealed that the two processes were in approximate equilibrium, amounting to 63 and 68 μmole of CO_2 change per liter per hour in one pond and 10.3 and 3.5 μmole per liter per hour in the other. Hourly photosynthetic rates per microliter of phytoplankton were 0.64 and 0.11 μmole, showing general agreement with data from the literature.

A limiting-factor equation derived from the Baule-Mitscherlich equation is presented, and levels of photosynthetic yield in aquatic environments are analyzed. This analysis suggests that several factors (light, CO_2, nitrogen, phosphorus, iron, algal population, etc.) may be limiting simultaneously, and that this is the rule in aquatic habitats resulting in yields representing only a small fraction of the yield obtainable if all factors were at optimal intensity.

The Lambert-Beer's law is applied to aquatic environments providing an equation from which a coefficient k relating light penetration to suspensoid concentration can be evaluated. The average value observed to date is 0.12 when light penetration is in meters and suspensoid concentration is in mg/liter, dry weight. Further application of this equation to aquatic environments is recommended.

LITERATURE CITED

Baule, B. 1918. *Landwirtsch. Jahrb.* 51:363.
Mitscherlich, E. A. 1909. *Landwirtsch. Jahrb.* 38:537.
Verduin, J. 1953. *Science* 117:392.
Verduin, J. 1954. *Ecology* 35:550.

Verduin, J. 1956. *Ecology* 37:40.
Verduin, J. 1960. *Limnol. and Oceanogr.* 5:372.
Whitney, L. V. 1938. *Trans. Wis. Acad. Sci.* 31:201.

Algae in Water Supplies of the United States

C. Mervin Palmer

U.S. Department of Health, Education, and Welfare
Public Health Service
Robert A. Taft Sanitary Engineering Center
Cincinnati, Ohio

In the United States the important sources of water for cities and towns are ground waters and surface waters. Algae may develop in water from underground sources after it is exposed to sunlight. Surface waters invariably are habitats for the growth of algae. The amount of growth may be small or very extensive depending upon the concentration of nutrients, turbidity, temperature, and length of exposure to light.

Preparation of surface waters for use generally requires treatment for algae and algal products in the water. The amounts of algae and algal products remaining in processed waters depend upon the raw water and the efficiency of the treatment process.

In recreational waters dense algal growths are detrimental. Water supplies for commercial fishing are affected by the amounts and kinds of algae present. In fishing areas, massive growths of attached algae interfere with harvesting, and late summer blooms can cause extensive fish kills if weather conditions or other factors cause oxygen release by the algae to be reduced while their oxygen consumption remains high or is increased. The flow of water in irrigation ditches in the western half of the United States is substantially reduced by dense growths of attached filamentous algae on the sides and bottom of the ditches and on the screens.

Marked regional differences in the nature and quality of available water supplies result from differences in the amount and distribution of rainfall, topography, hardness and turbidity of the water, depth of impoundments, water temperature, amount of pollution caused by agricultural or mining practices and by domestic as well as industrial wastes, and the extent of re-use.

Regional differences in the numbers and kinds of algae in water supplies cause differences in treatment practices. In some areas algal problems may be seasonal, but in others continuous. Portions of some American rivers have very large algal populations, whereas other portions and other streams have sparse populations. Algal growths in many surface waters are gradually increasing as greater populations are adding more nutrients to drainage areas.

A massive mixed growth of algae is always a potential problem because of its bulk and its tendency to produce malodorous products when a large amount of organic matter disintegrates. The number of genera or species of algae that individually are responsible for causing serious problems in water supplies is relatively small. Some genera of diatoms are significant in this respect and certain green algae, blue-green algae, and flagellates are also important. The algae referred to most frequently in water supply problems comprise 27 genera, seven of which are diatoms (Table I). *Asterionella, Melosira, Synedra,* and *Anabaena* are among the ones most often involved in filter clogging, odor production, and other problems. More than 135 genera of algae are recorded as common or important in American water supplies (Table II).

RIVER PROBLEMS

Many rivers are subject to occasional algal blooms which may come suddenly and persist long enough to cause a major crisis for communities using the water. The diatom *Synedra* is a notorious example. These long, slender diatoms effect short filter runs by forming an almost impervious mat on or in the top layer of sand, causing a rapid increase in head loss. This occurred at Laredo, Texas; water from the Rio Grande contained these diatoms in such numbers that filter runs were reduced to less than an hour. Almost as much water was then required to wash the filter as the filter would produce during the run.

The "diatom epidemics" in the Potomac River are a convincing example of the importance of diatoms in filter clogging. The first "epidemic" occurred in September 1934. Counts in the raw water ran as high as 2000 cells/ml. At the Dalecarlia treat-

TABLE I

Nuisance Algae Most Frequently Reported in American Water Supplies (Recorded by Geographic Regions)

	North-eastern	North Central	South Central	South-western
Diatoms				
Asterionella	×	×	×	×
Cyclotella	×	×	×	–
Fragilaria	×	×	×	×
Melosira	×	×	×	×
Stephanodiscus	–	×	–	×
Synedra	×	×	×	×
Tabellaria	×	×	×	–
Blue-Green Algae				
Anabaena	×	×	×	×
Anacystis	×	×	×	–
Aphanizomenon	×	×	×	×
Coelosphaerium	×	×	–	–
Oscillatoria	×	×	×	×
Green Algae – Nonfilamentous				
Pediastrum	×	×	×	×
Scenedesmus	–	×	×	–
Staurastrum	×	×	–	×
Green Algae – Filamentous				
Chara	–	–	–	×
Cladophora	–	×	–	–
Pithophora	–	–	×	–
Spirogyra	×	×	×	–
Vaucheria	–	×	–	×
Flagellates – Pigmented				
Ceratium	×	×	–	×
Chlamydomonas	×	×	×	–
Dinobryon	×	×	–	–
Euglena	–	×	–	–
Peridinium	–	×	–	–
Synura	×	×	–	–
Uroglenopsis	×	×	–	–

TABLE II

Common and Significant Genera of Algae in American Water Supplies

Diatoms	Flagellates	Green Algae, Nonfilamentous	Green Algae, Filamentous	Blue-Green Algae
Achnanthes	*Carteria*	*Actinastrum*	*Audouinella*	*Anabaena*
Amphiprora	*Cephalomonas*	*Ankistrodesmus*	*Basicladia*	*Anacystis*
Amphora	*Ceratium*	*Botryococcus*	*Batrachospermum*	*Aphanizomenon*
Asterionella	*Chlamydobotrys*	*Chlorella*	*Botrydium*	*Aphanocapsa*
Biddulphia	*Chlamydomonas*	*Chlorococcum*	*Bulbochaete*	*Aphanothece*
Cocconeis	*Chrysococcus*	*Chodatella*	*Chaetophora*	*Aulosira*
Cyclotella	*Cryptoglena*	*Closteriopsis*	*Chara*	*Calothrix*
Cymatopleura	*Cryptomonas*	*Closterium*	*Cladophora*	*Chroococcus*
Cymbella	*Dinobryon*	*Coelastrum*	*Desmidium*	*Coelosphaerium*
Diatoma	*Eudorina*	*Cosmarium*	*Dichotomosiphon*	*Cylindrospermum*
Fragilaria	*Euglena*	*Crucigenia*	*Draparnaldia*	*Gloeocapsa*
Gomphonema	*Glenodinium*	*Dictyosphaerium*	*Microspora*	*Gloetrichia*
Gyrosigma	*Gonium*	*Gloeocystis*	*Monostroma*	*Gomphosphaeria*
Melosira	*Gymnodinium*	*Hydrodictyon*	*Mougeotia*	*Lyngbya*
Meridion	*Hemidinium*	*Kirchneriella*	*Nitella*	*Merismopedia*
Navicula	*Lepocinclis*	*Lagerheimia*	*Oedogonium*	*Nostoc*
Nitzschia	*Mallomonas*	*Micractinium*	*Pithophora*	*Oscillatoria*
Pinnularia	*Pandorina*	*Micrasterias*	*Rhizoclonium*	*Phormidium*
Pleurosigma	*Peridinium*	*Netrium*	*Spirogyra*	*Plectonema*
Rhizosolenia	*Phacotus*	*Oocystis*	*Stichococcus*	*Pleurocapsa*
Stephanodiscus	*Phacus*	*Ophiocytium*	*Stigeoclonium*	*Raphidiopsis*
Surirella	*Platydorina*	*Pediastrum*	*Tribonema*	*Rivularia*
Synedra	*Pyramimonas*	*Phytoconis*	*Ulothrix*	*Scytonema*
Tabellaria	*Rhodomonas*	*Quadrigula*	*Vaucheria*	*Spirulina*
	Spondylomorum	*Scenedesmus*	*Zygnema*	*Stigonema*
	Synura	*Schroederia*		
	Trachelomonas	*Selenastrum*		
	Uroglena	*Sphaerocystis*		
	Uroglenopsis	*Staurastrum*		
	Volvox	*Tetraedron*		
		Tetrastrum		
		Xanthidium		

ment plant in Washington, D. C., the diatoms blocked the rapid sand filters so completely that it was necessary to wash them every 15 min. For six previous years, no serious inroads of organisms had hampered filtration and the river had given no evidence that any unusual conditions might arise. During the month of September 1934, however, the diatoms came suddenly and without warning.

Two years later, on the morning of September 15, 1936, each of the rapid sand filter units at the Dalecarlia plant dropped within an hour from the normal 4 million gallon rate to 1 million. Filter runs were reduced from an average of 50 hr to less than 1 hr. The offending diatom was *Synedra delicatissima,* and counts reached as high as 4800 cells/ml. During this same period the filter plant for Hagerstown, Md., about 60 air miles upstream, using water from the same river, experienced no difficulty and had a maximum of only 20 *Synedra*/ml. The 1936 epidemic at Washington lasted for 20 days.

A third epidemic of diatoms reaching the Washington water supply occurred in July 1938. Early in the month there was no unusual evidence of organisms in the water. On July 26, there appeared large quantities of the diatom reported as *Synedra pulchella*, but from its description as a number of cells fastened together like a comb, it probably was a species of *Fragilaria*. This diatom reached a maximum of about 1700 cells/ml and was controlled by use of large amounts of copper sulfate. The increased diatom production was ascribed to an increase in the amount of carbon dioxide in the water, which already contained a large amount of silica.

One of the worst algal problems in the history of the Cincinnati water supply occurred in 1957. Beginning in July, there were two months of successive blooms, first of *Synedra*, next *Melosira*, followed by *Anabaena*, and finally *Oscillatoria*. Blooms of the diatom *Asterionella* accompanied by objectionable river odors have also occurred in the Ohio River at Cincinnati.

In the Hocking River in Ohio the presence of the following algae was recorded during autumn, winter, and spring: *Lyngbya, Closterium, Navicula, Melosira, Synedra,* and *Pinnularia*. There were 42 genera of algae in the autumn, 33 in winter, and 20 in the spring.

Beginning in October 1957, plankton analyses have been made of samples collected monthly to semimonthly from many stations on rivers and some lakes and impoundments throughout the United States (Fig. 1). The results are published by the National Water Quality Network of the Division of Water Supply and Pollution Control of the U. S. Public Health Service. The data for the first three years are now available.

The clump count procedure is used in recording the plankton organisms as numbers per ml. In this method of enumeration, isolated cells and colonies are recognized as the unit. Analysis is commonly made on unconcentrated samples at a magnification of

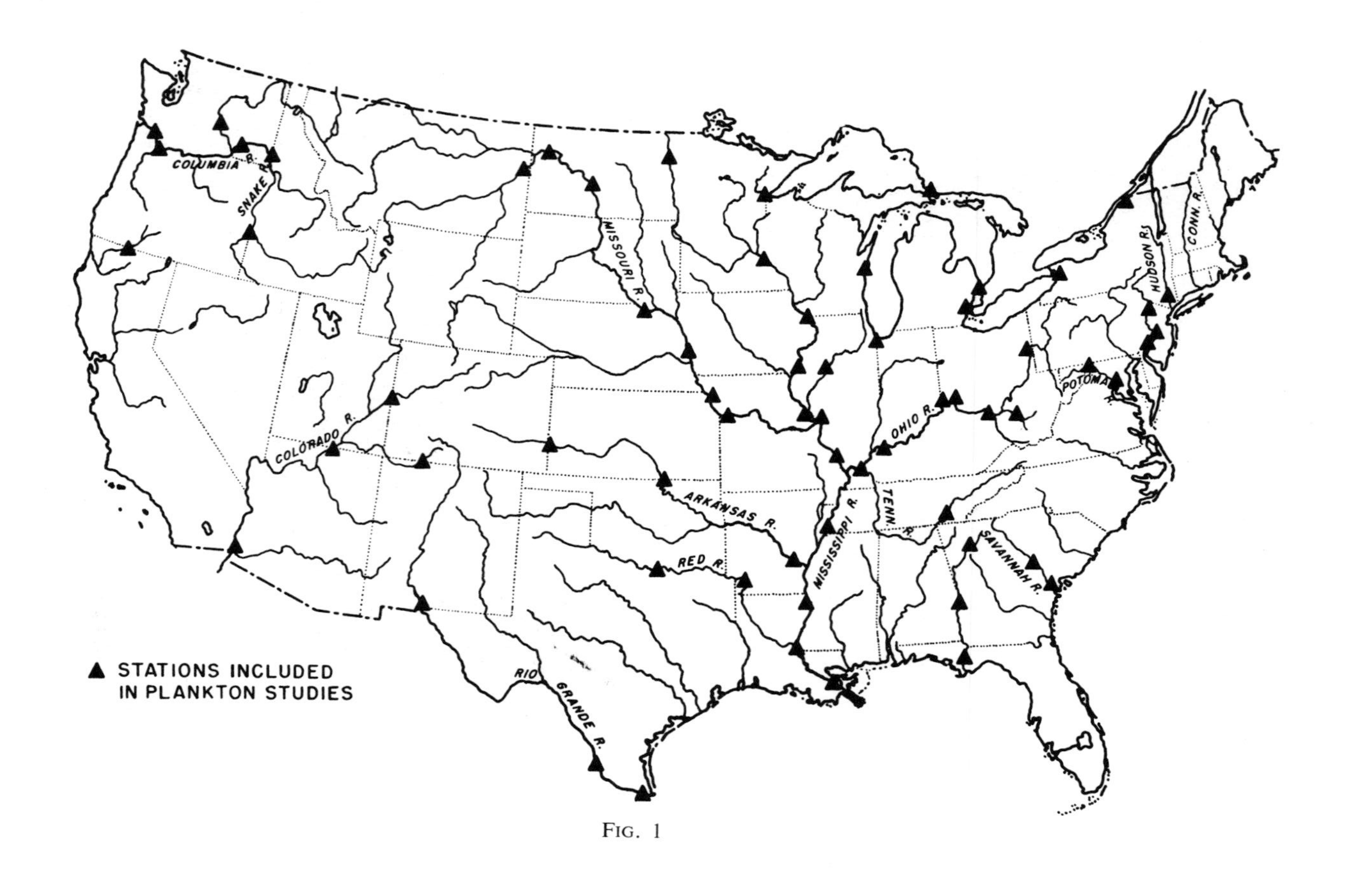
COLUMBIA R.
SNAKE R.
MISSOURI R.
COLORADO R.
ARKANSAS R.
RED R.
RIO GRANDE R.
MISSISSIPPI R.
TENN. R.
OHIO R.
SAVANNAH R.
POTOMAC
HUDSON R.
CONN. R.
▲ STATIONS INCLUDED
IN PLANKTON STUDIES

FIG. 1

200×. Two longitudinal strips (representing about 200 fields) of the Sedgwick-Rafter slide are examined, rather than only 10 fields as is often recommended.

Over the first two years the average count per monthly sample was 3625 algae/ml, the first year averaging 3460 and the second year 3850. The monthly average ranged from 1376 to 6745. November, December, and March had the lowest average counts; April, September, and October had the highest. Clearly, plankton algae are present in rivers in larger numbers than has generally been assumed. During the early winter the count may be lower, but seasonal fluctuation is less than had been suspected.

Occasional very high counts may raise the averages significantly. For example, the median for all rivers during the first year was only 1460/ml, in contrast to a mean of 3460/ml. Some individual counts exceeded 20,000, and one was above 50,000.

Over three years phytoplankton counts in samples from American rivers ranged from 10 to over 101,000/ml. Many counts are in the range of 1000 to 5000, and except for impounded areas, no river area tends to remain constantly below 1000. However, this is approached in the samples collected from the Missouri River at Bismark, N. D. At this station, of 40 samples obtained during a three-year period, only two had phytoplankton counts over 1000, these being 1110 and 1050. The other extreme is exemplified by the counts from the Mississippi River at Burlington, Iowa. Of 48 samples obtained during the same three-year period, only one had a count less than 1000 and the highest counts were 53,060, 25,930, and 23,940. From the Rio Grande at Laredo, Texas, 19 samples have had counts below 1000 and 20 have had counts above that figure. In one year the counts at this station ranged from 100 to 34,510/ml.

The maximum counts and the percentage of counts above and below 1000/ml for 15 rivers are given in Table III. Relative abundance of various algal groups is shown for the South Central States in Table IV. The five rivers with the highest average counts for the first year were the Mississippi, Arkansas, Merrimack, Missouri, and Columbia; the five with the lowest were the Red, Detroit, Colorado, Savannah, and Tennessee Rivers. High plankton counts are to be expected in rivers enriched by land drainage from productive soils, or by treated or untreated sewage from cities and towns, particularly where toxic industrial wastes are not abundant.

TABLE III

Abundance of Algae in American Rivers (October 1957-September 1960)

River	State	Sampling Point	Counts above 1000/ml (%)	Maximum Count
Mississippi	Iowa	Burlington	98	53,060
Arkansas	Oklahoma	Ponca City	78	101,200
Delaware	Pennsylvania	Philadelphia	75	4,750
Ohio	Ohio	Cincinnati	74	30,010
Missouri	Kansas	Kansas City	63	44,300
Potomac	Maryland	Great Falls	62	43,160
Red	Louisiana	Alexandria	61	29,070
Yellowstone	Montana	Sidney	61	21,480
Rio Grande	Texas	Laredo	51	34,510
Columbia	Washington	Pasco	47	4,370
Merrimack	Massachusetts	Lowell	41	16,250
Hudson	New York	Poughkeepsie	41	3,730
Colorado	Arizona	Yuma	37	3,050
Savannah	Georgia	Port Wentworth	28	10,860
Tennessee	Tennessee	Chattanooga	28	4,400

TABLE IV

Plankton Algae from Four Rivers of the South Central States

	Average No. per ml	Maximum No. per ml
Total algae	4,190	103,370
Centric diatoms	2,210	49,780
Pennate diatoms	870	12,240
Blue-green algae	450	20,830
Green algae	430	16,620
All other algae	230	7,020

TABLE V

Algae in Massachusetts Waters*

Habitat	Areal Standard Units per ml				
	Green algae	Blue-green algae	Diatoms	Flagellate algae (+Protozoa)	Total algae (+Protozoa)
Well waters	0	0	1	0	1
Infiltration galleries	1	1	13	6	21
Rivers	11	6	32	11	60
Spring waters	18	0	49	0	67
Reservoirs	83	86	713	83	965
Ponds and lakes	99	70	702	117	988

*Data given are averages of 10 samples from each type of habitat.

The same genera of algae tend to be dominant in all of the rivers of the country, and only a limited number of genera comprise the dominant forms generally encountered. Leading the list are several diatoms, the first six being *Cyclotella, Synedra, Melosira, Navicula, Asterionella,* and *Stephanodiscus.* Other algae high on the list are *Anacystis, Chlorella, Chlamydomonas,* and *Ankistrodesmus.*

An indication of what the plankton counts may reveal is found when, for 1959, four-month averages of the planktonic green algae at eight stations on the Mississippi River are compared. The average count decreases from the uppermost station at Red Wing, Minn., to the lowermost station at New Orleans, La. The average count per ml for the whole year was 2087 at Red Wing and 151 at New Orleans and the count at each intermediate station was less than that at the one above.

LAKES AND RESERVOIRS

Although it is commonly assumed that algae are more abundant in lakes and reservoirs than in rivers, the figures on the numbers of algae in the former often are misleading since they refer to the abundance of algae in the upper layer of water rather than the average number per total volume of water present. It is often possible to obtain a supply of water from the impoundment at a depth

TABLE VI

Abundance of Algae in American Lakes (October 1957-September 1960)

Lake	State	Sampling Point	Counts above 1000/ml (%)	Maximum Count (per ml)
Michigan	Indiana	Gary	60.0	9740
Havasu	California	Parker Dam	5.0	1130
Mead	Nevada	Hoover Dam	4.0	1110
Erie	New York	Buffalo	2.5	1690
Superior	Minnesota	Duluth	0	780

TABLE VII

Occurrence of Algal Groups in Massachusetts Lakes and Reservoirs

	Number of Lakes and Reservoirs			
Algal group	Often above 1000/ml	Occasionally above 1000/ml	Usually between 100-500/ml	Below 100/ml
Blue-green algae	7	10	18	22
Green algae	5	11	29	12
Diatoms	24	8	19	6
Flagellate algae and Protozoa	8	7	35	7

selected for its small algal population.

The comparative abundance of algae in waters from various types of habitat is shown in Table V. These again emphasize the large numbers in reservoirs and lakes compared to those in other habitats. More recent studies reported by the Basic Data Program are shown in Table VI. With the exception of the highly enriched water from Lake Michigan at Gary, Ind., the percentage of algal counts above 1000/ml is much lower than those for the rivers listed in Table III.

Diatoms are the algal group that most frequently occur in large populations in impoundments. The occurrence of algal groups in Massachusetts lakes and reservoirs is indicative of this (Table VII). Diatoms occurred at 1000 or more per ml in three times as many lakes as the next most abundant algal group.

Records for Indiana lakes and reservoirs indicate that a large number of kinds of plankton organisms may be present in abundance, with the many plants and animals in the shallow margins which can break away and add to the amount of living or dead organic matter dispersed in the water. All four types of algae are well represented in the lakes and reservoirs, and in addition, the Protozoa, Crustacea, sponges, worms, and other groups are abundant. Only a few of the more common genera are listed here. The diatoms are *Asterionella, Fragilaria, Melosira,* and *Synedra*; the green algae, *Pediastrum, Scenedesmus, Ankistrodesmus, Coelastrum,* and *Hydrodictyon;* the blue-green algae, *Anabaena, Aphanizomenon, Microcystis, Coelosphaerium,* and *Lyngbya;* and the flagellate algae, *Dinobryon, Ceratium, Chlamydomonas, Peridinium,* and *Mallomonas.* The tendency for a number of the algae to occur as extensive visible surface blankets or "blooms" is common.

RESERVOIR PROBLEMS

Water in reservoirs and lakes generally is held long enough to permit significant changes in its phytoplankton population. Reduction in turbidity due to silt permits light to penetrate and cause algal growth. The slower movement of the water permits organisms to accommodate themselves to environmental conditions. Localized benthic growths develop; deep water and shallow water plankton communities form; and surface mats of filamentous algae may occur. Any of these may increase to amounts that upset the balance of flora and fauna. At these times, critical problems that interfere with the use of the water often occur. Only a few examples of these problems can be considered here.

The amount of nutrients in water is an important factor in determining the amount and kind of algal growth. As a control in many fishing lakes in the southern part of the United States, a few hundred pounds of commercial fertilizer per acre of water surface is added. This stimulates growth of phytoplankton, which, in turn, reduces the growth of attached algae and other plants, and thus improves the lake for fishing.

Smaller ponds often are high in nutrients due to drainage from surrounding fields and may support dense growths of algae. A pond in a quarry at Bowling Green, Ohio, had an abundance of the flagellate algae *Pandorina, Chlamydomonas,* and *Cryptomonas* even during the winter.

Cities using water from the smaller lakes or impoundments report difficulties due to the plankton. Hobart, Ind., had so much difficulty from tastes and odors due to algae in its small supply lake that a new source was obtained by combining with Gary to use water from Lake Michigan.

The presence of a high proportion of diatoms in the plankton of Lake Michigan, together with the occasional development of flagellated algae such as *Dinobryon,* indicates that treatment plants obtaining water from this lake can expect frequent taste and odor problems and reduced filter runs. These problems do exist. Michigan City, Ind., reports that its chief source of tastes and odors has been algal growths or the reaction of chlorine with organic matter in the raw water. Filamentous algae appear in increasing numbers in April or May, depending upon the weather, but do not produce much taste and odor until later when counts reach 850 or more microorganisms per ml and tax the treatment plant's capacity to remove the odors and tastes. The Gary-Hobart plant reported that odors in the raw water from the lake were predominantly of natural origin, the most common odor being musty.

In June and July of 1951 the alga *Dinobryon* developed in the southern end of Lake Michigan in numbers sufficient to impart a pronounced fishy odor to the water and produce a threshold odor up to 13. A threshold odor of 10 is considered high for algal odors in the lake water at Chicago. Concurrently a range of 140 to 700 areal standard units of *Dinobryon* was reported in the raw water at the south district in Chicago and represented up to 47% of the total algal count. The fishy odor produced by *Dinobryon* has the reputation of being readily adsorbed by activated carbon. In two months approximately $70,500 worth of activated carbon was used to control the odor due to this alga. *Dinobryon* causes this type of nuisance almost every year at Chicago.

At Chicago ten species of diatoms may occur in Lake Michigan in large numbers at any time during the year. *Tabellaria* occurs in greatest abundance most frequently and is considered to cause more short filter runs than any other microorganism. *Fragilaria* and *Asterionella* are also important in filter clogging. In general, the length of the filter run decreases as the number of microorganisms increases. Representative filter runs in July ranged from 41 hr down to 3.5 hr; plankton collected on the filter ranged from 60,500 to 4,572,700 organisms per square inch of sand surface.

The new Hoover Reservoir that supplies part of the water for Columbus, Ohio, has had an algal population, during its first four years, principally of Diatomaceae and Chrysophyceae. *Mallomonas* was exceptionally abundant in the reservoir in June 1960, and Chrysophyceae were present throughout the summer. In 1958 *Anabaena* caused concern because of its high numbers, and in 1959 *Dinobryon* and *Mallomonas* were abundant in the upper end of the reservoir. Evidence indicates that Hoover Reservoir will support periodic and dangerous populations of algal species that cause odors and flavors.

For the treatment plant operator, one of the most elusive problems is the effective control of these tastes and odors in the water. Reservoirs and rivers are both able to support growths of taste and odor algae, but the reservoirs have a combination of factors that makes them constantly susceptible to the development of these organisms.

The appearance of large growths may interfere with various recreational uses of lakes and reservoirs and is objectionable to persons with cottages along the shores. The accumulation of large quantities of decomposing algae and water weeds that have been washed ashore constitutes a major nuisance. The growths also attract waterfowl, which contribute a heavy pollutional load. Attached algae may produce odors in the water or along the shore, clog inlet screens, and serve as nesting areas for the development of many kinds or organisms.

Attached algae and water weeds can become a problem in the shallow margins of lakes and reservoirs. One common control procedure is to cut them or pull them out by the roots. One Connecticut utility riprapped the shore of the reservoir in an attempt to eliminate their growth. *Chara, Nitella, Ulothrix,* and *Calothrix* are some of the attached algae which have been encountered. *Dichotomosiphon* has developed in the reservoir at the Bastrop, La., water works in July. *Pithophora* grows in profusion in Alabama during the warmer months of the year; it develops first on the bottom in the shallow areas and later forms floating mats. *Cladophora* is attached to the riprap and periphery of Lake Hefner at Oklahoma City, Okla. It is estimated that by weight sessile algae of Lake Hefner contribute three times as much organic matter as plankton for odor production.

ALGAL BLOOMS AND MATS

Many different genera of algae that visibly color water occur frequently. In the North Central states, blue-green algae that form blooms are *Anabaena, Anacystis (Microcystis), Aphanizomenon, Coelosphaerium, Gomphosphaeria, Oscillatoria,* and *Raphidiopsis.* Flagellate algae that have produced blooms are *Cephalomonas, Chlamydomonas, Euglena, Peridinium, Pandorina, Platydorina, Uroglena,* and *Volvox*. Two green algae that form blooms are *Pediastrum* and *Staurastrum.* In Lake Erie diatoms have sometimes made the water appear brownish because of their abundance; *Asterionella* and *Melosira* generally are the dominant genera. A bloom of *Melosira* was present in a reservoir at Columbus, Ohio, in early summer.

Some of the larger impoundments in central Texas have reported the dominance and even blooms of diatoms, including *Melosira, Synedra, Tabellaria, Fragilaria,* and *Asterionella.* Lake Overholser in Oklahoma contained more diatoms than other algae, with a relatively uniform population throughout the year, two of the dominant genera being *Cyclotella* and *Melosira.* In Pickwick Reservoir in Alabama diatoms predominated during the early part of the year followed by green and blue-green algae in the summer. An increase in the supply of phosphates in the water when silicates are already present from the colloidal clay of the surrounding soil often results in a diatom bloom. In turbid water the growth is likely to be *Melosira* or Synedra; in clear water it may be *Asterionella, Fragilaria,* or *Tabellaria.*

Blooms of *Melosira* have occurred in Lakes Amarillo, Bridgeport, Caddo, Dallas, Eagle, Waco, and Worth in Texas, and at Wilson Dam and Decatur, Ala. Blooms of *Anabaena* have been observed in Lakes Hamlin, Ranger, Possan Kingdom, and Wichita, Texas, and in various reservoirs in Louisiana.

Color has been a persistent problem in many New England water supplies. Humus may cause a brown color in shallow areas or various plankton algae may cause various shades of green, brown, or red. Color from certain very minute flagellate algae such as *Chlamydomonas* and *Gonyostomum* is sometimes difficult to control when the odors develop in open reservoirs of a distribution system. A prodigious growth of the filamentous alga *Oscillatoria prolifera* has produced a pink to copper-red color in Massachusetts water and *Aphanizomenon* frozen into the lake ice has given a distinct green color to New Hampshire water.

In Maryland, smaller ponds polluted with sewage have produced visible brilliant green blooms. The algae present in abundance were pigmented flagellates: *Pandorina morum, Eudorina elegans, Chlamydobotrys stellata,* and *Pyramimonas montana.*

Algae which tend to form blooms in reservoirs in the South-central area of the United States include the blue-green algae *Anabaena, Oscillatoria, Anacystis (Microcystis),* and *Aphanizomenon;* the green algae *Scenedesmus* and *Pediastrum;* the flagellates *Volvox, Chlamydomonas* and *Pandorina;* and the diatoms *Melosira, Synedra, Tabellaria, Fragilaria, Asterionella, Navicula,* and *Pinnularia.* In the Southwest, three of the algae that form blooms are *Anabaena, Oscillatoria,* and *Ceratium.* For small fishing lakes in Alabama and Mississippi, periodic applications of fertilizer sufficient to maintain a continuous bloom of algae at the surface are recommended. This bloom not only nourishes immense numbers of small animals upon which fish can feed, but it shades the bottom of the pond, thus preventing the growth of the larger attached aquatic plants.

SURFACE MATS

Surface mats formed by various filamentous genera of algae commonly occur. The ones reported for Ohio include *Cladophora, Hydrodictyon, Mougeotia, Oedogonium, Oscillatoria, Spirogyra, Stigeoclonium, Ulothrix,* and *Vaucheria.* Planktonic algae such as *Euglena* and diatoms are often present in large numbers associated with the filamentous algae in the mats.

In the Scioto River in Ohio, between Columbus and Piketon, masses of attached and floating algae have been observed throughout most of the year, particularly during early spring, early summer, and most of the autumn period.

TREATMENT PLANT PROBLEMS

Interference organisms cause problems in water treatment plants by shortening filter runs; clogging intake screens; forming slimy layers of growth on the walls of filters, settling basins, and intake pipes, or on the rough surfaces of aerators; and by increasing sludge deposits in settling basins. The organisms also change the pH, alkalinity, carbon dioxide and oxygen contents, color, and turbidity of the water. Tastes and odors may develop in water during treatment even when they are not evident in the raw water.

Growths on the basin and filter walls not only are unsightly, giving visitors a bad impression of plant sanitation, but they may also slough off and increase the organic material that has to be dealt with during sedimentation and filtration. When algae change the physical and chemical characteristics of water, frequent adjustments in the treatment process are required to prevent poor coagulation and insufficient settling, and to compensate for changes in chlorine demand, pH, and oxygen content of the water.

Treatments for interference organisms represent a very significant phase of the work at many water treatment plants. As a result of attempts to treat water during the *Synedra* "epidemics" at Washington, D. C., it was concluded that coagulation by excessive dosages of alum did more than any other single agent in dealing with the problem. Other promising agents were alum, activated carbon, and iron sulfate with lime and chlorine. Microscopic examination of the iron-chlorine-lime treatment floc revealed that more organisms were caught in this floc than in any of the others.

Ample sedimentation is required along with high coagulation for effective removal of organisms. The sediment which collects in a coagulation basin as a sludge has been responsible for tastes and odors in treated water. This can be stopped or prevented by more frequent removal of the sludge from the bottom of the basin. Prechlorination of raw water has been reported to prevent the accumulating deposits from becoming septic for a period of one or more months between seasonal cleaning schedules. Open reservoirs in the distribution system at Baltimore, Md., have also had to be cleaned to remove the heavy algal growths on the sides.

A report from Danbury, Conn., gives a vivid description of the effect of algae on coagulation and chlorination. The felting and dyeing processes of the city's major industries demand a water free of color, turbidity, iron, and aluminum, and with low constant pH and hardness. A weak floc, induced by large-scale algal concentrations, will pass appreciable amounts of these interfering substances through the filters. As algae increase in the raw water, small daily adjustments in chemical feed, equipment, and operational methods maintain the predetermined optimum conditions, such as maximum turbidity of 2.0 on the filters, maximum pH of 6.7 in the treated water, minimum filter runs of 23 hr, minimum rates of flow of 70,000 gallons per foot head loss per filter, minimum floc size in the last flocculator of 2.0 mm. Normally a large, fragile floc particle is obtained quickly, without agitation. Flocculators (agitators) are used, however, with certain growths, particularly *Coelosphaerium, Anabaena,* and other free floaters; although a smaller, poorer floc is formed, settled water with lower turbidity and color is obtained.

At Cleveland, Ohio, adjustment of the procedures for coagulation and sedimentation made it possible to settle out 88.7% of the organisms in the water. The algae that did not settle out readily were *Melosira* and *Nitzschia. Melosira* appeared to be the chief offender in filter clogging. At Lorain, Ohio, daily algal counts were found to be an aid in adjusting alum dosages and in accounting for sudden changes in chlorine demands.

Toledo, Ohio, abandoned the Maumee River for Lake Erie, as a water supply. Growths of algae in the lake water presented problems never encountered in the river water. Turbidity of the river water had helped produce a floc that precipitated the algae without much difficulty. The lake was found to have very heavy algal growths, especially in the late summer and in winter, but low turbidity. The organic matter from the algae made coagulation difficult and also shortened filter runs.

Penetration of sand filters by algae has also been a problem in treatment plants. Sand shrinkage in the filters at Akron, Ohio, was enough to permit water and algae to pass between the side walls of the basin and the sand.

At Cincinnati, Ohio, floating algae had been present in the settling basins and had to be skimmed at the outlet until control was obtained by almost continuous use of copper sulfate and chlorine. Prior to the time that prechlorination was started at the treatment plant in Warren, Ohio, algae grew abundantly at the surface of the water in the settling basins. Chlorination of the raw water eliminated the growth entirely and also increased the length of the filter runs.

At Indianapolis, Indiana, slime and biologic deposits were

eliminated from the filter sand and gravel when free chlorine residuals were maintained in filter effluents. Previously, it had been standard procedure to remove the sand and gravel from the rapid filters at about two-year intervals. The gravel was always slimy and the bottom layers of sand sticky. Although a new surface wash probably contributed to the better condition, it did not seem that this alone should be credited with eliminating biological growths.

In the South Central region, practices for reducing interference by algae in the water treatment or for reducing the tastes and odors in the treated water have been numerous. These range from the "bumping" of a filter to the covering of a reservoir at Fayetteville, Ark. It was sometimes practical to scrape the *schmutzdecke* from the surface of a filter in small treatment plants. Also, correct placement of the foot valve in the sedimentation basin, painting the walls of the basin with whitewash from the surface to 3 to 4 feet below the water level, and frequent and regular removal of the sludge from the bottom of the basin have been considered important in the control of algae and tastes and odors.

Improvements and frequent adjustments in the method of coagulation are advocated, since coagulation, as practiced in many plants, frequently reduces the organic content of the water by only about 50%.

CHEMICAL CONTROL OF ALGAE

In reservoirs and lakes, and occasionally in streams, control of planktonic algae has been attempted by use of algicides, commonly, copper sulfate. Various results have been reported. At 0.25 to 0.35 ppm copper sulfate the reservoir water at Norwalk, Ohio, was kept in good condition, except for occasional growths of some of the more resistant algae. At Youngstown, Ohio, a heavy growth of *Coelosphaerium* in a reservoir was treated effectively with copper sulfate at about 0.2 ppm. In the same area Lake Meander was first marked off into sections, each holding not less than 200 million gallons. Treatment then was carried out by section, with slightly higher concentrations being used in the deeper portions of the lake. Less than 0.5 ppm was found to be sufficient copper sulfate for most planktonic organisms, but a few required up to 9.5 ppm.

Achnanthes is a diatom reported in Indianapolis to be resistant to copper sulfate (1 ppm) but sensitive to chlorine (0.25 ppm). In contrast, an unidentified filamentous microorganism withstood chlorine (4 ppm) but was eliminated by copper sulfate (2 ppm).

At Tiffin, Ohio, where water was obtained from the Scioto River, copper sulfate at 1.8 ppm completely eliminated two kinds of algae, *Synura* and *Dinobryon.* The experience was repeated two years later. On other occasions the Tiffin plant has used this method on *Anabaena, Euglena,* and *Melosira.* Copper sulfate was reported to be ineffective against large numbers of the diatom *Synedra* at both Wilmington, Del., and Washington, D. C., where it was applied continuously at a concentration of 1 ppm. Copper sulfate applied for general control of algae at Cumberland, Hyattsville, and Laurel, Md., was effective at 0.35 to 0.5 ppm.

In California, copper sulfate has been used for many years and in large quantities for the control of algae in reservoirs. San Francisco has used it since 1912, the treatment at that time consisting of dragging bags of copper sulfate back and forth behind a boat. The procedure was later modified, the algicide being dissolved in a small amount of water and sprayed from nozzles under a pressure of 30 to 50 pounds over a zone up to 50 feet wide. With this method 8000 pounds can be applied to 250 acres each day. Approximately 80,000 pounds has been used annually for control of algae in San Francisco's local storage and impounding reservoirs, where copper sulfate appears to settle out slowly, remaining effective as an algicide for as long as 48 days.

Los Angeles has more than 25 open reservoirs ranging from about 10 to 180,000 acre-feet. They are subject to widely diverse climatic and topographic conditions, with elevations from about 375 feet to over 7000. Therefore, the great variety of biological activity could produce water of unsatisfactory quality. Since the only treatment before the water enters the distribution system is simple chlorination, the quality of the raw water supplies has to be rigidly controlled. The principal problems encountered are pond weed growths, plankton activity, and taste and odor control.

One method of applying copper sulfate to the Los Angeles, Calif., reservoirs has involved the use of a scattering machine on a fast boat to spread the copper sulfate crystals uniformly over the water. The crystals dissolve as they fall through the water, thus giving a uniform vertical distribution; the larger the crystals, the greater the depth affected. Up to 140 acres can be treated in an hour, and there are savings both in labor and the amount of copper sulfate required.

In the South Central region, copper sulfate is a commonly used algicide for treating impoundments ranging from small ponds to large reservoirs. Less than 0.2 ppm has been indicated as sufficient to destroy *Asterionella*; however, as much as 1.0 ppm has been recommended to control all algae and yet leave the water safe for livestock, humans, and fish. For Texas waters the range might be from 0.12 to 4.0 ppm, depending upon the types of algae present. For this area the optimum procedure would have to be determined for each situation. At Tyler, Texas, the recommendation involved treating the entire lake with 0.2 ppm of copper sulfate during early spring to retard growth, and adding copper sulfate around the margin of the lake once each month during the summer. Shreveport, La., arranged for seasonal treatment of Cross Lake with copper sulfate, as required by the concentration of algal growths. In May and June of one year, a total of 12,500 pounds was applied to the lake but it resulted in practically no reduction of odor due to algae. On Lake Dallas in Texas, when copper sulfate was used to control *Melosira,* the major offending diatom, the organism was destroyed but a bloom of other algae greener than before developed. As early as 1933, it was emphasized that prevention of maximum algal growth rather than its destruction later should be the operator's motto.

In New England, the only common algicide used has been copper sulfate, which appears to be effective if a sufficient dosage is applied before the algae become abundant. No fixed dosage of copper sulfate is suitable for any one reservoir or any one organism since several factors, including temperature and pH, must be considered in determinations of proper dosage.

Algicides used in limited concentrations tend to be selective in their toxicity. Destruction of one kind of alga sometimes causes rapid growth of a more resistant variety. This occurred at Rockport, Mass., when the elimination of *Tabellaria* by copper sulfate was followed by a bloom of *Chlamydomonas.* Indiscriminate use of an algicide may upset the balance of existing aquatic life and make possible unrestricted growth and dominance of a single kind of alga, causing more serious problems than an inoffensive balanced, mixed growth.

Chlorination has been used in New England for control of taste and odor and algal growths, for longer filter runs, better floc formation, and for maintenance of pipeline capacities. Prechlorination for better floc formation and longer filter runs does not necessarily improve the taste, odor, or appearance of the water.

In the Southwest, chlorine has been applied to open canals used either for irrigation or as a part of a public water supply system. Continuous treatment with low concentrations is more effective than occasional applications of high concentrations. Chlorine has been used for control of algae, especially when the particular forms were resistant to copper sulfate. An alga reported as *"Protococcus"* was destroyed in a Los Angeles reservoir with a chlorine dosage of up to 1 ppm. More recently, chlorination of the hypolimnion has been carried out to improve water quality, including reduction of tastes and odors. Baltimore, Md., reported heavy growths of algae in a small clear well and in a large reservoir for filtered water, in spite of a high equivalent residual chlorine content.

Tests with chlorophenyl dimethyl urea (CMU) at a fish hatchery in Newtown, Ohio, revealed that CMU is effective in eliminating floating algae. *Cladophora* was more resistant than *Spirogyra, Mougeotia,* and *Hydrodictyon.* CMU at 2 ppm prevented all species of blue-green algae and diatoms tested from developing.

CONTROL BY PREVENTIVE MEASURES

Attention has been given to the pre-impoundment preparation of areas to minimize subsequent development of biological growths and tastes and odors. At Little Rock, Ark., detailed specifications for clearing the new reservoir site of trees, stumps, brush, vines, and any other vegetation more than 6 inches in height were formulated. Biological nutrient materials, such as manure, garbage, and sawdust piles, were also removed or buried. In Alabama regulations require that the basin be completely cleared prior to impounding.

COVERING OF RESERVOIRS AND CANALS

Although the initial cost is high, the covering of canals and of smaller storage reservoirs has been found effective both in preventing the growth of algae and in preventing infestations of caddis flies, *Chironomus,* and insects. At Glendale, Calif., the covering was removed from the service reservoir; however, the abundant growth of algae that followed forced officials to construct a new roof because of public complaints about the green water. Pasadena and Los Angeles have had similar experiences. The covering of storage reservoirs to eliminate algae was urged by Forbes in Massachusetts as early as 1888.

Control of lands surrounding reservoirs makes it possible to reduce contamination, fertilization, and siltation in impounded water. Akron, Ohio, purchased several thousand acres around Lake Rockwell, an impounding reservoir, and substituted forest areas for cultivated areas. Limnological studies in Connecticut indicate that knowledge of the phosphorus, nitrates, and other nutrients in the water may be used to predict the kinds, abundance, and periodicity of plankton organisms in lakes and reservoirs.

One of the important sources of phosphorus in western Lake Erie is the discharge from rivers. The Maumee and Detroit Rivers together have been estimated to discharge 530 metric tons/year of bottom sediments rich in phosphorus.

As a result of raising the water level more than 50 feet in the Loch Raven reservoir in Maryland, the plankton count jumped from an average of less than 300 organisms/ml to almost 5000. The organisms represented in this great increase included several important taste and odor forms, such as the diatoms *Asterionella* and *Fragilaria;* the desmid *Staurastrum;* and the flagellate algae *Uroglenopsis, Dinobryon,* and *Peridinium.* One of the causes of increased growth of algae in the reservoir is the reduced turbidity, which dropped from an average of approximately 85 to 8.5 after impoundment. The maximum turbidity dropped from over 450 to 11. Light exclusion is one of the most important effects of suspended matter on algae, since it interferes with their photosynthetic action, thus depriving them of active growth.

STATEMENT ON ALGAL MANAGEMENT

The two statements that follow were made by men in the water treatment field. They indicate the increased emphasis which is being given to the detection and control of algae and other aquatic organisms in water supplies.

"The conviction is growing among sanitary engineers that the most successful method of combating algae will be finally by means of systematized biological control."

"Biological control is undoubtedly as important a factor in treatment plant operation as is the use of a coagulant, and the more carefully the biology of the various units in the plant and the impounding system is studied, the better and the more uniform will be the quality of water in service. . . . The ability to draw water from any desired stratum should be considered an indispensable element of biological control, and sufficient study of water at all levels should be made to enable intelligent selection of the point from which water for treatment is to be drawn."

LITERATURE CITED

Damann, K. E. 1951. Missouri River Basin plankton study. Fed. Security Agency, Pub. Health Serv., Environmental Health Center, Cincinnati, Ohio. 100 pp. (Mimeographed.)

National Water Quality Network. 1959. Annual compilation of data: Plankton population. October 1, 1957-September 30, 1958. *Pub. Health Serv. Publ.* No. 663 (1958 ed.):75-121.

National Water Quality Network. 1960. Annual compilation of data: Plankton population. October 1, 1958-September 30, 1959. *Pub. Health Serv. Publ.* No. 663 (1959 ed.):117-171.

National Water Quality Network. 1961. Annual compilation of data: Plankton population. October 1, 1959-September 30, 1960. *Pub. Health Serv. Publ.* No. 663 (1960 ed.):101-174.

Palmer, C. M. 1958. Algae and other interference organisms in New England water supplies. *J. New England Water Wks. Assoc.* 72:27-46.

Palmer, C. M. 1961. Algae and other interference organisms in water supplies of California. *J. A. Water Wks. Assoc.* 53:1297-1312.

Palmer, C. M. 1960. Algae and other interference organisms in the waters of the South Central United States. *J. A. Water Wks. Assoc.* 52:897-914.

Palmer, C. M. 1958. Algae and other organisms in waters of the Chesapeake area. *J. A. Water Wks. Assoc.* 50:938-950.

Palmer, C. M. 1961. Algae in rivers of the United States. Trans. 1960 Seminar on Algae and Metropolitan Wastes. Robert A. Taft San. Eng. Center, Tech. Rept. W61-3, pp. 34-38.

Palmer, C. M. 1962. Algae in water supplies of Ohio. *Ohio J. Sci.* 62:225-244.

Palmer, C. M. 1959. Algae in water supplies, an illustrated manual on the identification, significance, and control of algae in water supplies. *Pub. Health Serv. Publ.* No. 657. 88 pp.

Palmer, C. M. 1961. Interference organisms in water supplies. Proc. Oklahoma Water, Sewage and Indus. Wastes Assoc. for 1960, pp. 157-171.

Palmer, C. M., and Poston, H. W. 1956. Algae and other interference organisms in Indiana water supplies. *J. A. Water Wks. Assoc.* 48:1335-1346.

Williams, L. G. 1962. Plankton population dynamics. *Pub. Health Serv. Publ.* No. 663, Suppl. 2, 1962. 90 pp.

Williams, L. G., and Scott, Carol. 1962. Principal diatoms of major waterways of the United States. *Limnol. and Oceanog.* 7:365-379.

Algal Problems Related to the Eutrophication of European Water Supplies, and a Bio-Assay Method to Assess Fertilizing Influences of Pollution on Inland Waters

Olav M. Skulberg

Norwegian Institute for Water Research
Blindern, Norway

GENERAL BACKGROUND

It has been said that Europe is almost as complicated physically as it is politically. Although the European continent is the smallest principal division of the land surface of the globe distinguished by this term, the area still amounts to more than 10 million km^2. Situated between 71°N (Norway) and 36°N (Spain) and from 9°W (Portugal) to 66°E (Ural), deeply penetrated by the sea and with several great mountain ranges, the consequence is an outstanding variation in physical features and climatic contrasts. The hydrographic relations are correspondingly varied. The openness of Europe to oceanic influences and the topography give the landscape a multiplicity of rivers and lakes. But there are also great areas with small amounts of annual rainfall and with surface waters of sparse occurrence. Generalizations are misleading, but it is not wrong to state that almost all types of aquatic biotopes are represented, giving support to the most varied types of algal communities.

The problems of algae in water supplies are connected with the status of pollution in the water resources. A few remarks about the concentration of population in the European countries, and about the proportion of population provided with piped water, will give an idea of the severity of the pollution problems (World Health Organization, 1956). The average density of population in the whole of Europe is 46/km². The corresponding figure for the United States of America is 15. The Netherlands have the highest density of population with 324/km². Norway has in this respect the lowest figure with 10/km². Other examples are:

United Kingdom	199/km²
Italy	156/km²
Germany	137/km²
Greece	58/km²
Sweden	16/km²

One of the causes of the great variation in population density is the degree of industrialization in the different countries. The fact that industry and high population density follow each other makes the problem of pollution more severe.

The provision of drinking water by public water supplies is also accomplished in a varying degree throughout the European countries. In the United Kingdom approximately 95% of the population have piped water; the corresponding figure for Yugoslavia is 10%. The other countries of Europe fall between these extremes.

The general increase in the standard of living includes requirements for adequate water supply of high quality to the greater part of the people, and involves at the same time more potential pollution of the water resources. In the years to come, the algal problems in water supplies will concern a growing number of people in the European countries.

HISTORICAL SKETCH

Even if only a very superficial review of the historical background can be given in this connection, a short recapitulation of some steps in the development of algal studies in water supply practice in Europe might be of interest. To consider this relationship by itself is artificial. A deeper understanding presupposes a contemporary estimation of the history of biology and the history of limnology. The use of phycology in water supply practice is an example of applied ecology. For the development of this branch of science, refer to Allee et al. (1949).

The study of phycology has long traditions in Europe, as do the problems of algae in water supplies. In this connection we may point out Leeuwenhoek's observations in 1674 of "very little animalcules" discovered in fresh water (Dobell, 1960), and the first scientific classification of microorganisms in *Animalcula infusoria fluviatilia et marina* by Otto Mueller in 1786.

Early descriptions of algal troubles in European reservoirs of drinking water are cited by Baker (1949).

These records of algae infesting public water supplies go back to the beginning of the nineteenth century. The practical application of phycology in connection with water supply management originates as far back in time as this. The purity of drinking water was given more serious attention in the middle of the nineteenth century, and biological observations were now used in describing water quality. The microscopical examination of water soon proved to be a valuable support to the chemical analysis, particularly as a means of demonstrating the status of pollution. From now on hydrobiology and phycology in particular gradually gained technical importance. Historical accounts covering this period are presented in Kolkwitz and Marsson (1902) and Helfer (1916). The period may be represented by mentioning some European publications in chronological order:

1850. Hassall, A. H. A microscopic examination of the water supplied to the inhabitants of London and the suburban districts.
1853. Cohn, F. Über lebendige Organismen im Trinkwasser.
1865. Radlkofer, L. Mikroskopische Untersuchung der organischen Substanzen im Brunnenwasser.
1875. MacDonald, J. O. A guide to the microscopical examination of drinking water.
1876. Harz, C. O. Mikroskopische Untersuchungen des Brunnenwassers für hygienische Zwecke.
1883. Certes, A. Analyse microscopique des Eaux.
1894. Beijerinck, M. W. Notiz über den Nachweis von Protozoen und Spirillen im Trinkwasser.
1894. Zune, A. J. Traité d'analyse chimique micrographique et microbiologique des eaux potables.
1898. Mez, C. Mikroskopische Wasseranalyse.
1902. Kolkwitz, R. and Marsson, M. Grundsätze für die biologische Beurteilung des Wassers nach seiner Flora und Fauna.
1903. Lauterborn, R. Die Verunreinigung des Gewässer und die biologische Methode ihrer Untersuchung.
1908. Kolkwitz, R. and Marsson, M. Ökologie der Pflanzlichen Saprobien.

With these first formulations of method an increasing use of phycology was introduced. Water calamities occurred where water was rendered undrinkable because of excessive growth of algae, and the imperative need for knowledge to control such phenomena was realized.

In the first decades after the turn of the century there was increasing activity in the purely scientific field of hydrobiology. The influential publications by E. Naumann, A. Thienemann, and F. Ruttner are among the many contributions at that time. The modern taxonomy of algae was introduced, and monographs of the separate classes were published in the well-known series of L. Rabenhorst (*Kryptogamen-Flora von Deutschland, Österreich und der Schweiz*) and A. Pascher (*Die Süsswasserflora Deutschlands, Österreichs und der Schweiz*). These works on algae gave the necessary basis for ecological investigations of water resources. Numerous surveys of variations in standing stock of plankton with the seasons and in different biotopes were performed. Fundamental knowledge about the life of algae in lakes and rivers was the result of this work.

These extensive regional surveys of algal populations in lakes and rivers used for public water supply demonstrated the importance of algal metabolism for the chemical and physical relations of drinking water quality, and the effects of growths of algae for waterworks practice.

The quantity of literature published during the twenties and the thirties shows the growing interest in phycological investigations of water sources used for water-supply purposes. The methods used for such inquiries are formulated in the comprehensive manuals for examination of water and for water-supply practice (Olszewski and Spitta, 1931; Thresh, Beale, and Suckling, 1933).

The development up to the present may be characterized by an increasing understanding of the importance of biological considerations in sound waterworks practice (Hobbs, 1954). The European efforts to provide pure and adequate water supplies to the communities, and the research concerning purification of water, include important tasks for applied phycology.

ALGAE AND WATER SUPPLY IN EUROPE

The problems associated with algae in water supplies are of both direct and indirect nature. The algae may exert influence on the water quality by their metabolic activities. If the algae are present in sufficiently large numbers they may interfere with the operation of treatment plants. Algal troubles are of considerable economic importance to the cost of treatment.

The main difficulties caused by algae in water-supplies practice are: 1) Algae are filter-blocking; 2) algae impart objectionable tastes and odors to the water; 3) filter-penetrating species cause turbidity and discoloration of the water; 4) algae represent food for heterotrophic organisms developing in the mains and the service pipes; 5) algae form sediments in service reservoirs.

Most waterworks have one or several of these problems to deal with. High algal productivity is generally to be expected before problems become severe. But in some cases difficulties with algae are reported where water of very low fertility was used, and there has been no excess of algal growth. The type and magnitude of the algal vegetation are obviously of importance for the interference problems which arise, but it should be emphasized that the method of treatment as well as the way the waterworks is operated have considerable consequences.

Underground Supplies

Water from subterranean supplies generally contains only small populations of microorganisms, and the operators of waterworks of this type are not ordinarily faced with algal problems. Underground water is in several districts likely to be rich in plant nutrients, and if water of such quality is stored and exposed to the sun, heavy growth of algae may occur. These problems are of the same kind as those met with in water reservoirs, and will be mentioned later in this paper.

Lakes

Many cities and villages all over the European continent derive their water supply from lakes. This is generally the case within regions formerly covered by Pleistocene ice sheets, where moraines still dominate the topography. The Scandinavian lakes, the lakes of the Alps and their forelands, and the lakes of the English Lake District may serve as examples. Classified according to their productivity, these lakes range from the pioneer phase of oligotrophy through all stages of succession to eutrophy. Figure 1 shows an oligotrophic lake, Lake Maridalsvannet, Norway. Protected by forests in the drainage basin of igneous rocks and moraines, this lake has still kept its original character with small amounts of dissolved nutrients in the lake water.

FIG. 1. Lake Maridalsvannet, Norway.

FIG. 2. Lake Lugano, Italy.

Figure 2 is a photograph of Lake Lugano, also a glacial lake, but influenced by human activities in quite a different way. Ever since the first waterside settlement was established, this lake has been enriched from pollution, and is now of a eutrophic type.

Stagnant surface water supply is of importance in regions with lake systems on waterproof boulder clay, as occurring in East Prussia, in Denmark, and in France. These lakes are generally shallow and rather productive. Agricultural practice in the catchment areas supply drainage with fertile water.

The algal conditions of European lakes have been the object of studies since the beginning of this century. Information has accumulated about systematics of the species in the algal communities and their ecological relationships. The Central European lakes are perhaps at present among the best investigated of aquatic biotopes (Maerki, 1949; Tonolli, 1961). Scientific investigations of lakes in the United Kingdom (Macan and Worthington, 1951) and Scandinavia (Nygaard, 1949; Järnefelt, 1956; Skuja, 1956; Hauge, 1957) have contributed to the knowledge of hydrography and phycology of the northern part of the continent.

The algae problems associated with water supplies of the lake type are mainly caused by planktonic species. They will be considered in the following discussion.

A result of the investigations of phytoplankton in the lakes of Europe is the demonstration of the regionally uniform distribution of species over the whole continent. That means: lakes with similar conditions of chemistry potentially have the same selection of phytoplankton species developing without regard to geographical position. Comparative observations from lakes in Spain (Willén, 1960), Denmark (Nygaard, 1945), and northern Norway (Strøm 1926) show that the same algae can proliferate in these widely remote localities with different conditions of climate and geologic substratum. The important species of algae, active as interference organisms in water-supply practice, occur all over the continent.

Descriptive investigations of qualitative and quantitative variations in composition of phytoplankton seasonally and with depth are necessary for proper biological control and operation of a water supply. The object of such studies is to get information about the normal periodicity and enable the forecasting of periods when heavy algal growth are to be expected. Numerous investigations of this kind have been carried out in Europe.

To illustrate typical changes of standing stock of phytoplankton in temperate lakes of eutrophic and oligotrophic nature in Europe, examples of results obtained from two Norwegian investigations will be given. The two lakes considered are Maridalsvannet, an oligotrophic lake providing the city of Oslo (450,000 inhabitants) with drinking water, and Borrevannet, a eutrophic lake 55 km southwest of Oslo used as water supply for the city of Horten (13,200 inhabitants).

The phytoplankton development during one year was followed at intervals of one month. The water samples for quantitative plankton analysis were taken with a Ruttner sampler. The samples were fixed in formalin, and the counting performed by using an inverted Utermöhl plankton microscope (Utermöhl, 1931, 1958). Physical and chemical determinations were carried out on water samples from the respective depths of the lakes where the biological samples were collected.

Lake Maridalsvannet has a surface area of 3.9 km^2, the total water volume is 70 million m^3, and the greatest depth is 45 m. In winter the lake is ice-covered from December to April. The vernal full-circulation occurs in the beginning of May, the summer stagnation lasting to the beginning of August. The autumn full-circulation is observed during November.

The data for the chemical conditions of the water demonstrate only little variation throughout the year. With respect to the contents of plant nutrients the water is classified as oligotrophic. The average values of chemical data characterizing the nature of the water in the lake are listed in Table I.

TABLE I

Chemical Characteristics of Lake Maridalsvannet (1959-1960)

H_3O^+	pH	6.5
Electrolytic conductivity	κ_{20}	32×10^{-6}
Chloride	mg Cl/liter	1.4
Sulfate	mg SO_4/liter	5.1
Calcium	mg CaO/liter	6.6

The composition of the phytoplankton in Maridalsvannet was well correlated with the trophic stage of the lake. Many species of algae were present, and during the taxonomic analysis approximately 100 species were observed. They were distributed among the classes of algae as shown in Table II.

TABLE II

Number of Species of Phytoplankton in Lake Maridalsvannet (1959-1960)

Algal Class	Number of Species Identified
Schizophyceae	6
Chlorophyceae	37
Chrysophyceae	15
Bacillariophyceae	18
Dinophyceae	5

Tabellaria flocculosa, Crucigenia rectangularis, and *Diceras Chodati* had occurrence in the plankton communities. These algae are considered good indicators for oligotrophic conditions. The chrysophycean species which dominated in the early summer plankton were also those typical for oligotrophic lakes of northern Europe. The phytoplankton of Lake Maridalsvannet is described as a variant of the polymictic type, several species being in approximately the same abundance with other species less prevalent.

The outline of the sequence of phytoplankton development in Lake Maridalsvannet is illustrated in Fig. 3. The variations of standing stock of phytoplankton at a depth of 4 m are shown. The fluctuations of the dominating algal classes are characterized by small populations during the winter period, a gradual increase to the maximum occurrence observed during the late summer, and then subsequently decreasing population numbers until winter conditions were established. The major species of the phytoplankton communities and their quantitative occurrence during the period of investigation are listed in Table III. Seasonal maxima of particular species succeeded each other within the general pattern of development of their respective taxonomic class.

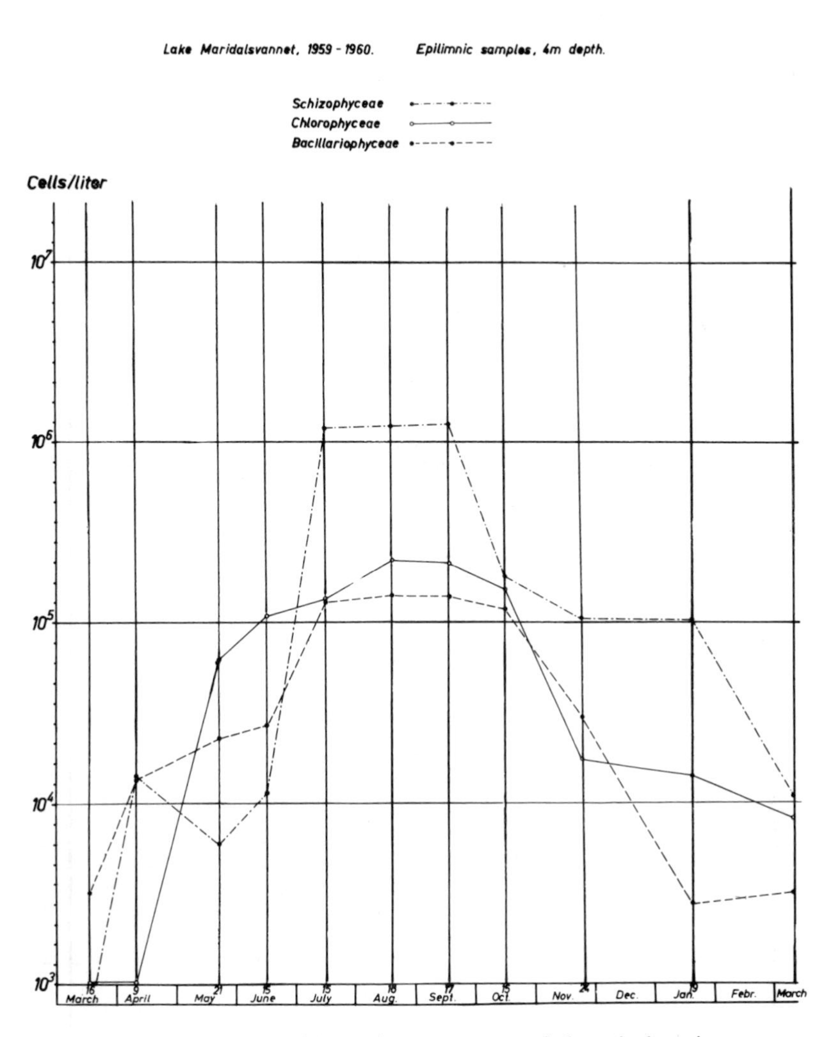

FIG. 3. Variations of the major components of phytoplankton in an oligotrophic lake.

Lake Borrevannet has a surface area of 1.8 km², the total water volume is 12 million m³, and the greatest depth is 16 m. The pronounced changes in the overall thermal structure and dynamics of the seasonal variations in this lake are in general the same as those described for Lake Maridalsvannet. But the lesser depth of Lake

TABLE III

Major Species of Phytoplankton in Lake Maridalsvannet at a Depth of 4 m (1959-1960)*

Taxon	Mar. 16, 1959	Apr. 9, 1959	May 21, 1959	June 15, 1959	July 15, 1959	Aug. 18, 1959	Sept. 17, 1959	Oct. 15, 1959	Nov. 24, 1959	Jan. 19, 1960	Mar. 10, 1960
Schizophyceae	–	23.5	–	15.2	1781.0	1878.0	1962.5	333.6	121.7	114.0	24.0
Chroococcus sp.	–	–	–	–	–	2.5	–	–	–	114.0	12.8
Merismopedia tenuissima	–	23.5	–	13.8	1778.0	1875.0	1962.5	333.6	121.6	–	11.2
Chlorophyceae	0.8	2.9	84.2	142.9	215.9	408.8	398.6	264.5	32.0	24.0	9.3
Ankistrodesmus falcatus	–	–	14.6	96.8	58.0	163.5	182.8	157.2	13.0	4.0	1.7
Chlorococcum sp.	–	–	37.0	1.0	9.0	3.5	–	–	–	–	–
Crucigenia rectangularis	–	0.2	0.3	–	13.9	13.9	3.0	0.3	–	–	–
Crucigenia tetrapedia	–	–	–	4.8	15.6	55.5	61.0	5.4	–	–	–
Dispora sp. (?)	0.04	1.0	12.0	25.8	33.4	47.0	35.6	42.8	10.8	–	–
Scenedesmus spp.	–	–	–	–	10.4	28.5	23.2	15.0	–	–	–
Bacillariophyceae	5.6	21.4	42.4	49.9	204.3	248.0	238.4	166.2	53.3	5.0	6.6
Cyclotella sp.	–	11.0	12.0	3.2	20.6	21.0	21.6	36.4	8.8	–	–
Melosira ambigua	3.0	1.5	3.6	4.6	25.8	69.5	99.0	79.0	22.4	1.5	2.6
Chrysophyceae	10.0	14.4	308.2	129.8	118.2	111.0	21.0	15.2	0.04	–	–
Kephyrion Rubri-claustri	–	13.0	12.2	0.4	0.6	–	–	–	–	–	–
Kephyrion spirale	–	1.4	31.8	17.2	22.4	18.0	7.8	7.0	–	–	–
Kephyrion sp.	–	–	225.0	64.8	25.6	22.0	6.0	4.6	–	–	–

* Figures indicate cells in thousands per liter of water.

Borrevannet means that the vernal and the autumnal circulation and mixing of water are more rapidly accomplished. In the autumn, for example, a uniform temperature from surface to bottom was observed in the beginning of October.

The chemical nature of the water masses of Lake Borrevannet is characterized by the average values of the factors listed in Table IV.

TABLE IV

Chemical Characteristics of Lake Borrevannet (1954, 1955, and 1959)

H_3O^+	pH	7.2
Electrolytic conductivity	κ_{20}	155×10^{-6}
Chloride	mg Cl/liter	16
Sulphate	mg SO_4/liter	15
Calcium	mg CaO/liter	38

During the stagnation periods oxygen deficients evolved in the hypolimnion.

Lake Borrevannet is classified as an edaphic and morphometric conditioned eutrophic lake.

The composition of the phytoplankton during the period of investigation typically demonstrated the eutrophic condition. The approximately 90 species identified were distributed among the taxonomic classes as listed in Table V.

TABLE V

Number of Species of Phytoplankton in Lake Borrevannet (1954-1955)

Algal Class	Number of Species Identified
Schizophyceae	18
Chlorophyceae	32
Chrysophyceae	3
Bacillariophyceae	32
Dinophyceae	4

The general pattern of seasonal variation in standing stock of phytoplankton at a depth of 3 m in Lake Borrevannet is illustrated in Fig. 4.

During the summer months the species of *Schizophyceae* made up a considerable part of the phytoplankton. The class was represented at a maximum in August and September. In the surface level of the lake algal blooms with *Anabaena spiroides, Aphanizomenon flos-aquae*, and *Microcystis aeruginosa* occurred. The chlorophyceans presented two maxima during the period of investigation, in June and in September. A few species of the order *Chlorococcales* dominated among the chlorophycean algae. Diatoms formed maxima when the ice cover had disappeared in the beginning of May and at the end of the summer stagnation in August. Two species of diatoms dominated the phytoplankton during these periods, *Stephanodiscus Hantzschii* in the spring outburst, and *Fragilaria crotonensis* in the autumn outburst.

Details of the phytoplankton development in Lake Borrevannet are given in Table VI. The numbers indicate algal cells, eventually higher morphological units, times thousand, per liter of water. The table reveals some characteristic features about the phytoplankton development. The algae respond quickly to changes in environmental factors, the culmination of their growth coinciding with the circulation periods of the water masses. The change in community structure by replacement of species is rapid. The phytoplankton consists for most of the year essentially of one or a few species in dominance, and is of a monomictic type.

A comparison of the data obtained on the development of phytoplankton in oligotrophic Lake Maridalsvannet and eutrophic Lake Borrevannet is of some interest. Figure 5 shows the monthly variations of gross phytoplankton in the epilimnion of the lakes; the data plotted represent cell numbers, eventually higher morphological units, of standing stock of algae per liter. The number of cells is not the best criterion to assess the magnitude of algal stock in a lake, but when the species composition of the population is known, it is justifiable to obtain an approximate idea using such data.

As might be expected from the greater concentration of nutrients in the water masses of Borrevannet, the algal vegetation is prolific in this lake. Population pulses of algae responsible

TABLE VI

Major Species of Phytoplankton in Lake Borrevannet at a Depth of 3m (1954-1955)*

Taxon	Dec. 12, 1954	Jan. 24, 1955	Feb. 28, 1955	Mar. 19, 1955	Apr. 16, 1955	May 7, 1955	June 2, 1955	July 12, 1955	Aug. 2, 1955	Sept. 8, 1955	Oct. 12, 1955	Nov. 5, 1955
Schizophyceae	0.3	0.04	8.0	14.0	0.04	11.0	3.5	38.9	34.4	117.5	33.6	0.3
Anabaena spiroides	–	–	–	–	–	–	–	1.4	–	27.0	–	–
Aphanizomenon flos-aquae	–	–	–	–	–	–	1.2	12.5	2.0	85.0	18.0	–
Chroococcus dispersus	–	–	–	–	–	11.0	–	–	3.0	–	–	–
Microcystis pulvera var. *inserta*	–	–	8.0	14.0	0.04	–	–	–	–	–	–	–
Chlorophyceae	11.0	4.1	1.0	9.7	3.9	28.1	783.5	334.5	268.3	3055.0	462.0	187.7
Ankistrodesmus falcatus	0.5	0.5	0.5	4.0	0.5	12.5	87.0	30.0	13.0	1.5	55.0	27.0
Dispora sp. (?)	–	1.5	–	5.5	1.5	13.0	426.0	170.0	6.0	1.0	57.0	85.5
Scenedesmus bijugatus	2.0	2.0	–	–	0.2	0.2	4.5	34.0	82.0	2930.0	280.0	16.5
Scenedesmus quadricauda	1.0	0.08	–	0.2	0.2	0.4	8.0	32.0	94.0	79.5	60.0	9.0
Tetraëdron minimum (?)	7.5	–	–	–	1.5	2.0	258.0	40.5	–	–	3.5	48.0
Bacillariophyceae	16.9	1.6	0.2	0.2	1.0	9945.5	98.3	351.0	585.7	271.4	45.8	8.0
Fragilaria crotonensis	14.0	1.5	–	–	–	2.0	54.0	222.5	430.0	246.0	14.1	3.5
Stephanodiscus Hantschii	–	–	–	–	–	9940.0	36.0	120.0	150.0	19.0	27.0	2.0
Dinophyceae	1.5	0.02	–	0.5	7.7	152.0	0.2	6.8	11.9	38.7	7.5	0.3
Peridinium aciculiferum	–	0.02	–	0.5	7.7	152.0	–	–	–	–	–	–
Peridinium cinctum	1.5	–	–	–	–	–	0.04	6.4	6.4	38.5	7.5	0.3

* Figures indicate cells in thousands per liter of water. *Anabaena* and *Aphanizomenon* are counted as trichomes: *Microcystis* is counted as colonies.

for difficulties in the treatment of water, and creating unpleasant conditions in the supply mains, occur at intervals. The spring and autumn outbursts of plankton are particularly troublesome.

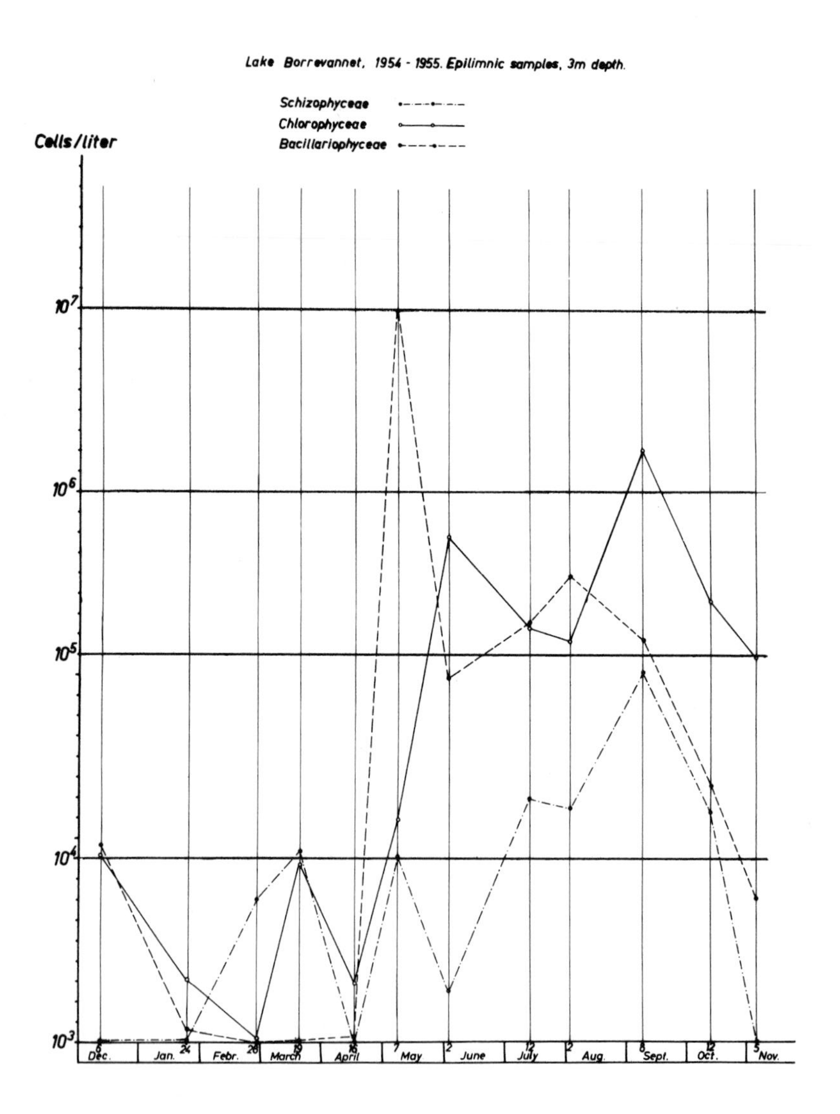

Fig. 4. Variations of the major components of phytoplankton in a eutrophic lake.

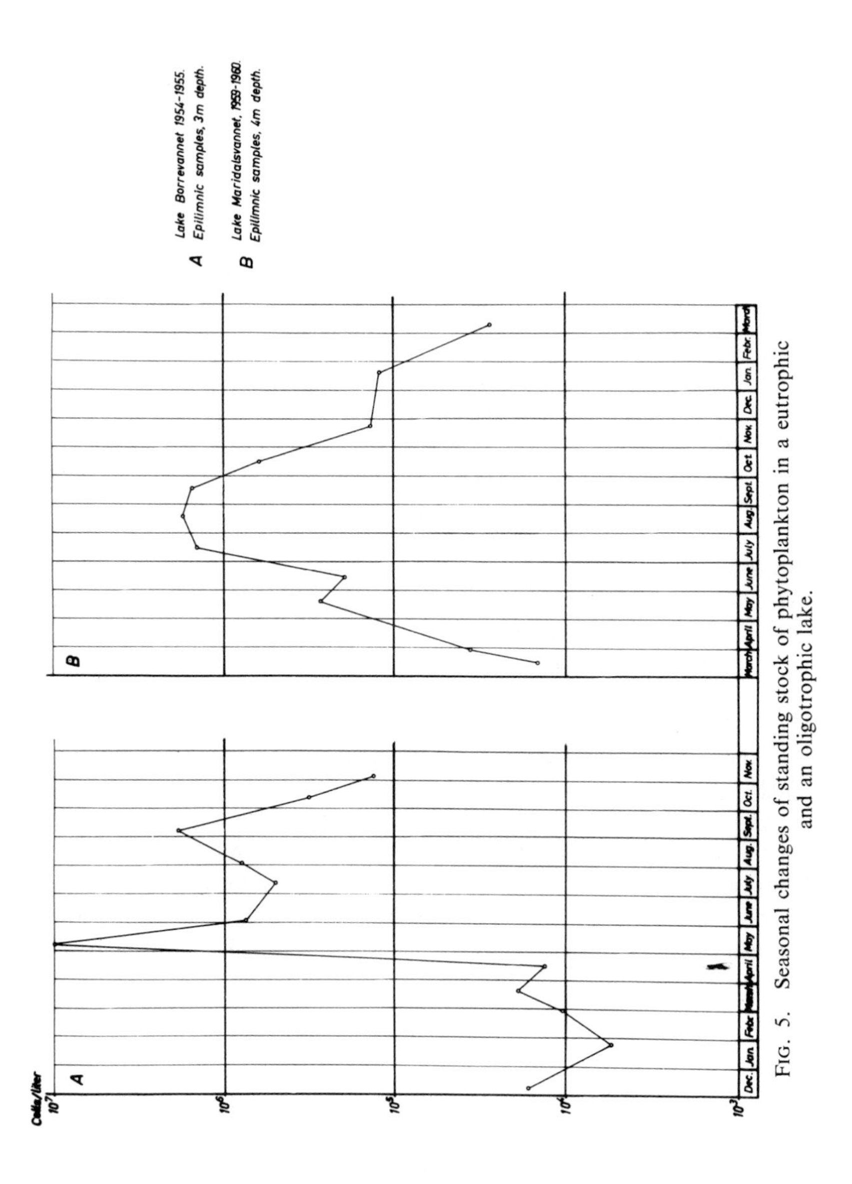

FIG. 5. Seasonal changes of standing stock of phytoplankton in a eutrophic and an oligotrophic lake.

In Lake Maridalsvannet the algal population is also of considerable amount with respect to number of cells, but in this case the major species belong to the nannoplankton. The algae give rise to minor troubles only for the city water supply, including the forming of sediments in the mains and service reservoirs and giving turbidity to the water.

These examples of algal development in the water supply of two cities of Norway are more or less typical of the situation for Northern Europe. Some special problems for the southern part of the continent will be mentioned briefly. Here the climatic conditions create other types of lakes with respect to thermal and dynamic relations. Warm monomictic lakes are typical for lower latitudes. In these lakes the water temperature never falls below 4°C at any depth. Stratification is only present during the summer, the winter being characterized by circulation of the water masses. There are also in Europe several lakes of transitional types between dimictic and warm monomictic lakes. Lake Zurich is of this category, and provides an example of algal problems related to eutrophic lakes with circulation during the whole of the winter (Thomas and Maerki, 1949; Zimmermann, 1961). The phycology of Lake Zurich is well described (Messikommer, 1954). The lake has been in a state of progressive eutrophy for several years, and today a heavy crop of algae is typical for Lake Zurich. A mass development of *Oscillatoria rubescens* together with small populations of *Fragilaria crotonensis, Synedra acus, Asterionella formosa,* and *Melosira islandica* are the cause of considerable difficulties for the water works of the city of Zurich. The water is taken from the lake below the thermocline and is treated by slow sand filtration. During the circulation period in winter the algae are transported down to the intake, and difficulties arise that are of considerable economic importance to the cost of filtration (Thomas, 1957).

RIVERS

Rivers form the sources of water supply to cities and concentrations of population in all countries of Europe. Among the best-known river works are those supplying London and Paris with drinking water from the Thames River and the Seine River, respectively. Reservoirs and impounding reservoirs are important constructions of water supplies derived from rivers. Figure 6 gives an example of a large reservoir in the water supply of London. The amphibious vehicle is a device adapted for the chemical treatment of the reservoir.

River water which has to be used for water-supply purposes is in many places rich in plant nutrients. The drainage from intensively cultivated fields and effluents from sewage works provide the water with fertility. When water is brought into a reservoir with a rather great surface area, suitable conditions for algal growth are created and prolific growth is the result. The engineer is faced with the problems of removing the algae from the water and re-establishing its quality. The engineer and the phycologist have to cooperate to master this task.

Investigations of algal problems of this kind are carried out in several European countries. The contributions from scientists in the United Kingdom, Czechoslovakia, Austria, and Poland will serve as examples.

The storage of water from the Thames and the Lee in the reservoirs of the Metropolitan Water Board promotes the growth

FIG. 6. River reservoir. (From Metropolitan Water Board, London.)

of algae to a great extent (Chevalier, 1953). The river water entering the reservoirs contains a primary load of algae, but the new environmental factors introduced by the reservoir give advantages to other species, and a second development of algal communities is supported. *Asterionella formosa, Fragilaria crotonensis, Stephanodiscus astrea, Synedra acus, Ankistrodesmus falcatus, Pediastrum* sp., *Anabaena circinalis, Aphanizomenon flos-aquae, Microcystis aeruginosa, Ceratium hirundinella,* and *Peridinium cinctum* are among the common species in the algal communities in the reservoirs (Taylor, 1961).

In the first year after the construction of an impounding reservoir, the biological and chemical changes in stored water are of considerable practical interest. A detailed investigation, including the study of algae, contributing knowledge about these problems was carried out at the Chew Stoke impounding reservoir, Bristol, England (Hammerton, 1959). *Asterionella formosa, Tribonema bombycinum, Microcystis* sp., *Anabaena* sp., and *Aphanizomenon flos-aquae* were the most important species, producing heavy growths. Only the representatives of *Schizophyceae* caused serious difficulties in the treatment for the water supply, and their mass occurrences took place during late summer and autumn.

The practical conclusions of phycological investigations on reservoir problems in England are treated in several publications (Pearsall, Gardiner, and Greenshields, 1946; Mackenzie and Greenshields, 1949; Hobbs, 1954; Ives, 1957; Chancellor, 1958).

Several impounding reservoirs in Central Europe, both at high altitudes and in the lowlands, have been objects of phycological investigations in recent years. A review of phytoplankton studies in mountain lakes and impounding reservoirs in the Alps contains twenty references (Pechlaner, 1961). In the industrial districts of Poland the population is frequently dependent on the use of water from impounding reservoirs. Hydrobiological investigations of these reservoirs (Starmach, 1958) conclude that phytoplankton plays the most important role among the organisms in the utilization of water reservoirs. *Fragilaria crotonensis, Asterionella formosa, Sphaerocystis planctonica, Microcystis aeruginosa,* and species of the order *Volvocales* are among the interference algae.

Prague, the capital of Czechoslovakia, derives its drinking water from the Želinka River. The Sedlice reservoir has been used as a study object in order to obtain knowledge necessary for controlling and mastering the processes taking place in a reservoir during the year (Štěpánek, 1960). Phycological studies are incorporated in the research project (Chalupa and Štěpánek, 1960; Štěpánek and Pokorný, 1960; and Štěpánek et al., 1960). Species of the genera *Aphanizomenon, Chlamydomonas, Kirchneriella,* and *Scenedesmus* are reported having mass occurrence in the reservoir.

Algal problems are more common for water supplies taken from rivers than from lakes. Rivers as sources for water supplies will be of increased importance in the years to come, because other sources of suitable water with required capacity are becoming scanty in many European countries. The mastering of algal problems concerned with river water supplies will also in the future be an important aspect of applied phycology.

PHYCOLOGY AND THE STUDY OF EUTROPHICATION

In the following portion of this paper attention will be concentrated upon some aspects of the relationship between phycology and the study of eutrophication phenomena.

The gradual changes in productivity of inland waters from oligotrophy through mesotrophy to eutrophy are study objects of scientific interest (Lindeman, 1942), and the results have considerable practical and economic importance (Jaag, 1956). The process of eutrophication is primarily caused by the supply of plant nutrients to the aquatic environment, even though other factors are involved, such as morphometric relations and the degree of biological utilization of the nutrient substances. The manifestation of eutrophication is the increasing, primary productivity; in extreme cases the water-bloom phenomenon develops, and in stagnant water oxygen supply becomes depleted. The consequence of eutrophication is reduced fitness for use of the water. The primary cost of industrial and domestic water supply is considerably increased. Damage is done to fisheries by preventing the development of salmonids and coregonids, and coarse-fish are favored. The landscape is aesthetically deteriorated, and the recreational value of the water for public health is adversely affected. In almost all European countries it is recognized that eutrophication phenomena require attention. Some countries have worked out programs for the conservation of natural resources, and in these the protection of water plays an important part.

The influence of the eutrophication process on lake succession has been investigated intensively in the United Kingdom (Pearsall, 1921; Pennington, 1943) and in Switzerland (Nipkow, 1920; Thomas and Maerki, 1949; Thomas, 1951; and Züllig, 1956). One important result of those investigations has been the demonstration of phases of lake development. There is no slow and steady change from oligotrophic to eutrophic conditions, but periods of rapid changes alternating with rather long periods of stability. This steplike character of the evolution is perhaps best illustrated by the corresponding change having taken place in the communities of algae, and which are recorded in the bottom deposits of the lakes. Although well known, some of the observations concerning Lake Zurich, Switzerland, will for the purpose of continuity be briefly recapitulated.

Figure 7 is the classical picture of a core of the deposits from Lake Zurich (Nipkow, 1920). The layers made up of finer and coarser particles of clay and the overlying column with characteristic zones of black sediments are shown. Evidently some radical changes took place in the lake corresponding to the first formation of the black sediment. These changes are ascribed to the transition from oligotrophic to eutrophic conditions in the lake. During 1896 a mass occurrence of *Tabellaria fenestrata* was reported, and the first black zone in the core coincides with the invasion of *Oscillatoria rubescens* in Lake Zurich.

The history of inland waters of the European continent through the last hundred years yields examples of lakes changing from oligotrophic to eutrophic conditions (Thienemann, 1928; Jaag, 1956). The phycological surveys of the Swiss lakes enable the reconstruction of time for the mass invasion of *Oscillatoria rubescens,* which generally paralleled the alteration in lake metabolism. Well-known examples are listed in Table VII (Jaag, 1952, 1956).

There are fundamental problems to be investigated before an understanding of the eutrophication process and the water-bloom phenomenon is possible. Research along autecological and synecological lines has to be performed. But work is in progress, and promising results are obtained (Ambühl, 1960; Kliffmüller, 1960; and Staub, 1961).

Practical application of the knowledge gained about the process of eutrophication is still a problem. The state of eutrophication

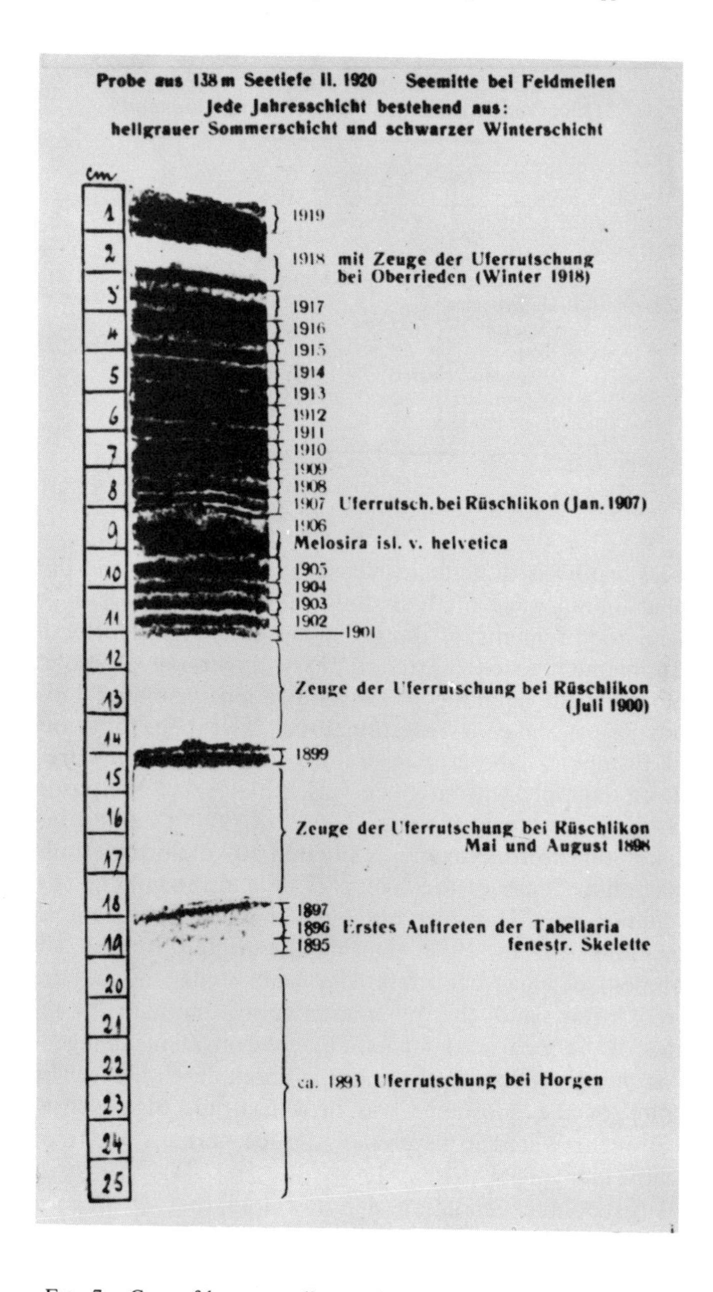

FIG. 7. Core of bottom sediments from Lake Zurich, Switzerland.

TABLE VII

Observed Invasions of *Oscillatoria rubescens*

Locality	Year
Murtensee	1825
Baldeggersee	1870
Zürichsee	1898
Hallwilersee	1898
Zugersee	1898
Rotsee	1910
Vierwaldstättersee	1930
Wäggitalersee	1946
Lago Lugano	1956

of inland waters is difficult to measure and evaluate, and the conventional limnological methods only seldom give information that meet the requirements of the designing engineers (Hörler, 1950). As a supplement to chemical and physical methods of importance for the practical handling of eutrophication problems, bio-assay methods using algae as test organisms have been developed in several European laboratories. Before these procedures are mentioned, an example will be given which may serve the purpose of illustrating the practical aspects.

Lake Steinsfjord, Norway, is situated 30 km northwest of Oslo. The maximum depth of the lake is 22 m, and the volume of water is 45 million m^3. The lake has been the object of limnological investigations (Strøm, 1932). During the winter of 1961 a heavy development of algae occurred. The water under the ice turned a faint red color, and the phenomenon was noticed by the inhabitants of the local settlements. The concentrations of algae were found as shapeless flocs in the water in boreholes in the ice (Fig. 8). The color of the substance was brownish-red. Algal substances had also accumulated in fissures of the ice perhaps due to phototactic movements.

Microscopic examination demonstrated that *Oscillatoria rubescens* was the cause of the phenomenon. The habitus of the alga from Lake Steinsfjord is very close to the taxonomical description of the species (Figs. 9 and 10). This is to my knowledge the first

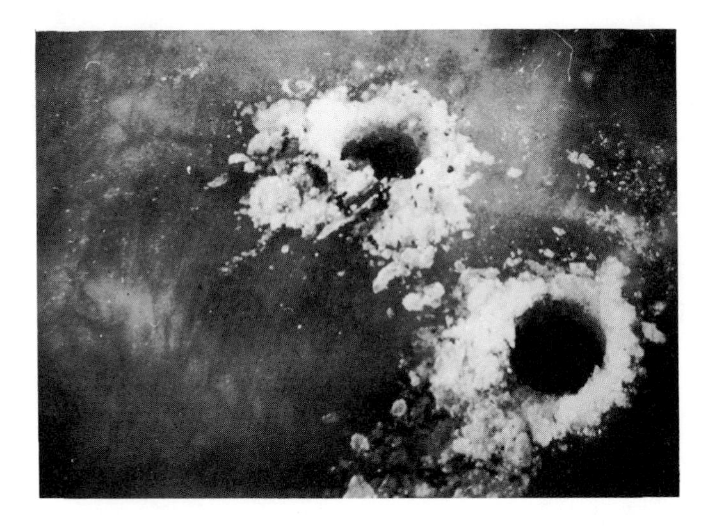

FIG. 8. Flocs of *Oscillatoria rubescens* in boreholes in the ice.

invasion of *Oscillatoria rubescens* reported in a Scandinavian lake.

The development of *Oscillatoria rubescens* in Lake Steinsfjord was observed during the winter 1962. However, this winter the population density of the alga was not of the same magnitude. A comparison of the observations from two characteristic days in the winters of 1961 and 1962, respectively, is shown in Fig. 11. The phytoplankton consisted mainly of *Oscillatoria rubescens* on March 4, 1961, with other species (*Asterionella formosa, Melosira* cf. *ambigua,* etc.) less prevalent. On February 27, 1962, *Asterionella formosa* was the major species of the phytoplankton community, with *Oscillatoria rubescens* less abundantly represented. The hydrographical data plotted on the diagram are not sufficient to explain the difference in phytoplankton development of these two years. Very many physical and chemical factors have to be determined before an understanding of such an invasion of an alga may be possible.

Some questions present themselves: Is this occurrence of *Oscillatoria rubescens* an evidence of the increasing eutrophy of the lake? Will the development of Lake Steinsfjord follow the course

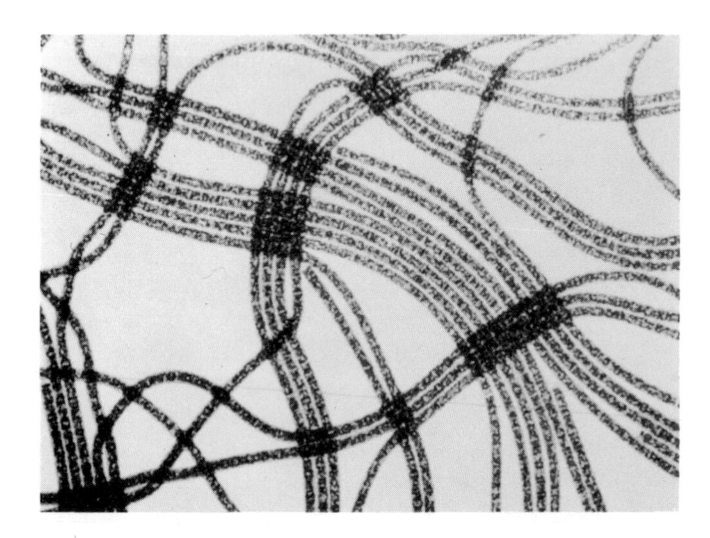

FIG. 9. Habitus of *Oscillatoria rubescens* from Lake Steinsfjord.

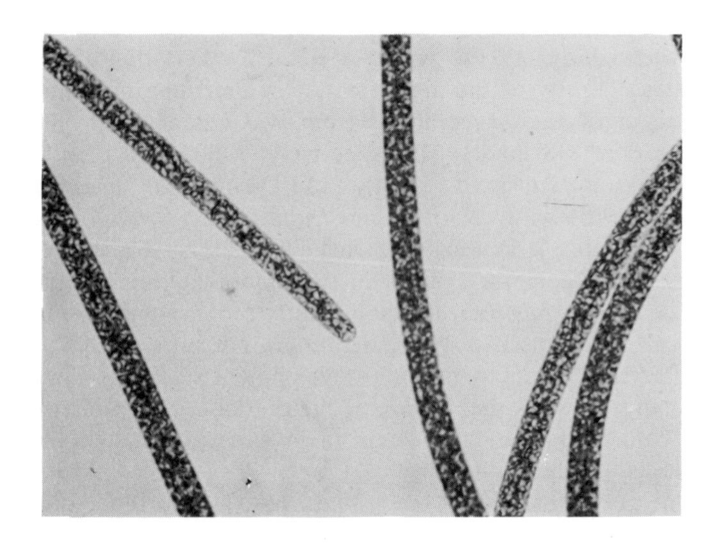

FIG. 10. Details of trichomes of *Oscillatoria rubescens*.

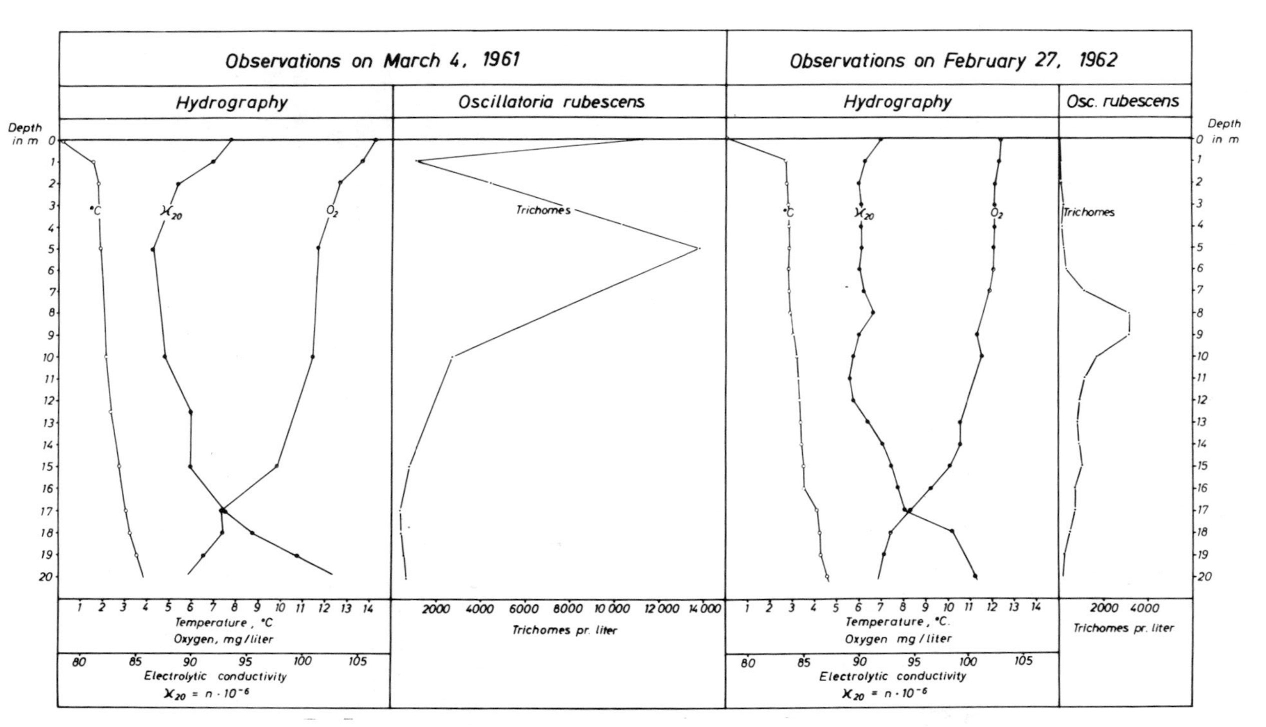

FIG. 11. Occurrence of *Oscillatoria rubescens*, and hydrographical data from Lake Steinsfjord.

represented by the Central European lakes containing *Oscillatoria rubescens* blooms? What about allowing the sewage from the population to be discharged into the lake, the catchment area being in a place of increased settlement? These problems are of far-reaching consequences, and it does not seem likely that the ordinary limnological methods are sufficient to give background knowledge for deciding them.

One particularity in this connection is the difficulty of interpreting the results gained from physical and chemical analysis with respect to what they indicate about the biological conditions of the recipient. There is good reason to believe that the use of bioassay methods might be a supplement to the physical and chemical methods of importance for a better assessment of the effect of eutrophication on inland waters.

LABORATORY CULTURING OF ALGAE AS A BIO-ASSAY METHOD FOR AN ASSESSMENT OF POLLUTION

The application of laboratory cultures of algae for the study of the potential fertility of water is not a new procedure. Some important publications connected with the development of these techniques are listed below:

1910. Allen, E. J., and Nelson, E. W. On the artificial culture of marine plankton organisms.
1923. Atkins, W. R. G. The phosphate contents of fresh and salt waters in its relationship to the growth of algal plankton.
1927. Schreiber, E. Die Reinkultur von marinen Phytoplankton und deren Bedeutung für die Erforschung der Produktionsfähigkeit des Meerwassers.
1929. Naumann, E. Grundlinien der experimentellen Planktonforschung.

Perhaps most influential of these were the investigations of Schreiber. He worked at the biological station on Helgoland. In his treatise dated 1927 he describes a procedure for what he calls a "physiological analysis of sea water." By using a species of *Carteria* as test organism, and observing the growth of the flagellate in water samples made free of organisms by ultrafiltration, he was able to determine the contents of phosphorus and nitrogen in the actual water mass. From the results of these culture experiments Schreiber was also able to determine which salt, eventually what combination of salts, was the limiting nutrient factor for the growth of phytoplankton in the sea around Helgoland during the season.

Schreiber's method resulted in knowledge which completed the information obtained by chemical analysis.

Some investigations of recent times include the following:

1948. Rohde, W. Environmental requirements of fresh-water plankton algae. Experimental studies in the ecology of phytoplankton.

1956. Bringmann, G., and Kühn, R. Der Algen-Titer als Mass-stab der Eutrophierung von Wasser und Schlam.

1956. Potash, M. A biological test for determining the potential productivity of water.

1959. Lund, J. W. G. Biological tests on the fertility of an English reservoir water.

The publication of Rohde cited is different from the others mentioned. It is included to illustrate the type of work on which the culture technique methods depend. The other inquiries are all based on the same philosophy: that the potential fertility of a water can be measured and expressed in terms of algal growth.

Examples follow of work along such lines carried out by the Norwegian Institute for Water Research. There is no standard method described for procedures of bio-assay of this kind, and an original technique has been developed and practiced.

The species of alga used as the test organism is *Selenastrum capricornutum.* Taxonomically the species belongs to the order *Chlorococcales* and the family *Selenastraceae.* The position of the species in the genus *Selenastrum* or the genus *Ankistrodesmus* is questionable (Skuja, 1948). The cells are characteristically lunate with apices acutely pointed, about 2 μ broad and 20 μ long (Fig. 12). *Selenastrum capricornutum* is solitary. As a matter of curiosity it may be mentioned that the species was first described from a Norwegian locality by Printz (1914, p. 92). The distribution in Norway is wide, but in no localities are large populations observed. The alga is found in oligotrophic as well as in eutrophic waters.

Selenastrum capricornutum has proved to have several superior qualities as a laboratory organism, among which are: 1) The alga is morphologically easy to identify; 2) the form variations with changing growth conditions are small; 3) the alga is solitary except during cell division; 4) the alga is, as far as has been investigated, obligatorily autotroph, and has a minimum of growth requirements.

A clone culture of *Selenastrum capricornutum* is kept in the laboratory. The nutrient solution used for the cultivation is described in the literature (Staub, 1961; Hughes, Gorham, and Zehnder, 1958). The composition of the medium is presented in Table VIII.

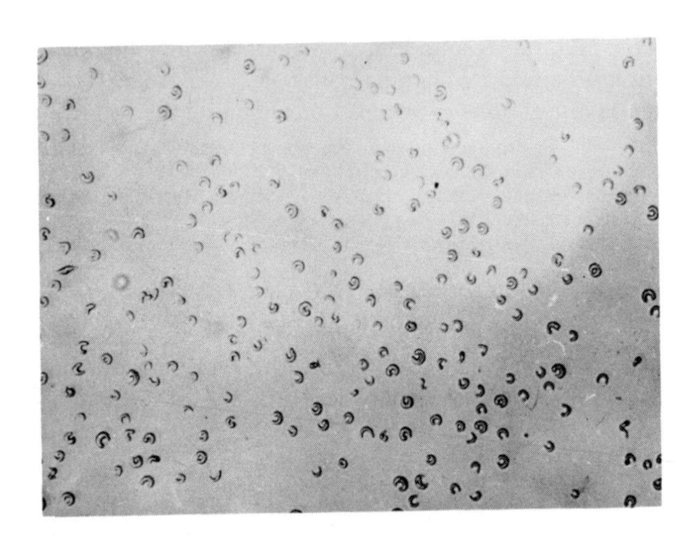

FIG. 12. Habitus of *Selenastrum capricornutum.*

From the stock solution dilutions of the actual concentrations used for the culture experiments are made.

The method involves the following procedure:

1. Water samples from the locality in nature are autoclaved and inoculated from the clone culture of *Selenastrum capricornutum.*

TABLE VIII

Composition of the Nutrient Solution

$NaNO_3$	467 mg/liter
$Ca(NO_3)_2 \cdot 4H_2O$	59 mg/liter
K_2HPO_4	31 mg/liter
$MgSO_4 \cdot 7H_2O$	25 mg/liter
Na_2CO_3	21 mg/liter
Fe-EDTA*	10 ml/liter
Trace-element solution†	0.8 ml/liter

* Five ml of a 0.1 N solution of $FeCl_3 \cdot 6H_2O$ in 0.1 N HCl and 5 ml of a 0.1 N solution of Komplexon III are mixed and diluted to 500 ml. Corresponds to 0.5 mg Fe^{3+}/liter.
† After Gaffron (Hughes *et al.*, 1958).

2. The algae are grown in Pyrex-glass flasks which are shaken slowly at 30°C and illuminated by fluorescent lamps.

3. The extent of growth of the test alga is determined by measuring the red fluorescence due to chlorophyll.

Direct counting of algal cells by the inverted microscope technique, and absorptiometric and turbidimetric determinations, have been found less suitable than the alcoholic extraction procedure and chlorophyll measurements for describing the growth of the algae. The primary cause of this is the varying amount of particles contained in the different water samples from localities in nature.

The procedure results in growth curves which are discussed and compared with those obtained using the standard culture solution of known chemical composition. The growth curves express a measure of the amount of nutrients available for *Selenastrum capricornutum* in the actual water sample. They indicate the amount of fertilizing substances contained in the water.

In the following an investigation in which the *Selenastrum* method was applied will be briefly commented on. The recipient system studied was the Nitelv River, Norway (Fig. 13), flowing from Lake Harestuvannet, a distance of 43 km, to Lake Øyeren. The five localities of Stryken, Åneby, Slattum, Kjellerholen, and Nybrua were the sampling stations. The water accumulating to make the river is a dilute solution of alkaline earth and alkali bicarbonate with a variable quantity of minor inorganic constituents in true solution. Some ionic constituents which determine the chemical nature of the water are listed in Table IX. Concentrations of all the components mentioned vary with locality and fluctuate with the season and the prevailing weather conditions.

TABLE IX

Chemical Characteristics of Nitelv River Water

Components	Average Value
Ca^{++}	5.4 mg/liter
Mg^{++}	1.3 mg/liter
Fe^{++}, Fe^{+++}	0.3 mg/liter
SO_4^{--}	7.0 mg/liter
Cl^-	4.3 mg/liter
H_3O^+	7.1 pH

The changing properties of the river water as it flows down its course are demonstrated in Fig. 14. Data of electrolytic conductivity and permanganate value are plotted. In the lower river region the water carries a certain load of pollution from drainage of cultivated soil and sewage from populated areas, and the observations of permanganate value and the electrolytic conductivity describe the relations. But these observations reflect no

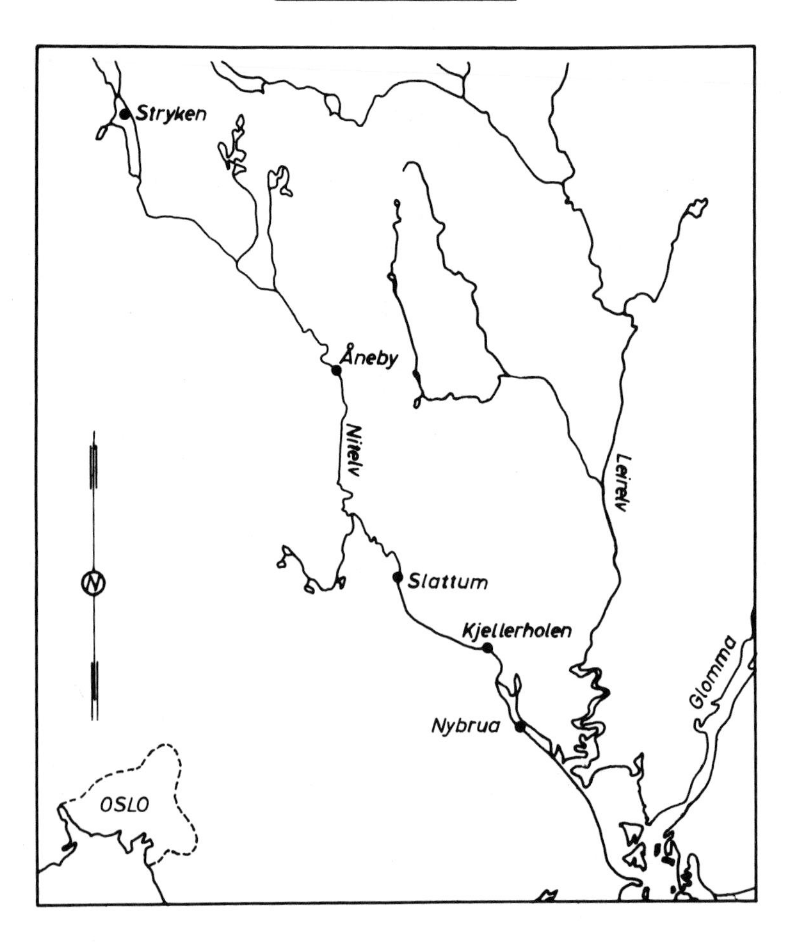

FIG. 13. Nitelv River sampling stations.

marked changes in eutrophic direction manifested in the biological conditions between the stations Aneby and Slattum.

Let us compare these facts with the results obtained from the *Selenastrum* method. The growth curves representing water samples from the five locations are given in Fig. 15. The diverse possibilities of the river water for supporting algal growth are described by the curves. The radical change in water quality in

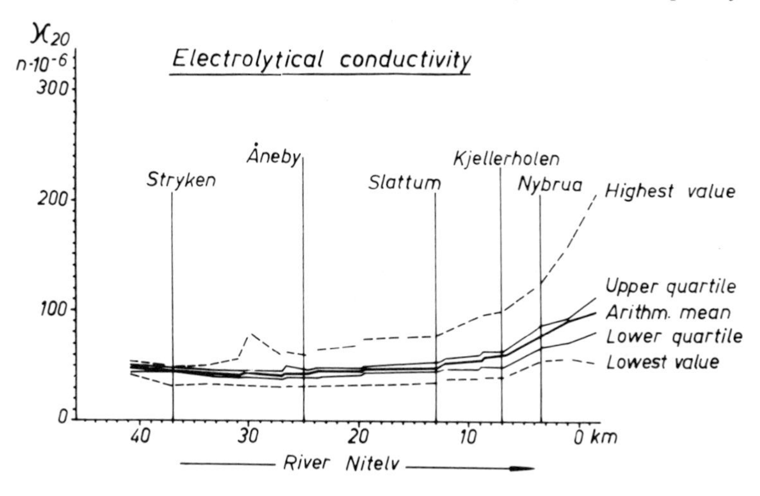

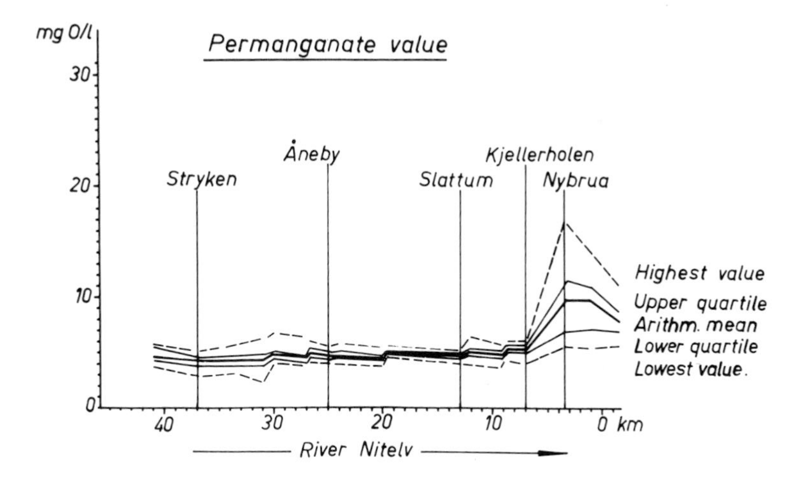

FIG. 14. Physiographical factors of the Nitelv River.

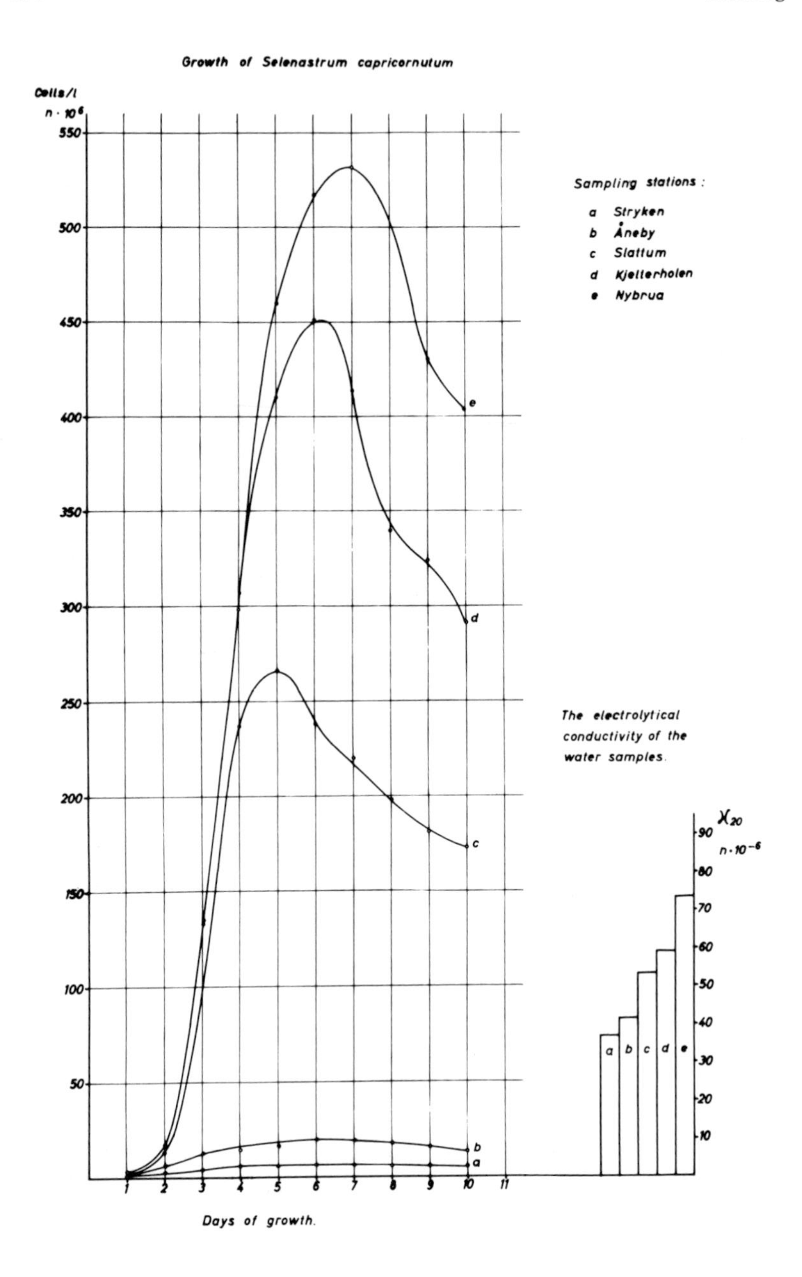

FIG. 15. River water as a substrate for algal growth.

eutrophic direction has taken place between the stations Åneby and Slattum. This corresponds with the results obtained from the field biological survey. Further biological testing indicated that in the upper river region phosphorus is generally the limiting factor in the water for the growth of *Selenastrum capricornutum*; in the lower river region primarily nitrogen and secondarily phosphorus are the minimum factors among the elements. In this example information was obtained by the *Selenastrum* method in addition to that collected by other methods.

Growth curves obtained in lake water collected from a variety of localities ranging from oligotrophic to extreme eutrophic conditions give evidence that the *Selenastrum* method may be applied for characterizing their potential fertility for algal development (Fig. 16).

The difficulties and limitations of the *Selenastrum* method may be briefly pointed out. All algae differ to some extent in their physiology. It is therefore a question of whether the information obtained by using this method is relevant to the problems of the locality studied. The effects of biotic factors of importance for the development of algae in nature are excluded. The sterilization process involved can change the chemical-physical properties of the water, which may have consequences for the resulting growth of the test alga. The water of the locality studied may contain poisonous substances which are growth-inhibiting. Nevertheless, if practiced with care, the application of this bio-assay method has the following possibilities:

1. The procedure may result in data being of additional help concerning the assessment of the effects of pollution on eutrophication.

2. It may result in a useful parameter for comparative limnology.

3. The method may be a supplement to chemical and physical methods of importance for the practical handling of eutrophication problems. Population equivalents for the fertilizing influences of pollution may be defined. The effect of sewage-treatment plants in reducing the component of mineral nutrients in the effluents may be tested.

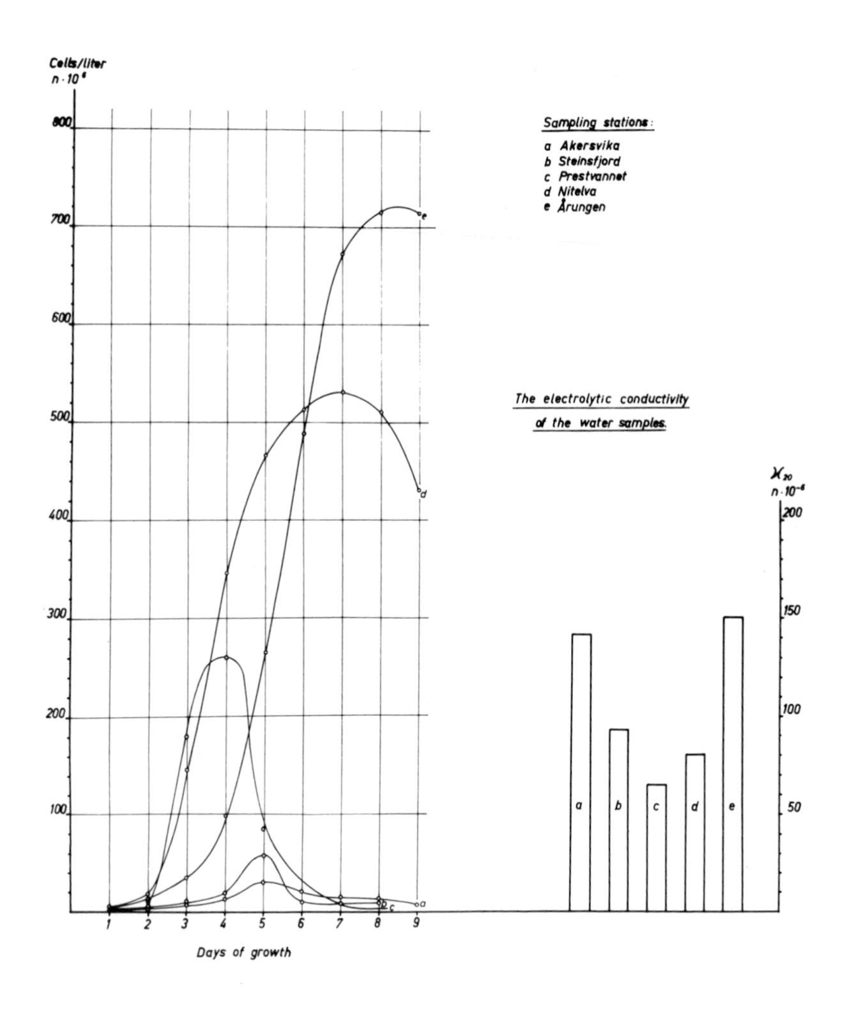

FIG. 16. Growth of *Selenastrum capricornutum* in water samples from various localities.

LITERATURE CITED

Allee, W. C., et al. 1949. Principles of animal ecology. London.
Allen, E. J., and Nelson, E. W. 1910. On the artificial culture of marine plankton organisms. *J. Mar. Biol. Assoc.*, Vol. 8.
Ambühl, H. 1960. Die Nährstoffzufuhr zum Hallwilersee. *Schweiz. Z. f. Hydrologie,* Vol. 22, Fasc. 2.
Atkins, W. R. G. 1923. The phosphate content of fresh and salt waters in its relationship to the growth of algal plankton. *J. Mar. Biol. Assoc.*, Vol. 13.
Baker, M. N. 1949. The quest for pure water. New York.
Beijerinck, M. W. 1894. Notiz über den Nachweis von Protozoen und Sperillen in Trinkwasser. *Centralbl. f. Bakt.* 15:10.
Bringmann, G., and Kühn, R. 1956. Der Algen-Titer als Mass-stab der Eutrophierung von Wasser und Schlamm. *Gesundheits-Ingenieur,* 77(23/24).
Certes, A. 1883. Analyse microscopique des eaux. Paris.
Chalupa, J., and Štěpánek, M. 1960. Limnological study of the reservoir Sedlice near Želiv. XIII Permanganate, C.O.D. and plankton. Scientific paper from Inst. of Chem. Tech., Faculty of Technology of Fuel and Water 4(2). Prague.
Chancellor, A. P. 1958. The control of aquatic weeds and algae. Her Majesty's Stationery Office, London.
Chevalier, W. S. 1953. London's water supply. 1903-1953. A review of the work of the Metropolitan Water Board. London.
Cohn, F. 1853. Über lebendige Organismen im Trinkwasser. *Z. f. klin. Med. v. Dr. Günsberg.* Vol. IV.
Dobell, C. 1960. Antony van Leeuwenhoek and his "Little Animals." New York.
Forel, F. A. 1895. Le Léman, Monographie Limnologique. Tome Second. Lausanne.
Hammerton, D. 1959. A biological and chemical study of Chew Valley Lake. Proceedings of the Society for Water Treatment and Examination. Vol. 8, Part 2.
Harz, C. O. 1876. Mikroskopische Untersuchungen des Brunnenwassers für hygienische Zwecke. *Z. f. Biol.* 12:75.
Hassall, A. H. 1850. A microscopic examination of the water supplied to the inhabitants of London and the suburban districts. London.
Hauge, H. V. 1957. Vangsvatn and some other lakes near Voss. A limnological survey in Western Norway. *Folia Limnol. Scand.,* No. 9.
Helfer, H. 1916. Geschichte der biologischen Analyse des Wassers. *Arch. f. Hydrol. und Planktonkunde,* Vol. 11.
Hobbs, A. T. 1954. Manual of British water supply practice. Cambridge.
Hughes, E. O., Gorham, P. R., and Zehnder, A. 1958. Toxicity of a unialgal culture of *Microcystis aeruginosa. Canad. J. Microbiol.* 4.
Hörler, A. 1950. Was erwartet der Abwasseringenieur von der hydrobiologisch-limnologischen Forschung? *Schweiz. Z. f. Hydrol.,* Vol. 12, Fasc. 2.
Ives, K. J. 1957. Algae and water supplies. *Water and Water Engineering.*
Jaag, O. 1952. Die Notwendigkeit des Gewässerschutzes und unser Ziel der Abwasserreinigung in der Schweiz. *Schweizer Baublatt,* No. 38.
Jaag, O. 1956. The pollution of surface and ground water in Switzerland. Fourth European Seminar for Sanitary Engineers. Geneva.
Järnefelt, H. 1956. Zur Limnologie einiger Gewässer Finnlands. XVI. Mit besonderer Berücksichtigung des Planktons. *Ann. Zool. Soc. Bot. Fenn. "Vanamo,"* 17(1).
Kliffmüller, R. 1960. Beiträge zum Stoffhaushalt des Bodensees. *Internationale Revue der gesamten Hydrobiologie,* 45(3).
Kolkwitz, R., and Marsson, M. 1902. Grundsätze für die biologische Beurteilung

des Wassers nach seiner Flora und Fauna. *Mitteilungen d. Kgl. Prüfungsanstalt f. Wasserversorgung u. Abwasserbeseitigung.* No. 1.

Kolkwitz, R., and Marsson, M. 1908. Ökologie der pflanzlichen Saprobien. *Ber. d. Deut. Bot. Gesell.* Jahrgang 1908, 26a(7).

Lauterborn, R. 1903. Die Verunreinigung der Gewässer und die biologische Methode ihrer Untersuchung. Ludwigshafen.

Lindeman, R. L. 1942. The trophic-dynamic aspect of ecology. *Ecology* 23(4).

Lund, J. W. G. 1959. Biological tests on the fertility of an English reservoir water (Stocks Reservoir, Bowland Forest). *J. Inst. of Water Engineers* 13(6).

Macan, T. T., and Worthington, E. B. 1951. Life in lakes and rivers. *The New Naturalist.* London.

MacDonald, J. O. 1875. A guide to the microscopical examination of drinking water. London.

Mackenzie, E. F. W., and Greenshields, F. 1949. The influence of algal growth on certain aspects of civil engineering design. The Institution of Civil Engineers. Excerpt from the proceedings of the conference on biology and civil engineering, September 21-23, 1948. London.

Maerki, E. Die Limnologie der schweizerischen Seen und Flüsse. *Schweiz. Z. f. Hydrol.,* Vol. 11, Fasc. 3/4.

Messikommer, E. 1954. Die Algenflora des Zürichsees bei Zürich. *Schweiz. Z. f. Hydrol.* Vol. 15, Fasc. 1.

Mez, C. 1898. Mikroskopische Wasseranalyse. Berlin.

Naumann, E. 1929. Grundlinien der experimentellen Planktonforschung. Die Binnengewässer, Vol. 6. Stuttgart.

Nipkow, F. 1920. Vorläufige Mitteilungen über Untersuchungen des Schlammabsatzen im Zürichsee. *Z. f. Hydrol.* 1.

Nygaard, G. 1945. Dansk Planteplankton. Copenhagen.

Nygaard, G. 1949. Hydrobiological studies on some Danish ponds and lakes. *Det Kongelige Danske Videnskabernes Selskab, Biologiske Skrifter* 7(1).

Olszewski, W., and Spitta, O. 1931. Untersuchung und Beurteilung des Wassers und des Abwassers. Berlin.

Pearsall, W. H. 1921. The development of vegetation in the English lakes, considered in relation to the general evolution of glacial lakes and rock basins. *Proc. Roy. Soc.*

Pearsall, W. H., Gardiner, A. C., and Greenshields, F. 1946. Freshwater biology and water supply in Britain. Freshwater Biol. Assoc. Scientific Publication No. 11.

Pechlaner, R. 1961. Lebenswelt in alpinen Speicherseen. Wasser und Abwasser. Zur Limnologie der Speicherseen und Fluss-staue. Vol. 1961. Vienna.

Pennington, W. 1943. Lake sediments: the bottom deposits of the north basin of Windermere, with special reference to the diatom succession. *New Phytologist* 42.

Potash, M. 1956. A biological test for determining the potential productivity of water. *Ecology,* Vol. 37.

Printz, H. 1914. Kristianiatraktens Protococcoideer. *Videnskapsselskapets Skrifter. I Mat.-Naturv. Klasse.* 1913, No. 6.

Radlkofer, L. 1865. Mikroskopische Untersuchung der organischen Substanzen im Brunnenwasser. *Z. f. Biol.* 1:26.

Rohde, W. 1948. Environmental requirements of fresh-water plankton algae. Experimental studies in the ecology of phytoplankton. *Symbolae Botanicae Upsaliensis* 10:1.

Schreiber, E. 1927. Die Reinkultur von marinen Phytoplankton und deren Bedeutung für die Erforschung der Produktionsfähigkeit des Meerwassers. *Wissenschaftliche Meeresuntersuchungen, aus der Biologischen Anstalt auf Helgoland* 16(10). Kiel.

Skuja, H. 1948. Taxonomie des Phytoplanktons einiger Seen in Uppland, Schweden. *Symbolae Botanicae Upsalienses* 9:3.

Skuja, H. 1956. Taxonomische und Biologische Studien über das Phytoplankton

Schwedischer Binnengewässer. *Nova Acta Regiae Societatis Scientiarum Upsaliensis*, Ser. IV, 16(3).
Starmach, K. 1958. Hydrobiological basis of the water utilization by waterworks. *Polskie Arch. Hydrobiol.* 4(17).
Staub, R. 1961. Ernährungsphysiologisch-autökologische Untersuchungen an der planktischen Blaualge *Oscillatoria rubescens* D. C. *Schweiz. Z. f. Hydrol.*, Vol. 23, Fasc. 1.
Štěpánek, M. 1960. Limnological study of the reservoir Sedlice near Želiv. X. Hydrobioclimatological part. The relation of the sun radiation to the primary production of nannoplankton. *Scientific Papers from Inst. of Chem. Tech., Faculty of Tech. of Fuel and Water* 4(2). Prague.
Štěpánek, M., and Pokorný, J. 1960. Practical knowledge of the bacteriomass and phytonannoplankton development in the reservoir Sedlice during 1957/1958. *Scientific Papers from Inst. of Chem. Tech., Faculty of Tech. of Fuel and Water* 4(1). Prague.
Štěpánek, M., et al. 1960. Application of algicide CA 350 in reservoirs. *Scientific Papers from Inst. of Chem. Tech., Faculty of Tech. of Fuel and Water* 4(2). Prague.
Strøm, K. M. 1926. Plankton from Finmark lakes. *Tromsoe Museums Aarshefter* 49(1).
Strøm, K. M. 1932. Tyrifjord, a limnological study. *Skrifter utgitt av Det Norske Videnskaps-Akademi i Oslo, I. Mat.-Naturv. Klasse* 1932, No. 3.
Taylor, E. W. 1961. Thirty-ninth report on the results of bacteriological, chemical and biological examination of the London waters for the years 1959-1960. Metropolitan Water Board.
Thienemann, A. 1928. Der Sauerstoff im eutrophen und oligotrophen See. Ein Beitrag zur Seetypenlehre. Die Binnengewässer, Vol. 4. Stuttgart.
Thomas, E. A., and Maerki, E. 1949. Der heutigen Zustand des Zürichsees. Verhandlungen der Internationalen Vereinigung für theoretische und angewandte Limnologie, Vol. 10.
Thomas, E. A. 1951. Produktionsforschungen auf Grund der Sedimente im Pfäffikersee und Zürichsee. Verhandlungen der Internationalen Vereinigung für theoretische und angewandte Limnologie, Vol. 11.
Thomas, E. A. 1957. Der Zürichsee, sein Wasser und sein Boden. Jahrbuch vom Zürichsee, Vol. 17.
Thresh, Beale, and Suckling. 1933. The examination of waters and water supplies. London.
Tonolli, V. 1961. Pubblicazioni dell Istituto Italiano di Idrobiologia. Pallanza.
United Nations. 1957. Water and the world today. A United Nations Review Reprint. New York.
Utermöhl, H. 1931. Neue Wege in der quantitativen Erfassung des Planktons. Verhandlungen der Internationalen Vereinigung für theoretische und angewandte Limnologie, Vol. 5.
Utermöhl, H. 1958. Zur Vervollkommung der quantitativen Phytoplankton-Methodik. Mitteilungen der Internationalen Vereinigung für theoretische und angewandte Limnologie, No. 9.
Willén, T. 1960. Phytoplankton algae from three Spanish lakes. *Svensk Botanisk Tidsskrift* 54(4).
World Health Organization. 1956. Water pollution in Europe. Geneva.
Zimmermann, P. 1961. Chemische und bakteriologische Untersuchungen im unteren Zürichsee während der Jahre 1948-1957. *Schweiz. Z. f. Hydrol.*, Vol. 23, Fasc. 2.
Züllig, H. 1956. Sedimente als Ausdruck des Zustandes eines Gewässers. *Schweiz. Z. f. Hydrol.*, Vol. 18, Fasc. 1.
Zune, A. J. 1894. Traité d'analyse chimique, micrographique et microbiologique des eaux potable. Paris and Brussels.

The Biotic Relationships within Water Blooms

George P. Fitzgerald

Hydraulic and Sanitary Laboratories
University of Wisconsin
Madison, Wisconsin

ABSTRACT

The sequence of algae blooms in an aquatic environment with periods of nearly unispecific conditions at certain times is discussed. It is pointed out that even in sewage stabilization ponds where there is an excess of the usually considered nutrients present such sequences of algae also occur. The rapid replacement of the dominant species by minor species when the former is removed by toxic chemical treatment of the water is considered to indicate the controlling factor for dominance of one species over others is very short-lived in the absence of the particular alga.

The effect of fertilization and organic pollution on the type of algae in an aquatic environment is discussed. Other factors controlling the species of algae in a bloom that are reviewed are predation by animals, attacks by bacteria or fungi, and the effects of extracellular products of certain algae in limiting the development of other species.

The many factors that might influence the development of an algal bloom, the duration of a bloom, and the ultimate decline of a bloom have been the subject of many research investigations, reviews, and symposia. The subject is so diverse that it cannot be dealt with on a simple fact-to-fact basis, but requires information from a great many fields of interest in order that a very general understanding of a very complex system can be evolved. Complete understanding may never be achieved, but by accumulating and digesting the information available to us from the fields of ecology, physiology, and chemistry, the search for specific items of fact in each separate field can be directed toward a common understanding rather than hard theories based only upon a very narrow field of study.

Reviews by McCombie (1), Rice (2), Brook (3), Tucker (4), and

Saunders (5) have pointed out the many fields of study that have been applied to this problem. The general factors that influence the development and degree of development of algae blooms will be reviewed here. Emphasis will be placed on the natural cycle of algal blooms and the factors which bring about changes in the biotic composition of a bloom.

The sequence of algae in an aquatic environment is well documented: the general spring bloom of diatoms, the early summer growth of green algae, and the blue-green algal blooms in the late summer (3, 4, and 6-9). A great deal of information has been collected with the view that such changes in plankton populations are dependent upon the available supply of nutrients. However, one must also be aware of sequences of algal blooms that occur under conditions of nutrient supply far in excess of those in natural lakes. Such a situation has been recorded in studies of the use of sewage stabilization ponds in stripping sewage treatment plant effluents of algal nutrients (10). During the period from April 1956 through August 1956, a series of blooms of algae in a half-acre pond were separated by periods of low algal activity that varied from a few days to three weeks (the latter part of June 1956) despite a continuous supply of nutrients in the form of secondary sewage effluent. The order of algal blooms was *Euglena, Chlamydomonas, Chlamydomonas* again, *Closteridium,* and finally *Eudorina.* Thus, nutrient supply, as we normally consider the level of phosphorus, nitrogen, and various other essential elements, cannot be the only factor controlling the dominance of one species over others.

Studies on the application of chemicals toxic to algae for the purpose of controlling obnoxious quantities of algae in lakes have demonstrated that frequently the predominant algal form would be greatly reduced in numbers and then some of the previously minor forms would rapidly multiply and become the excessive algal growths of the future (7 and 11-13). Such a sequence was recorded for a 27-acre lake in Wisconsin, Spaulding's Pond, that received a series of treatments with 2,3-dichloro-naphthoquinone for the control of blue-green algae. The sequence of dominant algae was: *Anabaena, Microcystis, Aphanizomenon, Microcystis, Aphanizomenon,* and finally a mixture of *Aphanizomenon* and *Microcystis.* The similarity of such artificially induced sequences of algal dominance to the natural sequences mentioned earlier and the sometimes very short interval of time between dominant species demonstrated that, while the combination of factors that allow one species of algae to hold dominance over other species may be extremely successful in that almost unispecific conditions may exist, this situation must diminish very rapidly under certain conditions, as has been pointed out by Lefèvre, Jakob, and Nisbet (14).

In studies of nature and nature as man affects it one must always be prepared for the exception to the frequent course of events. Ganapati (15) has recorded the fact that in many ponds in Madras, India, there are continuous blooms throughout the year of *Microcystis* of such abundance as to exclude almost all other species of algae. Studies of the algae of Lake Waubesa, Wis., have also shown that the dominance of one species of algae over all others may last throughout the summer growing period (July through October). This situation has not always been the case in this Madison lake. Early studies by Birge and Juday (6) showed that in 1916 the numbers of algae contributing to the algal bloom of Lake Waubesa increased from 10 species on May 10, 15 species on July 25, 20 species on September 19, to 25 species by October 30. In 1925, Domogalla (7) recorded blooms of *Aphanizomenon, Anabaena,* and a mixture of *Lyngbya* and *Microcystis.* Lackey (9) measured 19 blooms in 1942-1943 that were caused by at least 10 species of green, flagellate, diatom, and red algae plus at least one bloom of *Microcystis*. In contrast, during the period of 1951 and 1952, net samples of algae of Lake Waubesa indicated only the presence of *Microcystis* and *Aphanizomenon*. Similar samples collected in 1955 and 1957 (16) showed more than 99% of the algal population to consist of *Microcystis* for at least two months of the summer. During the period of time from 1951 through 1957, Lake Waubesa was the only lake of the Madison area directly receiving the secondary effluent of the Madison Sewage Treatment Plant.

Since December 1958, the effluent from the sewage treatment plant has been diverted around the Madison lakes rather than through Lakes Waubesa and Kegonsa. Algal samples of the Madison lakes collected during the summer of 1959 indicated that Lake Waubesa no longer was unispecific with *Microcystis*. Significant numbers of *Melosira, Oscillatoria, Ceratium,* and other algae were present, as had been the case in Lakes Monona and Kegonsa of the Madison chain of lakes during 1955, 1957, and 1959. Thus, the effect of fertilizing this lake with treated sewage appears to have been a factor in the growth of bloom-forming blue-green algae, as was predicted by Sawyer and Lackey in 1943 (17).

One cannot generalize too far from these data, however, since nutrition alone has not been shown to be the cause of the sequences of algae one frequently observes, as was pointed out in the discussion of sewage stabilization ponds. There is no question that pollution of an aquatic environment by organic chemicals will affect the type and numbers of algae present, but the question of why only certain organisms survive or thrive so as to exclude others has obviously not been completely answered (18 and 19).

A review of the factors other than nutrition that might influence the algal population of an environment would of course bring out the fact that algae are the prime source of organic material in an aquatic environment and as such could be included in the food chain of many organisms. A number of papers have presented evidence that algae are eaten by certain zooplankters (20-26) and the theories of the effect of such predation have been discussed (5 and 27-29).

There is evidence that an inverse relationship frequently exists between the density of phytoplankton and zooplankton. This might be the result of the plankton animals migrating into patches of phytoplankton and, when present in sufficient numbers, grazing these down very quickly. In the meantime fresh growths of phytoplankton will have occurred in neighboring areas now devoid of animals and an inverse relationship will be established (28).

Another theory of the reason for the inverse zooplankton-phytoplankton relationship mentioned suggests that certain phytoplankton may have an "exclusion" effect on zooplankton so that they migrate from high concentrations of phytoplankton (5, 27, and 29), the "exclusive" relationship being mediated by external products of the plants (27).

Since there is evidence (21 and 30) that algae can at times pass through the zooplankton without being affected by digestive processes, it is possible that the presence of other microorganisms, either bacteria or fungi, is required to make the organic matter of algae available. Canter and Lund (31) have studied the relation of a parasitic fungus to the numbers of *Asterionella* in English lakes, but it was not decided if the fungus caused a decrease in the diatom numbers or if the fungus became epidemic because the algae were made susceptible to it by other factors. Evidence by other authors on the relationship of algae and bacteria (32 and 33) would tend to support the theory that healthy algae were not subject to degradation by bacteria or fungi, while dead or weakened algae are rapidly decomposed by bacteria. Because of this resistance to bacteria, Waksman, Stokes, and Butler (32) have suggested that animal forms may be largely responsible for the destruction of living diatoms in the plankton and the role of bacteria consists of the destruction of dead diatoms as well as of the animal residues.

In situations where the algae are so abundant as to require their control by chemical means it appears that animal predation or attacks by microorganisms are not enough to cause a shift in the dominant species present, but once the dominant species is eliminated one or more other species increase in numbers and become the dominant forms to the exclusion of other species. The reason for only one species being present in an environment that will support another species as soon as the original inhabitant has been destroyed is thought by some authors to be due to the presence of controlling chemicals excreted by the algae (2, 14, and 34-39).

The large amount of evidence showing that certain algal cultures produce extra-cellular products that inhibit the development of other species of algae leaves no doubt that such products can have an influence in the ecology of blooms. However, the products which are so effective in some instances as to allow a unispecific condition to exist in a medium of excess nutrients (2 and 10) must be very short-lived in nature because the development of succeeding species of algae or aquatic plants is very rapid once the original dominant species disappears. It can be noted that 40% of the organic matter of a bacteria-free culture of *Chlorella* was found to be present as extracellular products, whereas in the presence of sewage bacteria (1 ml of sewage per liter of culture) only 8% of the organic matter of the culture was found to be extracellular (33). Thus, if the alga is not continually producing an algistatic product it may be that bacteria and fungi can rapidly lower its concentration below the critical effective level.

The nature of the extracellular products of algal cultures that inhibit the same species or other species has been the subject of many investigations (2, 5, 35, 36, 38, 40, and 41). It has been found that the products are heat labile (2, 35, and 38) and can be removed from culture filtrates by adsorption onto carbon (2 and 36). With continued study of this interesting subject it may be that man will be able to make use for his own purposes products closely related to those the algae have been found to produce.

LITERATURE CITED

1. McCombie, A. M. 1953. Factors influencing the growth of phytoplankton. *J. Fish. Res. Bd. Canada* 10:253-282.
2. Rice, T. R. 1954. Biotic influences affecting population growth of planktonic algae. *Fish. Bull. U.S.* 54:227-245.
3. Brooke, A. J. 1957. Water blooms. *New Biology* 23:86-101.
4. Tucker, A. 1957. The relation of the phytoplankton periodicity to the nature of the physico-chemical environment with special reference to phosphorus. *Am. Midland Naturalist* 57:334-370.
5. Saunders, G. W. 1957. Interrelations of dissolved organic matter and phytoplankton. *Botan. Rev.* 23:389-409.
6. Birge, E. A., and Juday, C. 1922. The inland lakes of Wisconsin, the plankton. *Wis. Geol. and Nat. Hist. Survey, Bull.* No. 64, Sci. Series No. 13.
7. Domogalla, B. P. 1926. Treatment of algae and weeds in lakes at Madison, Wisconsin. *Eng. News Record* 97:950-954.
8. Pennak, R. W. 1949. Annual limnological cycles in some Colorado reservoir lakes. *Ecol. Monographs* 19:233-267.
9. Lackey, J. B. 1945. Plankton productivity of certain south eastern Wisconsin lakes as related to fertilization, II Productivity. *Sew. Works. J.* 17:795-802.
10. Fitzgerald, G. P. 1961. Stripping effluents of nutrients by biological means. Algae and Metropolitan Wastes. Trans. 1960 Seminar, R. A. Taft S.E.C., US P.H.S. Tech. Report W61-3.
11. Domogalla, B. 1935. Eleven years of chemical treatment of the Madison lakes: its effect on fish and fish foods. *Trans. Am. Fish Foc.* 65:115-121.
12. Rodhe, W. 1949. Die Bekämpfung einer Wasserblüte von *Microcystis* und die gleichzeitige Förderung einer neuen Hochproduktion von *Pediastrum* im See Norrviken bei Stockholm. *Proc. Intern. Assoc. Limnol.* 10:372-376.
13. Fitzgerald, G. P., and Skoog, F. 1954. Control of blue-green algae blooms with 2,3-Dichloronaphthoquinone. *Sew. and Ind. Wastes* 26:1136-1140.
14. Lefèvre, M., Jakob, H., and Nisbet, M. 1950. Compatabilités et antagonismes entre algues d'eau douce dans les collections d'eau naturelles. *Proc. Intern. Assoc. Limnol.* 11:224-229.
15. Ganapati, S. V. 1941. Ecology of a temple tank containing a permanent bloom of *Microcystis aeruginosa*. *J. Bombay Nat. Hist. Soc.* 42:65-77.
16. Lawton, G. W. 1961. The Madison lakes before and after diversion. pp. 108-117. Algae and Metropolitan Wastes. R. A. Taft S. E. C. Tech. Report W61-3.
17. Sawyer, C. N., and Lackey, J. B. 1943. Investigation of the odor nuisance occurring in the Madison lakes, particularly Monona, Waubesa, Kegonsa from July 1942 to July 1943. Governor's Committee.
18. Tarzwell, C. M. 1957. Biological problems in water pollution. Trans. 1956 Sem. R.A. Taft S.E.C., US P.H.S.
19. Tarzwell, C. M. 1960. Biological problems in water pollution. Trans. 2nd Sem. R.A. Taft S.E.C., US P.H.S. Tech. Report W60-3.
20. Wright, J. C. 1958. The limnology of Canyon Ferry Reservoir. I. Phytoplankton-zooplankton relationships in the euphotic zone during September and October, 1956. *Limnol. and Oceanog.* 3:150-159.
21. Gibor, A. 1957. Conversion of phytoplankton to zooplankton. *Nature* 179:1304.

22. Marshall, S. M., and Orr, A. P. 1955. On the biology of *Calanus finnmarchieus.* VIII Food uptake, assimilation and excretion in adult and stage V *Calanus. J. Mar. Biol. Assoc. U-K* 34:495-529.
23. Loosanoff, V. L., Hanks, J. E., and Ganaros, A. F. 1957. Control of certain forms of zooplankton in mass algal cultures. *Science* 125:1092-1093.
24. Lear, D. W., Jr., and Oppenheimer, C. H., Jr. 1962. Consumption of microorganisms by the copepod *Tigriopus Californicus. Limnol. and Oceanog.* VII (Suppl.):63-65.
25. Pennington, W. 1941. Control of the numbers of fresh water phytoplankton by small invertebrate animals. *J. Ecology* 29:204-211.
26. Fleming, R. H. 1939. The control of phytoplankton populations by grazing. *J. Conseil Exploration de la mer* 14:1-20.
27. Lucas, C.E. 1947. The ecological effects of external metabolites. *Biol. Rev. Cambr. Phil. Soc.* 22:270-295.
28. Bainbridge, R. 1953. Studies on the interrelationship of zooplankton and phytoplankton. *J. Mar. Biol. Assn. U-K.* 32:385-445.
29. Hardy, A. C. 1936. Plankton ecology and the hypothesis of animal exclusion. *Proc. Linn. Soc. Lond.* 148:64-70.
30. Welch, P. S. 1935. Limnology. McGraw-Hill Book Co., Inc., New York. 471 pp.
31. Canter, H. M., and Lund, J. W. G. 1948. Studies on plankton parasites. I. Fluctuations in the numbers of *Asterionella formosa* Hass. in relation to fungal epidemics. *New Phytol.* 47:238-261.
32. Waksman, S. A., Stokes, L. J., and Butler, M. R. 1937. Relation of bacteria to diatoms in sea water. *J. Mar. Biol. Assn.* 22:359-373.
33. Fitzgerald, G. P., The effect of algae on B.O.D. measurements. I. The factors affecting the measurement of algal B.O.D. (Submitted 1962.)
34. Akehurst, S. E. 1931. Observations on pond life, with special reference to the possible causation of swarming of phytoplankton. *Roy. Micro. Soc. J.* 51:237-265.
35. Swanson, C. A. 1943. The effect of culture filtrates on respiration in *Chlorella vulgaris. Am. J. Botany* 30:8-11.
36. Pratt, R. 1944. Influence on the growth of *Chlorella* of continuous removal of chlorellin from the solution. *Am. J. Botany* 31:418-421.
37. Rodhe, W. 1948. Environmental requirements of fresh-water plankton algae. *Symbolae, Botanicae Upsalienses* 10:1-149.
38. Lefèvre, M., Nisbet, M., and Jakob, E. 1948. Action des substances excrétées en culture, par certaines espèces d'algues sur le métabolisme d'autres espèces d'algues. *Proc. Intern. Assoc. Limnol.* 10:259-264.
39. Jørgensen, E. G. 1956. Growth inhibiting substances formed by algae. *Physiologia Plant.* 9:712-726.
40. Shilo, M., and Aschner, M. 1953. Factors governing the toxicity of cultures containing the phytoflagellate *Prymnesium parvum. J. Gen. Microbiol.* 8:333-343.
41. Fogg, G. E., and Westlake, D. F. 1955. The importance of extracellular products of algae in freshwater. *Proc. Intern. Assoc. of Theoretical and Applied Limnol.* 12:219-232.

Toxic Algae*

Paul R. Gorham

Division of Applied Biology
National Research Council
Ottawa, Canada

ABSTRACT

Laboratory study has provided a better understanding of the reasons for the variable toxicity of waterblooms of blue-green algae. At least two toxins, produced by strains of *Microcystis aeruginosa* and *Anabaena flos-aquae,* are responsible for acute poisonings of various animals. Bacteria associated with the algae also produce toxins that are responsible for less acute poisonings. The effects of algal and bacterial toxins may be superimposed. *Microcystis* fast-death factor is a cyclic polypeptide of moderate toxicity which kills livestock and other animals but not waterfowl. The structure of *Anabaena* very-fast-death factor is not yet known. It kills a variety of animals, including waterfowl. Variable toxicity of waterblooms is determined by: 1) dominance by toxic strains of algae or bacteria, or both; 2) concentration of toxic organisms; 3) release of toxin(s); and 4) consumption of toxin(s) in sufficient amounts by susceptible animals before appreciable dilution, adsorption or destruction occurs.

INTRODUCTION

In June 1962 I received a report of algal poisoning from J. G. O'Donoghue, D.V.M., Veterinary Services Branch, Department of Agriculture, Edmonton, Alberta. A farmer whose land bordered a shallow lake about two square miles in area had one horse and two cows die very suddenly. The veterinarian who was called found 66 head of cattle quite sick. Six of them were very sick, had algal stains on their feet and muzzles, and died within 24 hours. The veterinarian ordered the animals removed from access to the lake, which had a bloom of blue-green algae, and there was no further

* Issued as N.R.C. No. 7804.

trouble. He diagnosed this as a case of algal poisoning. It is a fairly typical case history of the sort which I am going to discuss. It illustrates the kind of circumstantial evidence that, until fairly recently, was all there was to indicate that a few species of planktonic blue-greens are, upon occasion, poisonous to livestock, horses, cattle, swine, chickens, waterfowl, and various other animals (see reviews by Fitch et al., 1934; Olson, 1951; Grant and Hughes, 1953; Vinberg, 1954; Ingram and Prescott, 1954; and Schwimmer and Schwimmer, 1955).

There are indications that algae are sometimes toxic to man. A suspicious connection between algal pollution and gastrointestinal disorders has been noted (Tisdale, 1931a,b; Veldee, 1931; Dillenberg and Dehnel, 1960; and Senior, 1960), but actual proof that the algae were directly responsible is lacking. Poisoning of fish by blue-green algae has also been reported (Carl, 1937; Prescott, 1948; Mackenthun, Herman, and Bartsch, 1948; Lefèvre, Jakob, and Nisbet, 1952; Shelubsky, 1951; and Astakhova, Kun, and Teplyi, 1960). In many cases of fish kill, death is caused by anoxia resulting from decomposition of large masses of algae under shallow water conditions. In other cases, fish have been killed while there was still an adequate oxygen supply so that poisonous substances have been implicated.

There are other kinds of algae besides blue-green algae that are poisonous. Fish kills are caused by the yellow-brown alga *Prymnesium parvum* (Otterstrom and Steemann-Nielsen, 1939; and Shilo and Aschner, 1953) and by the dinoflagellates *Gymnodinium veneficum* (Abbott and Ballantine, 1957), *Gymnodinium brevis* (Ray and Wilson, 1957), and *Gonyaulax monilata* (Gates and Wilson, 1960). Shellfish poisoning of humans is caused by the dinoflagellates *Gonyaulax tamarensis* (Medcof et al., 1947) and *Gonyaulax catenella* (Riegel et al., 1949). These latter organisms produce a poisonous substance which accumulates in the digestive glands of the shellfish that feed upon them and if man or other animal eats the shellfish, paralysis and quite frequently death ensues. Species of blue-green algae have been found to poison other algae (Lefèvre et al., 1952; Jakob, 1954; and Proctor, 1957a,b), and there are indications of toxicity to zooplankton as well (Braginskii, 1955; and Dillenberg and Dehnel, 1960). However, it is the toxicity of the blue-green algae to higher animals that has concerned me most and on which

I am going to concentrate in this paper. The Drs. Schwimmer are going to discuss the medical aspects.

One or two case histories will serve to indicate how serious intoxications or poisonings by waterblooms of blue-green algae can be. At Storm Lake, Iowa, in October and November 1952, one of the most serious outbreaks of algal poisoning on record occurred (Rose, 1953). A series of blooms of *Anabaena flos-aquae* caused an estimated loss of 5000 to 7000 Franklin's gulls, 560 ducks, 400 coots, 200 pheasants, 50 fox squirrels, 18 muskrats, 15 dogs, 4 cats, 2 hogs, 2 hawks, 1 skunk, and 1 mink. In South Africa, in 1943, thousands of cattle, sheep, and many other animals were killed along the Vaal dam in the Transvaal when the reservoir developed a poisonous bloom of *Microcystis* (Steyn, 1945; and Stephens, 1945). Louw (1950) carried out a detailed study of the toxic principle involved in these *Microcystis* poisonings. The workers in South Africa concluded that the poisonous substance was an unidentified alkaloid that affected the central nervous system and liver. They also identified a secondary poisoning caused by the phycobilin pigments of this alga. They found that these pigments accumulated in the skin of animals that regularly drank water polluted with this alga and caused an increase in photosensitivity. This, in turn, produced a sloughing of the skin, general weakening of condition, and occasional deaths.

Poisonings by waterblooms of blue-green algae are virtually worldwide in occurrence. They have been most frequently reported in central North America, especially from the states of North Dakota, South Dakota, Minnesota, Iowa, Wisconsin, Illinois, and Michigan and the provinces of Alberta, Saskatchewan, Manitoba, and Ontario. There is a good review by Vinberg (1954) on the situation as it has been found in Russia. It differs in no important respects from that which we observe here. Algal poisonings have also been reported from Argentina (Mullor, 1945); Australia (Francis, 1878); Israel (Shelubsky, 1951); Morocco (Lefèvre et al., 1952); Bermuda (Ingram and Prescott, 1954); Brazil (Branco, 1959), and Finland (Hinderson, 1933).

Six species of blue-green algae* have been incriminated in these intoxications (Fig. 1): *Nodularia spumigena* Mert. (not shown), *Microcystis aeruginosa* Kütz. *emend.* Elenkin, *Coelosphaerium Kützingianum* Nägeli, *Gloeotrichia echinulata* (J. E. Smith) Richter,

*The nomenclature adopted is that of Prescott (1961).

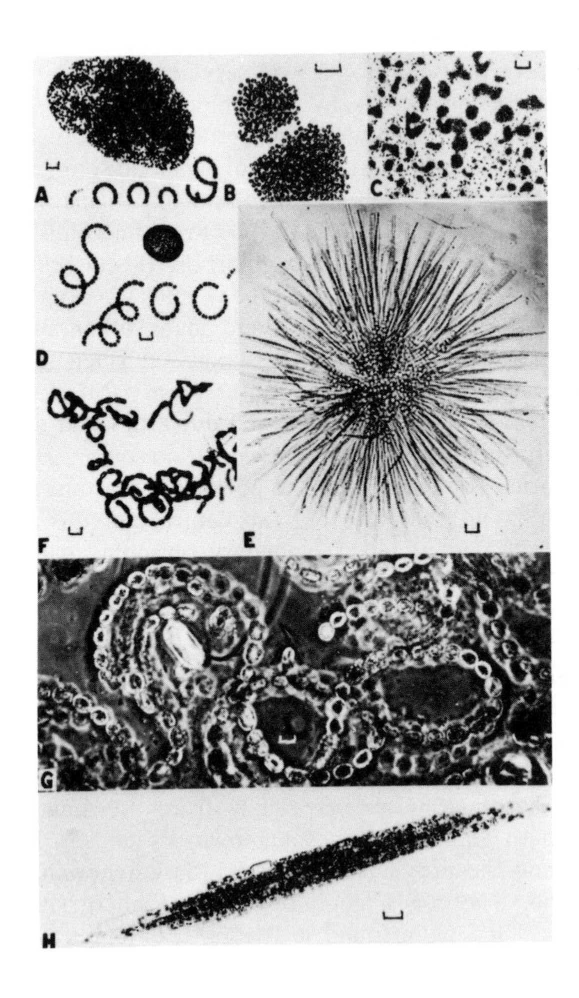

FIG. 1. Species of blue-green algae suspected of toxicity. A) *Microcystis aeruginosa* colony (top) with nontoxic *Anabaena spiroides* (bottom). Scale = 20 μ. B) *Microcystis aeruginosa* colony. Scale = 20 μ. C) *Microcystis aeruginosa* colony, with subcolonies in pronounced gelatinous sheath. Scale = 100 μ. D) *Coelosphaerium Kützingianum* colony (top) with nontoxic *Anabaena spiroides* (bottom). Scale = 20 μ. E) *Gloeotrichia echinulata* colony. Scale = 20 μ. F) *Anabaena flos-aquae* filaments. Scale = 20 μ. G) *Anabaena flos-aquae* filaments. Scale = 10 μ. H) *Aphanizomenon flos-aquae* colony. Scale = 10 μ. (From Gorham, 1960.)

Anabaena flos-aquae (Lyngb.) de Bréb. (including *An. Lemmermannii*), and *Aphanizomenon flos-aquae* (L.) Ralfs. Of these six species it is primarily *Microcystis* and *Anabaena* and, to a lesser extent, *Aphanizomenon* that have repeatedly been blamed for the most dramatic and serious poisonings. Waterblooms consisting predominantly of one or another of these species vary greatly in toxicity. Poisonings have occurred one day, one week, or one season and not the next. This variability is sometimes correlated with a change in species composition and sometimes it is not—especially if it occurs rapidly. The symptoms and survival times vary from one case to another. When one studies the records one finds that some animals, including a full-grown cow, have died in half an hour or less, while others have died in 24 to 48 hr. Algal poisoning is, therefore, an intrinsically interesting problem in toxicology since there are few poisons known that can kill a large animal in less than half an hour.

LABORATORY STUDY

When we began our investigations about eight years ago we decided that it would probably be most valuable to take this problem into the laboratory to see if we could isolate and grow the algae and prove that they were actually responsible for the poisonings. We also wished to find out how many toxins were involved, and, if possible, to try and identify them. Finally, we hoped to provide an explanation for variable toxicity.

Microcystis aeruginosa

We first had to learn how to isolate and grow cultures of planktonic blue-green algae. We made a number of colonial isolates of *Microcystis aeruginosa* in the medium of Fitzgerald, Gerloff, and Skoog (1952) and incubated them under continuous illumination from "white" fluorescent lamps. A few grew slowly and were unialgal. Since we were aiming at producing rather large quantities, we attempted to improve this medium, using the first unialgal strain we had isolated, called NRC-1 (Hughes, Gorham, and Zehnder, 1958). By a better balancing of the concentration of the major salts and a change of chelator and minor elements we arrived at Medium No. 11, which gave much higher yields.

Strain NRC-1 was toxic to mice when administered orally or

intraperitoneally. With a minimal lethal dose the survival time was 30 to 60 min. The symptoms were violent convulsions and pallor, followed by prostration and death. Immediate autopsy revealed a loss of peripheral circulation and an engorged, mottled-looking liver. Steyn (1945) and Louw (1950) in South Africa had indicated that *Microcystis* toxin was a liver poison. We made one rather interesting observation and that was that when healthy cells were injected into mice they caused no effect whatever. Only when cells were frozen and thawed, mechanically disintegrated, or decomposed by semianaerobic incubation to release the toxin did poisoning occur. As already mentioned, the toxin was active when given by either the oral or intraperitoneal route and it appeared to be, as judged by the symptoms produced in laboratory animals, the toxin we were looking for. The evidence just presented indicated that it is an endotoxin that is ordinarily released when the algae grow old or decompose.

We scaled up production of *M. aeruginosa* NRC-1 by stages until we were finally producing a continuous laboratory waterbloom. We first used 3.5-liter tower fermentors with fluorescent lamps on opposite sides (Fig. 2A; Zehnder and Gorham, 1960), adapting conventional microbiological methods to the production of these algae. We were aiming at getting enough material to do some meaningful chemistry on the isolation and identification of the toxin, and thought this scale would probably be adequate. We believed that we had *M. aeruginosa* NRC-1 bacteria-free, having used the purification procedures then current, and so the fermentors were hopefully designed for sterile culture. We could introduce sterile nutrient at the top and take samples of algae under sterile conditions from the bottom at suitable intervals, thereby maintaining the culture in vigorous growth for weeks at a time. Figure 2B is a schematic diagram of a fermentor assembly. The air was passed through a drying tower containing silica gel to make sure that no extraneous bacteria or molds had sufficient moisture to grow in the sterile cotton filter that came ahead of the fritted glass plate in the bottom of the fermentor. On the fermentor there was an input for nutrient at the top and a sampling outlet at the bottom. The condenser at the top served as a foam breaker. At first we plugged the top with cotton, but we were always bothered by foaming as the culture aged. This is caused, we believe, by secretion of materials such as polypeptides and polysaccharides

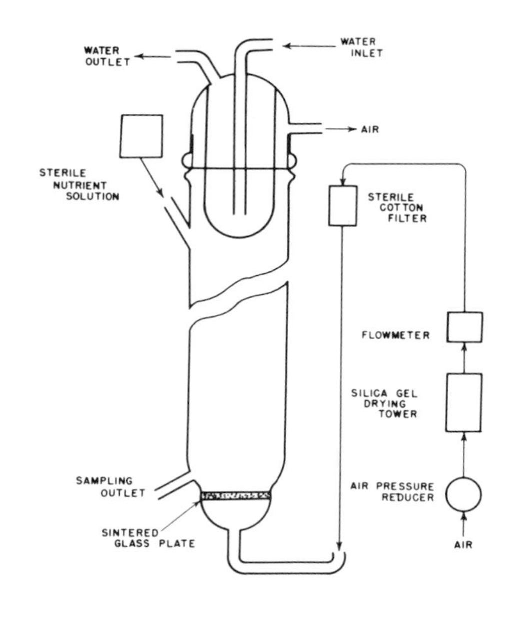

FIG. 2. A) Two cultures of *Microcystis aeruginosa* NRC-1, in 3.5-liter tower fermentors continuously illuminated by a pair of twin 40-w "white" fluorescent lamps. B) Diagram of tower fermentor with foam-breaking condenser top and auxiliary equipment. (From Zehnder and Gorham, 1960.)

into the medium. By putting the condenser on top, we discovered that the foam would break and we could grow cultures to maturity satisfactorily.

With 1-liter tower fermentors we investigated growth and toxin production by *M. aeruginosa* NRC-1 (Harris and Gorham, 1956). Figure 3 summarizes the effects of time and temperature on growth (measured as dry weight per liter corrected for nutrient salts) and toxin production (measured as mouse units—the minimal dose to kill a 25-g mouse). Under the conditions used, the alga had a thermal growth optimum of about 32.5°C. Growth was poor at 35°C. By contrast, the optimum temperature for the production and accumulation of toxin was 25°C, and this reached a peak in about 4 days. At 32.5°C, very little toxin, if any, was produced, or if it was produced it was rapidly destroyed. The bulk of the toxin was recovered from the cells; only traces were recovered from the medium. Even at 25°C, when there was a big decline in toxin concentration in the cells after 5 or 6 days, there was no increase in

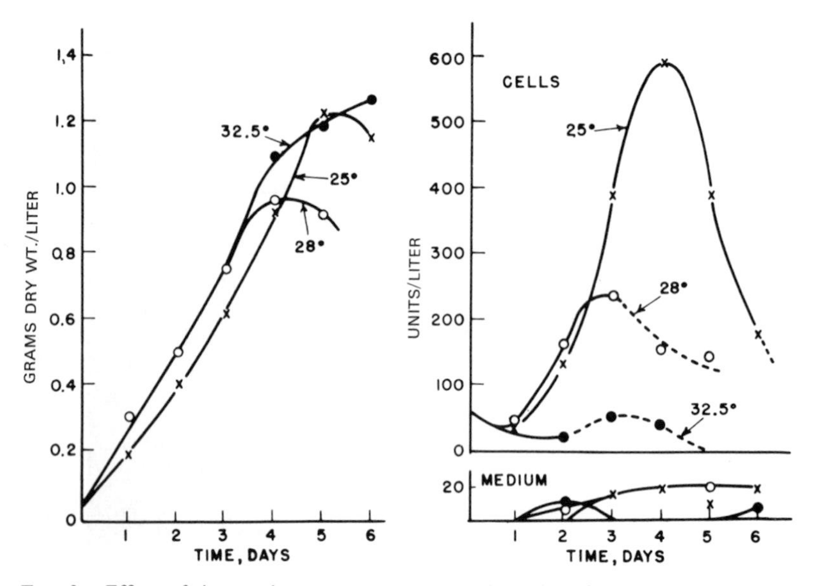

FIG. 3. Effect of time and temperature on growth (yield) of *Microcystis aeruginosa* NRC-1 in a tower fermentor and on the amount of the fast-death factor present in the cells and the medium. (From Harris and Gorham, 1956.)

toxin recovered from the medium. This was of practical significance later on, since it meant that we had only to work up the cell fraction and not the bulky medium to obtain most of the available toxin.

Figure 4 summarizes the effects of light intensity, temperature, and aeration on growth (3-day yields) and toxin accumulation by *M. aeruginosa* NRC-1. There was a distinct effect of light intensity on growth (top graphs) in these tower fermentors. The thermal growth optimum increased from about 30°C to 35°C at the higher of the two light intensities employed, regardless of aeration rate. Toxin accumulation, however, was greatest at 25°C with the higher aeration rate, regardless of light intensity. The fact that we could produce toxin in well-aerated cultures at 25°C just about as well at low light intensity as at high light intensity was of great practical significance when we later had to mass culture the alga on a much larger scale. We tried extra carbon dioxide to boost the yields and toxin production still higher but found that algal growth and,

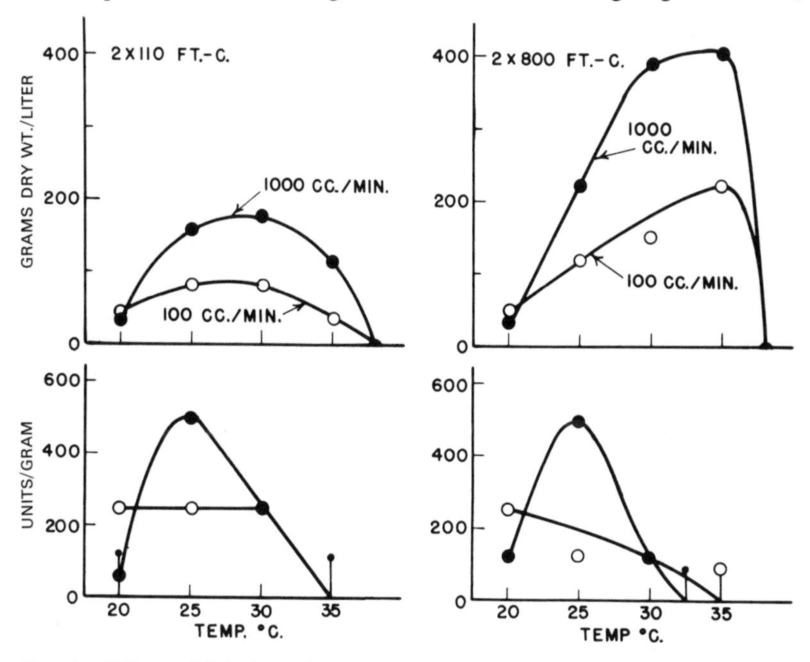

FIG. 4. Effects of light intensity, temperature and aeration on growth (3-day yield) of *Microcystis aeruginosa* NRC-1 in a tower fermentor (top) and on the amount of fast-death factor present in the cells (bottom). (From Harris and Gorham, 1956.)

indirectly, toxin production were inhibited by 1.0% carbon dioxide in the air stream. This was partly caused, perhaps, by a drop in pH to unfavorable levels. However, Medium No. 11 is moderately well-buffered with phosphate and silicate so the pH did not change very drastically with the extra carbon dioxide.

At about this time we realized that production of toxic *M. aeruginosa* NRC-1 by the tower fermentors was inadequate to meet our needs for the isolation and identification of the toxin. We therefore set up an "algal factory" (Bishop, Anet, and Gorham, 1959) based on the facts we had learned about toxin production in the tower fermentors. We used 9-liter Pyrex bottles as culture vessels. Batteries of these were placed in double rows on either side of three rows of "white" fluorescent lamps. Cultures were stirred as well as aerated by means of compressed air passed through sintered glass filter sticks at a rate of about 2 liters/min. Figure 5 is a picture of our "algal factory" when we had about 300 bottles growing at one time and were inoculating and harvesting 40 to 50 bottles a day. The inoculum was carefully grown under aseptic conditions, but during the final stage the algae were grown in unautoclaved medium. From this laboratory waterbloom we were able to harvest between one and two kilograms of freeze-dried cells per month.

With this amount of toxic alga on hand we were able to conduct some large animal tests to make sure that we were not working with an artifact. Although the symptoms in laboratory animals were right, we wanted to take it a step further to see what the effects would be on livestock. In collaboration with Drs. H. V. Konst and P. D. McKercher, P. R. Robertson, and J. Howe (1959) of the Animal Diseases Research Institute, Canada Department of Agriculture, Hull, P. Q., we were able to do this. We tested the freeze-dried alga on a sheep, a calf, guinea pigs, rabbits, chickens, and domestic ducks. We were interested in whether *Microcystis* poisoning had anything to do with the waterfowl sickness problem. The results with the sheep and the calf as well as with larger laboratory animals confirmed that this was, indeed, the toxin that we were looking for (Gorham, 1960). The symptoms were the same as those reported in case histories implicating *Microcystis.* To further clinch the matter, we sent a sample of freeze-dried alga to Professor D. G. Steyn in Pretoria, South Africa. He very kindly conducted tests on it and reported that the effects produced on laboratory animals were indistinguishable from those produced by

FIG. 5. "Algae factory" for large-scale culture of *Microcystis aeruginosa* NRC-1. Three hundred aerated cultures in 9-liter bottles continuously illuminated by three rows of fluorescent lamps situated between double rows of bottles. (From Gorham, 1960.)

the *Microcystis* that they had studied in the early 1940s. We concluded, therefore, that we were dealing with one of the major algal poisons. I say algal poison but, of course, we had not yet proven it was really the alga that was responsible for toxin production. I shall come to that a little later. Of all the animals tested only the ducks were resistant. This established that *Microcystis* poisoning is not connected with waterfowl sickness.

Proceeding on the supposition that it was an algal poison that we were dealing with, we set about isolating and identifying it. You may remember that I mentioned that *Microcystis* toxin had been thought to be an alkaloid. We proceeded on this assumption for quite a while. It is a nonvolatile substance that is irreversibly adsorbed on charcoal. It is also very strongly adsorbed on cellulose so that chromatography becomes very difficult. It is adsorbed on the acid form of resins but not on the hydroxyl form. We found that it actually exists in two forms which we later identified as the free acid and as the salt. It is water- and alcohol-soluble but not soluble in such solvents as acetone, ether, chloroform, or benzene. Toxic extracts had no antibiotic activity when tested by standard procedures against a number of different bacteria. It proved to be quite a difficult task to get the toxin out in any quantity and decide what class of compound it belonged to. To make a long story short, we eventually discovered that it is not an alkaloid but a polypeptide (Bishop et al., 1959). Moreover, it is one of five closely related polypeptides that are extracted together from the alga. We were able to resolve this mixture by paper electrophoresis in borate buffer. Figure 6 shows an electrophorogram on which the five polypeptides are clearly distinguished. Number 2 is the toxic polypeptide. The other four polypeptides are nontoxic.

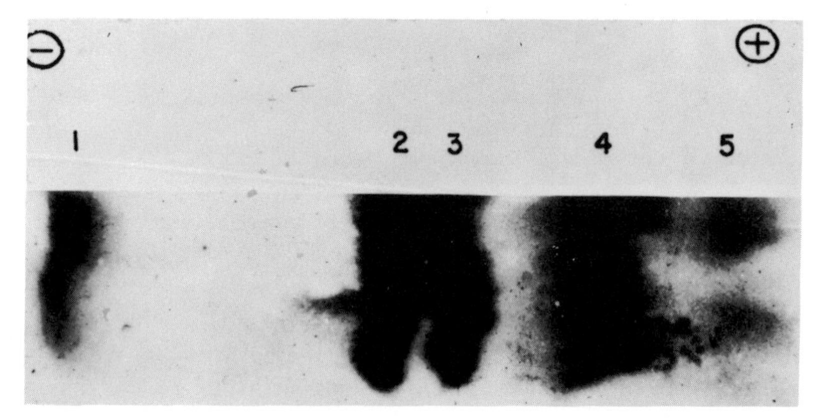

FIG. 6. Electrophorogram of toxic and nontoxic polypeptides from *Microcystis aeruginosa* NRC-1 on Whatman 3MM paper, 0.1M borate buffer, pH 9.1. (From Bishop, Anet, and Gorham, 1959.)

Upon hydrolysis, all five polypeptides were found to have five amino acids in common (Table I). Numbers 3, 4, and 5 also have a number of unknowns. The similarities in amino acid composition accounted for the great difficulty we had in finding a way to separate these peptides. Table II shows the proportional composition of the toxic polypeptide. This is what we have termed *Microcystis* fast-death factor (Hughes et al., 1958). It is composed of seven different amino acids. The molecular weight must be low since it dialyzes slowly through Visking casing. This means that there are probably no more than 10 or 20 amino acids in the molecule and that the molecular weight is of the order of 1300 to 2600. It has a D-serine instead of the normal L-serine. This was shown by D-amino acid oxidase on chromatograms of the hydrolyzed constituents. The dinitrofluorobenzene test for end-groups indicated that the toxin has a cyclic structure. We suspect, but do not know, that its toxicity is a consequence of possessing D- instead of L-serine, L-ornithine, or a cyclic structure, since these features have been correlated with biological activity in other peptides of natural occurrence (Bishop et al., 1959). We isolated enough of the purified toxin to conduct an adequate test of its intraperitoneal toxicity toward mice, and found that the LD_{50} (which, interestingly, is almost the same as the minimal lethal dose) was only 0.5 mg/kg body weight. *Microcystis* fast-death factor is, therefore, only a moderately toxic substance. This has some significance for the interpretation of case histories, as we shall see later.

TABLE I

Amino Acids Present in Five Polypeptides from *Microcystis aeruginosa* NRC-1. (From Bishop, Anet, and Gorham, 1959.)

Amino Acids	Polypeptide				
	1	2	3	4	5
Aspartic	+	+	+	+	+
Glutamic	+	+	+	+	+
Serine	+	+	+	+	+
Valine	+	+	+	+	+
Ornithine	+	+	+	+	+
Alanine	–	+	+	+	+
Leucine	–	+	–	–	–
No. of unknowns	0	0	4	4	3

TABLE II

Proportional Composition of the Toxic Polypeptide from *Microcystis aeruginosa* NRC-1. (From Bishop, Anet, and Gorham, 1959.)

Amino Acid	Proportion
Aspartic	1
Glutamic	2
D-Serine	1
Valine	1
Ornithine	1
Alanine	2
Leucine	2

We isolated other strains of *Microcystis* and tested them for toxicity. To our surprise, we discovered that there were toxic and nontoxic strains (Simpson and Gorham, 1958a; and Gorham, 1960, 1962). To date, we have made a total of 28 unialgal isolates and have found that fewer than one third of them produce the fast-death factor.

When we examined the purity of *M. aeruginosa* NRC-1 more critically we discovered that our so-called pure culture was contaminated by at least five different kinds of bacteria (Simpson and Gorham, 1958a). By devising appropriate bacteriological media and culture conditions, we managed to grow each of them. We attempted to purify the alga by a variety of means, including antibiotics, antiseptics, washes, sheath removal, bacteriostatic agents, ultraviolet light, and single-cell isolation. Although we found several ways to reduce the number of contaminants we could not eliminate them altogether. We were forced, therefore, to use roundabout methods to establish that it was the alga and not the bacteria that was primarily responsible for the production of the fast-death factor. We showed this by growing and testing each of the bacterial contaminants for toxicity. They were either nontoxic or else produced different symptoms or survival times than the fast-death factor. By differential centrifugation of a large batch of algae we were able to get an algae-rich and a bacteria-rich fraction. When tested on mice we found that all of the toxicity was associated with the algae-rich fraction. When a nontoxic strain, *Microcystis aeruginosa,* Wisc. 1036, was grown in the presence of the bacterial contaminants from strain NRC-1, it still remained nontoxic. As I have already mentioned, we tried single-cell isolations as one method of getting pure cultures (Simpson and Gorham, 1958b). This was done with a micromanipulator and a sterilized isolation chamber of the design shown in Fig. 7. Hanging from the under side of the coverglass were drops of sterile medium. A single *Microcystis*

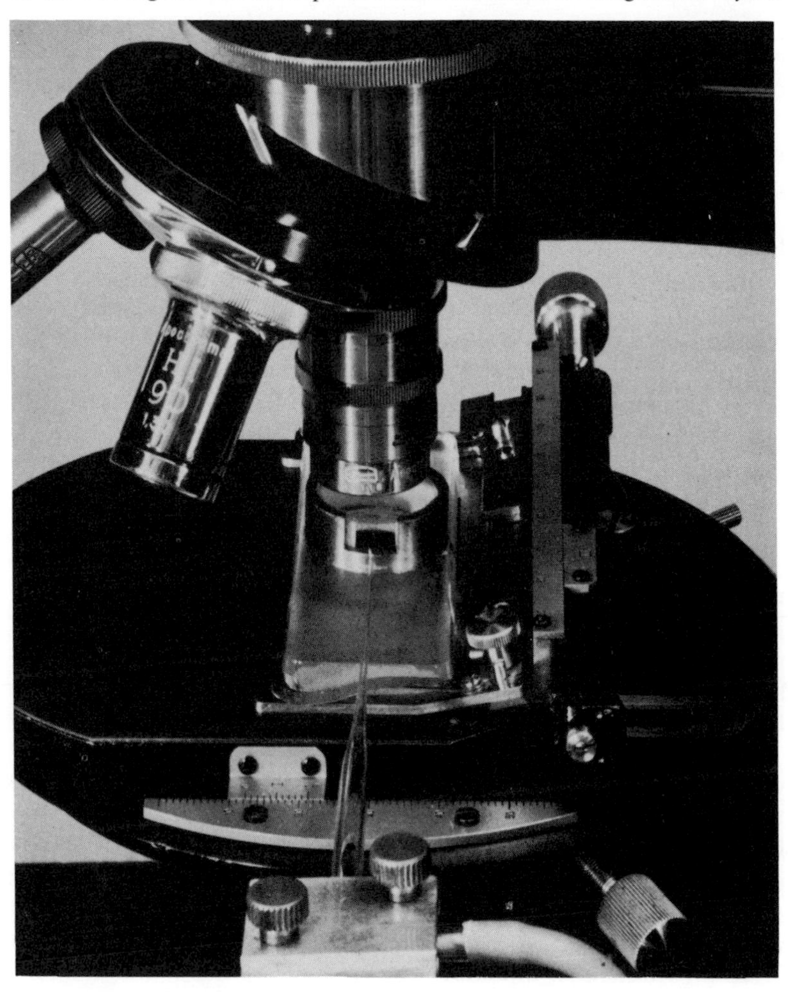

FIG. 7. Brass ring with coverglass sealed on top to form a moist chamber for isolating and washing single cells of *Microcystis aeruginosa* NRC-1. In the foreground is the micropipette clamped in a micromanipulator that is used for these operations. (From Simpson and Gorham, 1958b.)

cell was washed 8 to 14 times by transfer with a micropipette into these drops of sterile medium and then inoculated into shake-flasks and incubated for three to six months. We made 35 such isolates but managed to grow only two and these proved to be contaminated. The remaining 33 made either very limited or no growth in six months. Although the results were disappointing as far as purification was concerned, we were very fortunate in another respect: One of the two clones that did grow was toxic and the other was nontoxic! This established the fact that not only could toxic and nontoxic strains of *Microcystis* be derived by colony isolation from the same bloom but also that strain NRC-1 of *Microcystis* is genetically heterogeneous for toxin production.

The nature of the association of the bacteria with the alga looks suspiciously like symbiosis. Whenever we reduced the bacterial population to the point of elimination or got anywhere near it, the alga would either fail to grow or grow very poorly. We attempted to discover what growth factors the bacteria might be contributing to the algae, with the hope that we could then eliminate the bacteria successfully. Besides known vitamins and growth factors for higher plants we tested such things as peptones, caseitones, yeast extracts, corn-steep liquor and blood—either singly or in various combinations. In no case did we observe early growth stimulation of the alga. We reasoned that this might reveal a need for some growth factor even in the presence of bacteria that were supplying the same growth factor. More recently, a microbiologist colleague, Dr. C. Quadling, has again attempted purification of *M. aeruginosa* NRC-1 without success, using a wider selection of antibiotics and a variety of antimetabolites. The problem of how to purify this alga is still open.

We had long observed with Medium No. 11 (which we had devised for heavy yields of *M. aeruginosa* NRC-1 from large inocula) that there were pronounced lag phases and large replicate variability. We have recently turned our attention to balancing the mineral medium more carefully so that there will not be any lag phase or significant replicate variability from small inocula. We have devised ASM medium that accomplishes this for *M. aeruginosa* NRC-1 (McLachlan and Gorham, 1961, 1962). ASM medium contains all ingredients within the optimum concentration range and does not precipitate during autoclaving. We intend to use it in further attempts to purify these algae.

Aphanizomenon flos-aquae

Leaving *Microcystis* for the time being, let us turn our attention to some of the other species of blue-green algae that have been implicated in algal poisonings. With the knowledge gained from our experience with *Microcystis* we now knew that we could expect to find toxic strains of these algal species, too. We therefore changed our tactics somewhat and endeavored, first of all, to locate toxic blooms of these other species. From these we planned to grow as many isolates as possible to give us the best possible chance of finding any toxic strains that might be present.

Phinney and Peek (1961) had reported that blooms of *Aphanizomenon flos-aquae* in Klamath Lake, Oregon, were highly toxic at times. In the summer of 1960, with the cooperation of Professor Phinney and Mr. Peek, of the Department of Biology, University of Oregon, we arranged to have iced samples of bloom flown from Klamath Lake to Ottawa. They consisted of about 50% *Aphanizomenon* and 50% *Microcystis.* Professor Phinney and Mr. Peek kindly gave us a dried sample of toxic bloom of similar composition that they had collected from Klamath Lake in 1957. By the time we received the fresh samples they had begun to decompose and none of the many isolates that we made grew. Both the 1957 and 1960 samples were toxic when tested on mice. The symptoms and survival times were alike and indistinguishable from those produced by *Microcystis* fast-death factor. The amount of toxin present as well as the symptoms could be adequately explained by the presence of 50% *Microcystis* in the bloom. This strongly suggested that *Aphanizomenon* was not toxic after all, or if it was toxic, that it produced the same kind of toxin as *Microcystis.* We have succeeded in isolating 10 strains of *Aphanizomenon* from sources in Ontario and Saskatchewan (McLachlan, Hammer, and Gorham, 1963), but five of these that have been tested so far are nontoxic. Toxicity tests have been negative on a number of blooms consisting predominantly of *Aphanizomenon* that were collected in Saskatchewan (Hammer, 1962) as well as in Ontario. Olson (1960a,b) has tested 24 blooms composed predominantly of *Aphanizomenon* that he collected from Minnesota lakes during a three-year period from 1948 to 1950. Only two of these (containing 5 to 10% *Microcystis*) produced fast deaths with mice. The rest produced slow deaths (5 to 24 hr) or no deaths. It is conceivable that shortly before the two samples that produced fast deaths were taken there had been a greater proportion of *Microcystis* present which had decomposed and released its toxin. A decision as to whether fast-death-producing strains of *Aphanizomenon* exist must be left in abeyance, but the suspicion we now have is that the final answer may be negative.

To get isolates of *Aphanizomenon* and *Anabaena* to grow proved more difficult than *Microcystis.* We tried a variety of media. *Aphanizomenon* had been cultured before (Gerloff, Fitzgerald, and Skoog, 1950) but *Anabaena flos-aquae,* as far as we then knew, had not (cf. Guseva, 1941). We made many isolation attempts using different media and procedures before we finally succeeded. After we had developed ASM medium, we used it, with or without soil extract, and succeeded in growing isolates of both species. Some of the *Aphanizomenon* strains that we have isolated are shown in Fig. 8. This figure shows two interesting things: 1) that we still get akinete production in our cultures, although not in all of them; and 2) that there are two distinguishable forms—a large-diameter form with distinct septation and a small-diameter form with indistinct septation. Whether we have two distinct species, we are not prepared to say, but our cultures rather suggest it. After becoming aware of this, we re-examined the sources of our collections and observed both forms living side by side in nature. We feel, therefore, that the two forms are not just abnormal products of our cultural conditions. When the trichomes are aggregated in fascicles as they are usually found in nature, the two forms look much more alike. They gradually tend to lose the colonial habit in culture; it may take six months or so before it is entirely gone. We have found that the colonial habit can be experimentally induced, although not reliably, by the addition of certain preparations of soil extract to the medium. We have done it on two different occasions, but we cannot do it at will (McLachlan et al., 1963).

Anabaena flos-aquae

Some of the most dramatic cases of algal poisoning on record have been attributed to *Anabaena Lemmermannii* (= *An. flos-aquae*) and *Anabaena flos-aquae* (Fitch et al., 1934; Deem and Thorp, 1939; Olson, 1951, 1960a; Rose, 1953; and Firkins, 1953). Mice, chickens, rabbits, and guinea pigs have died within 2 to 20 min. after receiving minimal lethal doses of *Anabaena* bloom. These short survival times suggested to us that the toxin responsible was

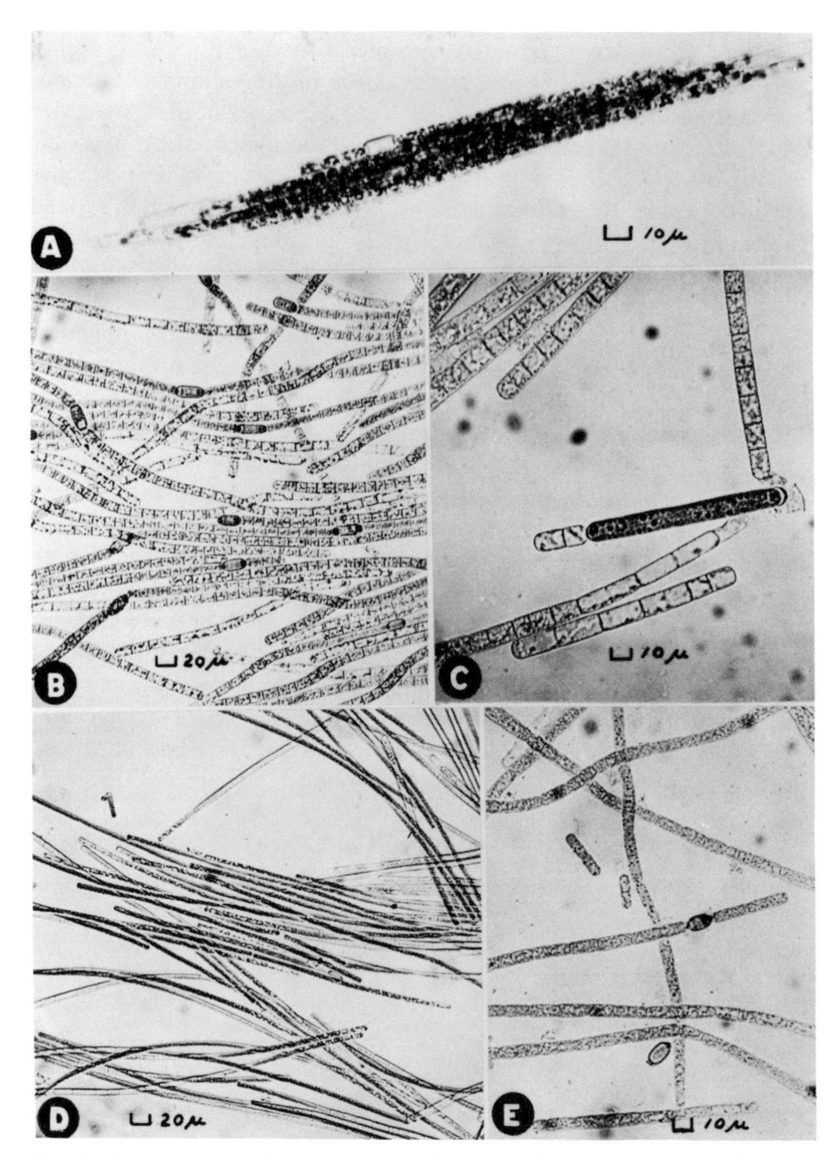

FIG. 8. *Aphanizomenon flos-aquae.* A) Small colony from Rideau River, Ottawa, Ont., with cellular dimensions typical of those given for the species. B,C) Trichomes of strains NRC-23, -24, -27, -29, or -30 in ASM medium with soil extract, showing heterocysts, akinetes, and distinct septation. D,E) Trichomes of strains NRC-19, -20, -28, -31, or -32 in ASM medium with soil extract, showing heterocysts and indistinct septation. (From McLachlan, Hammer, and Gorham, 1963.)

different from *Microcystis* toxin. Since we were unable to locate any toxic blooms of *Anabaena flos-aquae* in the vicinity of Ottawa, we enlisted the collaboration of Dr. U. T. Hammer of the University of Saskatchewan. He located a toxic bloom consisting predominantly of *An. flos-aquae* in Burton Lake, Saskatchewan, on August 3, 1960 (Table III; Gorham et al., 1962). A minimal lethal dose (intraperitoneal) killed mice in 1 to 2 min, preceded by symptoms of paralysis, tremors, and mild convulsions. Autopsy revealed no discernible changes. These effects were not at all like those produced in mice by *Microcystis* fast-death factor. Another sample was collected and shipped to Ottawa on September 14. It produced the same effects as before when administered to mice either orally or intraperitoneally. The very-fast-death factor was found to be about equally distributed between the cells and the water. A large collection was made on September 24, which proved to be more toxic than that made ten days earlier, and had most of the toxin located in the cells. Freeze-dried samples of this material were sent for comparative tests to Prof. T. A. Olson, University of Minnesota, and to Prof. Lloyd D. Jones, D.V.M., South Dakota State College (who had worked with Dr. Rose during the outbreak of *Anabaena* poisoning in Iowa in 1952). Both men kindly tested our material on laboratory animals and reported that the symptoms and survival times were indistinguishable from those which they had observed previously with blooms of *An. Lemmermannii* or *An. flos-aquae* in

TABLE III

Toxicity of Waterblooms of *Anabaena flos-aquae* collected from Burton Lake, Saskatchewan. (From Gorham et al., 1962.)

Intraperitoneal, 20-g male mice, dosage range 20-640 mg/kg body wt.

Collection Date	Fraction Tested	MLD mg/kg	Mouse Units* per Fraction
August 31, 1960	Algae + water†	160	—
September 14, 1960	Algae†	320	81
	Water‡	40	84
September 24, 1960	Algae	160	275
	Water	80	25
June 20, 1961	Algae + water	80	—
August 18, 1961	Algae + water	(1 ml/mouse)§	—

* Mouse unit = minimum dose to kill a 25-g mouse.
† Freeze-dried, weighed, suspended in water.
‡ Evaporated to dryness, 95% ethanol extract of residue dried, weighed, taken up in water.
§ Freshly-collected bloom.

Minnesota and Iowa, respectively. We concluded from this that the bloom of *An. flos-aquae* in Burton Lake was producing the same kind of poisoning as described in published case histories.

Figures 9A and 9B show the appearance of typical colonies

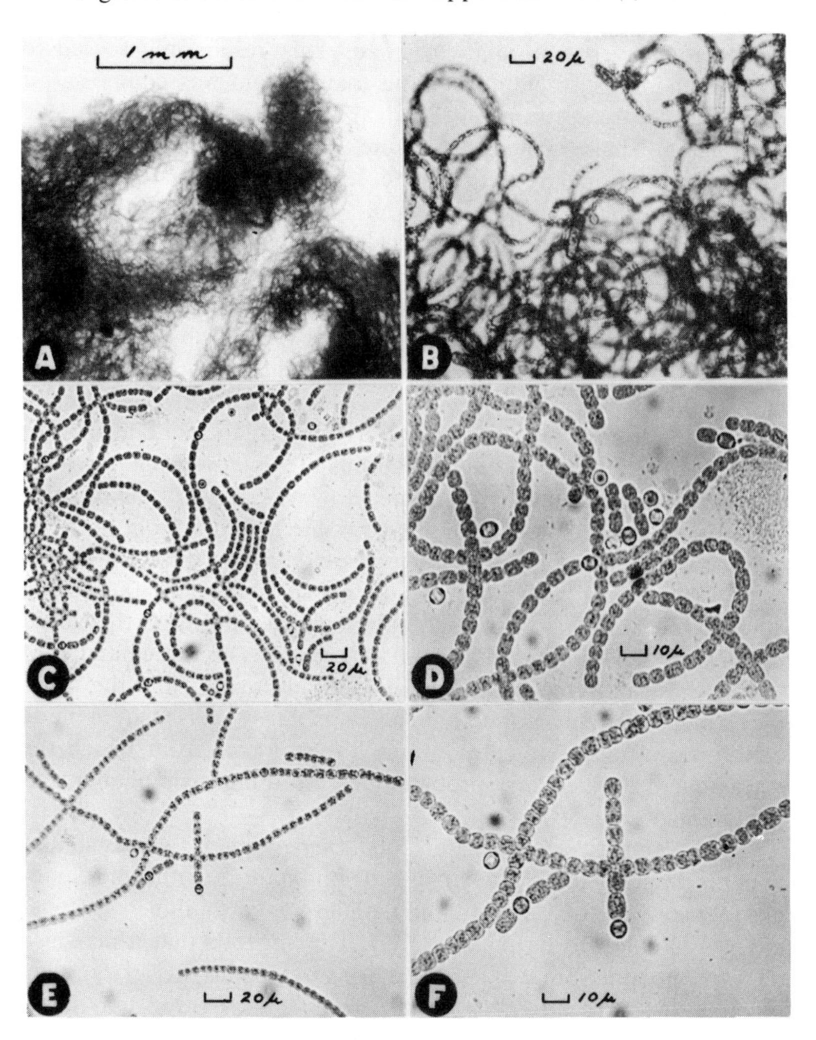

FIG. 9. *Anabaena flos-aquae* from Burton Lake, Saskatchewan. A) Colony from waterbloom collected September 14, 1960. B) Edge of colony from waterbloom collected September 24, 1960. C,D and E,F) Filaments of strains NRC-21 and -22, respectively, in ASM medium with soil extract. (From Gorham et al., 1962.)

of *An. flos-aquae* in the September 14 and 24 collections. Many such colonies were isolated from the September 14 collection, washed, and cultured in ASM medium containing soil extract. Only two strains were finally obtained in unialgal culture and these gradually lost the colonial habit (Figs. 9C-F). To our great disappointment, both strains were nontoxic to mice. This result only served to strengthen our suspicions that the toxic bloom was composed of toxic and nontoxic strains of the alga. The following year, 1961, we tried again. There occurred on June 20 a toxic bloom (Table III) from which we managed to isolate 12 unialgal strains of *An. flos-aquae*—8 toxic and 4 nontoxic. The symptoms and survival times (2 to 7 min) produced by minimal lethal doses of the toxic strains were the same as those produced by the parent bloom. Early in July, the lake was treated with copper sulfate, but on August 18 there was a further light bloom that was toxic. Subsequent treatment of the lake with copper sulfate prevented any further observations of blooms.

We have not had an opportunity, as yet, to make a detailed study of toxin production by our strains of *An. flos-aquae.* We have made a few interesting observations, however. As with the blooms, we have recovered toxin in highly variable amounts from both the culture filtrates and the cells. The toxicity of the different strains as they have been grown has been highly variable, and rather low in comparison with that of the parent bloom. That this is partly caused by physiological conditions was shown by tests on three batches of the same strain. One batch was moderately toxic while the other two were essentially nontoxic. That it is partly caused by genetic conditions is also indicated, since the colonies from which the strains were derived very probably possessed nontoxic filaments as contaminants.

The very-fast-death factor from *Anabaena flos-aquae* is similar in some respects to *Microcystis* fast-death factor. It withstands autoclaving and is soluble in water and ethanol but insoluble in acetone, ether, and chloroform like the latter. This suggests that it may also be a polypeptide. Since it seems to be secreted into the surrounding medium more readily than the *Microcystis* factor and acts more swiftly, it is possible that its molecular weight is lower. Of course we have no more rigorously proven that the toxin comes from the alga in the case of *Anabaena* than in the case of *Microcystis.* Since we were able to isolate toxic and nontoxic strains of *Anabaena* from the same sample of bloom, these strains probably have similar bacterial contaminants and it seems highly probable, therefore, that the alga is the source of the toxin.

BACTERIAL TOXINS

From the work I have described, it now seems reasonably clear that certain strains of certain species of blue-green algae produce the major algal poisons. What about the bacteria that are always present in blooms—do they produce toxins too? We have some evidence that they do. Early in our work with *Microcystis* we noticed that strains or blooms that did not produce fast deaths with mice sometimes produced slow deaths (4-48hr) preceded by symptoms of fur standing up, irregular breathing, and lethargy (Hughes et al., 1958). Slow deaths, or the symptoms produced by the slow-death factor or factors, were sometimes observed with survivors that had received sublethal doses of *Microcystis* containing the fast-death factor. These observations led us to suspect that slow-death factors were bacterial in origin. When the bacteria that we isolated from *M. aeruginosa* NRC-1 were grown in quantity and tested, we found that some produced slow deaths preceded by symptoms of the sort I have just described. One also produced fast deaths after semi-anaerobic incubation (Gorham et al., 1959) but the symptoms were entirely different from those produced by *Microcystis* fast-death factor. Thomson (1958) and his co-workers (Thomson, Laing, and Grant, 1957) have also described two different types of bacteria isolated from a strain of *Microcystis aeruginosa* and a strain called *Anacystis montana* f. *minor* (= *Microcystis aeruginosa*). These two types of bacteria produce different slow-death factors. One is heat-stable while the other is not. In a survey of bacteria from 25 algal collections Thomson found that the majority of the isolates were of the one type that produced the heat-stable factor. The importance of bacterial poisons as causes of death by toxic waterblooms is difficult to assess with any certainty at the present time. There are some cases on record, however, where the symptoms, survival times, or both suggest that the toxins involved may have been bacterial rather than algal in origin (e.g., McLeod and Bondar, 1952). The role of the bacteria needs further study, but from what we already know it is clear that their potential contribution must not be overlooked when trying to evaluate the symptoms of a suspected algal poisoning.

WATERBLOOMS AND STRAINS OF BLUE-GREEN ALGAE

In our studies to date, we have tested the toxicity of 20 predominantly unialgal waterblooms and 59 strains of blue-green algae, representing 11 species, by intraperitoneal injection into mice (Tables IV and V). So far, we have encountered very fast deaths only with *Anabaena flos-aquae,* fast deaths only with *Microcystis aeruginosa,* and either slow deaths, caused by bacteria, or no deaths with all the rest. It should be noted that the nontoxic group includes three of the five species most suspected of being poisonous; one bloom and five strains of *Aphanizomenon flos-aquae,* one strain of *Coelosphaerium Kützingianum* (Cambridge 1414/1), and one strain of *Gloeotrichia echinulata* (Cambridge 1432/1). We have made a number of unsuccessful attempts to isolate new strains of *Coelosphaerium* and *Gloeotrichia* in the hope of finding toxic strains of these species. Until this has been accomplished, and quite a few blooms have been tested as well, we cannot decide whether the poisonings attributed

TABLE IV

Toxic Response Produced by Freeze-Dried Cells from 24 Waterblooms of Blue-Green Algae. (From Gorham, 1962.)

Intraperitoneal, 20-g male mice, dosage range 40-640 mg/kg body wt.

Dominant Species	Blooms Producing:		
	Very fast deaths (1-10 min)	Fast deaths (1-2 hr)	No deaths or slow deaths (4-48 hr)
Anabaena flos-aquae	5	–	3
Aphanizomenon flos-aquae	–	–	1
Lyngbya Birgei	–	–	1
Microcystis aeruginosa	–	7	3
Aphanizomenon (50%) plus *Microcystis* (50%)	–	4	–
Total	5	11	8

TABLE V

Toxic Response Produced by 59 Strains of Blue-Green Algae (From Gorham, 1962.)

Intraperitoneal, 20-g male mice, dosage range 40-640 mg/kg body wt.

Species	Strains Producing:		
	Very fast deaths (1-10 min)	Fast deaths (1-2 hr)	No deaths or slow deaths (4-48 hr)
Anabaena flos-aquae	8	–	6
Anabaena limnetica	–	–	3
Anabaena spiroides	–	–	2
Anabaena Scheremetievi	–	–	2
Anacystis nidulans	–	–	2
Aphanizomenon flos-aquae	–	–	5
Coelosphaerium Kützingianum	–	–	1
Gloeotrichia echinulata	–	–	1
Microcystis aeruginosa	–	8	20*
Nodularia sphaerocarpa	–	–	1
Total	8	8	43

*Includes nine small-celled strains formerly classified (Gorham, 1962) as *Anacystis cyanea* f. *minor.*

to *Aphanizomenon, Coelosphaerium,* and *Gloeotrichia* in the past were caused by toxic strains of these algae or of bacteria associated with them. None of the other species that were tested and found to be nontoxic have been implicated in algal poisonings.

WATERFOWL SICKNESS

This brings me to a final point that I should like to mention. When we had enough toxic *Microcystis* to test a variety of animals we were surprised to find that massive doses administered orally did not kill domestic ducks although we found that they were susceptible when injected intraperitoneally. Ducks apparently have a detoxifying mechanism in their digestive tracts that can keep pace with all the *Microcystis* toxin they are likely to consume. Olson (1961) has found that *Anabaena* toxin behaves very differently. He has dosed ducks with toxic *Anabaena* bloom and found that they died in very short times with symptoms like those that have been reported for waterfowl sickness. Olson has observed the same limber neck symptom that is so characteristic of botulinus poisoning. It occurs rapidly, however, whereas with botulinus poisoning its appearance is slow. It takes 24 to 48 hr or even longer to bring about death by botulinus poisoning. When he dosed ducks with *Anabaena* bloom, they died within a matter of an hour or two. Polyvalent botulinus antiserum, types A, B, and C, failed to counteract the *Anabaena* poison, thereby providing another indication of difference. Waterfowl sickness is known to be caused by botulinus poisoning, but it has been suspected that this is not the whole story. *Anabaena* poisoning appears to be the other part of the story.

SUMMARY

The work I have described has furnished at least partial answers to the questions we asked at the outset: 1) Are the algae or the bacteria responsible for the toxicity of waterblooms? 2) Is more than one toxin involved? 3) Why do waterblooms vary in toxicity? From what we have learned we can say that the algae are almost certainly responsible for the acute poisonings produced by some waterblooms. Bacteria appear to play a secondary role, causing less acute poisonings of various sorts. The symptoms of sublethal poisoning produced by bacteria may sometimes be superimposed on those produced by algal toxins. We now know that more than one algal toxin is involved, and that more than one bacterial toxin is involved too. We have isolated and identified *Microcystis* toxin, which does not cause waterfowl sickness. Work is under way to isolate and identify *Anabaena* toxin, which appears to be partly responsible for waterfowl sickness. *Microcystis* toxin is a polypeptide of only moderate toxicity. Animals must therefore eat or drink quite a lot of a toxic waterbloom of this species at one time to be poisoned. Variable toxicity of waterblooms is determined first of all by the degree of dominance by toxic strains rather than species of algae and this, in turn, is governed by such environmental factors as light, temperature, and inorganic and organic nutrition. These factors also govern toxin production. Variable toxicity is also caused by the need for the algae and toxins to be collected in adequate quantity, released in suitable form, and sufficient amounts consumed by susceptible animals for deaths to occur before dilution, adsorption, or destruction render the toxins harmless. It is no longer difficult to understand why some animals may drink and not be killed by a particular toxic waterbloom. Animals that escape sickness or death have either consumed too little or detoxified what they have consumed too rapidly for any effects to occur.

ACKNOWLEDGMENTS

I wish to acknowledge the fact that the results described in this report represent the work of a number of colleagues and post-doctoral fellows with whom I have collaborated for varying periods of time and on various aspects of the problem. They are Dr. E. O. Hughes, who began the project, Dr. Alfons Zehnder, Dr. R. E. Harris, Mrs. M. McBride, Dr. Miriam Wolfe, Dr. C. T. Bishop, Dr. E. F. L. J. Anet, Dr. Beulah Simpson, Dr. H. V. Konst, Dr. P. D. McKercher, Dr. P. R. Robertson, Dr. J. Howe, Dr. J. McLachlan, Dr. U. T. Hammer, and Dr. W. K. Kim. In addition, I have had the excellent technical assistance of Mr. D. Wright throughout the investigations. I wish to acknowledge that without the careful and conscientious work of all these people there would probably have been little to report.

My colleagues and I wish to express our appreciation to Dr. J. G. O'Donoghue and to Professors Lloyd D. Jones, T. A. Olson, H. K. Phinney, C. A. Peek and D. G. Steyn for their valuable co-operation. We also wish to thank Mr. R. Whitehead, who took the photographs. Special thanks are due to NATO, the University of Louisville, and to Dr. and Mrs. D. F. Jackson for organizing this study conference.

LITERATURE CITED

Abbott, B. C., and Ballantine, D. 1957. The toxin from *Gymnodinium veneficum* Ballantine. *J. Mar. Biol. Assoc. U. K.* 36:169-189.

Astakhova, T. V., Kun, M. S., and Teplyi, D. L. 1960. Cause of carp disease in the lower Volga. *Dokl. Akad. Nauk. S.S.S.R.* 133:1205-1208. [A.I.B.S. Transl. *Doklady (Biol. Sci. Sect.)* 133:579-581. 1961.]

Bishop, C. T., Anet, E. F. L. J., and Gorham, P. R. 1959. Isolation and identification of the fast-death factor in *Microcystis aeruginosa* NRC-1. *Can. J. Biochem. Physiol.* 37:453-471.

Braginskii, L. P. 1955. O toksichnosti sinezelenykh vodoroslei (On the toxicity of blue-green water plants). *Priroda* 1955:117.

Branco, S. M. 1959. Algas toxicas—controle das toxinas em aguas de abasticimento. *Revista do Departmento de Aguas e Esgatos de São Paulo (Brasil)* 20(33):21-30; 20(34):29-42.

Carl, G. C. 1937. Flora and fauna of brackish water. *Ecol.* 18:446-453.

Deem, A. W., and Thorp, F. 1939. Toxic algae in Colorado. *J. Am. Vet. Med. Assoc.* 95:542-544.

Dillenberg, H. O., and Dehnel, M. K. 1960. Toxic waterbloom in Saskatchewan, 1959. *Can. Med. Assoc. J.* 83:1151-1154.

Firkins, G. S. 1953. Toxic algae poisoning. *Iowa State Coll. Veterinarian* 15:151-152.

Fitch, C. P., et al. 1934. "Waterbloom" as a cause of poisoning in domestic animals. *Cornell Veterinarian* 24:30-39.

Fitzgerald, G. P., Gerloff, G. C., and Skoog, F. 1952. Stream pollution. Studies on chemicals with selective toxicity to blue-green algae. *Sewage and Ind. Wastes* 24:888-896.

Francis, G. 1878. Poisonous Australian lake. *Nature* 18:11-12.

Gates, J. A., and Wilson, W. B. 1960. The toxicity of *Gonyaulax monilata* Howell to *Mugil cephalus. Limnol. and Oceanogr.* 5:171-174.

Gerloff, G. C., Fitzgerald, G. P., and Skoog, F. 1950. The isolation, purification, and culture of blue-green algae. *Am. J. Bot.* 37:216-218.

Gorham, P. R. 1960. Toxic waterblooms of blue-green algae. *Can. Vet. J.* 1:235-245.

Gorham, P. R. 1962. Laboratory studies on the toxins produced by waterblooms of blue-green algae. *Am. J. Public Health* 52:2100-2105.

Gorham, P. R., et al. 1959. Toxic waterblooms of blue-green algae. Proc. IX Int. Bot. Congr. 2:137 (abstract).

Gorham, P. R., et al. 1962. Isolation and culture of toxic strains of *Anabaena flos-aquae* (Lyngb.) de Bréb. XV Int. Cong. Limnol. Abstracts: 91. [Verh. Int. Ver. Limnol. 15: in press.]

Grant, G. A., and Hughes, E. O. 1953. Development of toxicity in blue-green algae. *Can. J. Public Health* 44:334-339.

Guseva, K. A. 1941. Tsvetenie Uchinskogo Vodokhranilischa. *Trudy Zool. Inst. Akad. Nauk, S.S.S.R.* 7:89-121. (Bloom on the Ucha reservoir. N.R.C. Tech. Transl. TT-939, 1961.)

Hammer, U. T. 1962. An ecological study of certain blue-green algae (Cyanophyta) in Saskatchewan lakes (doctorate dissertation). Univ. of Saskatchewan, Saskatoon.

Harris, R. E., and Gorham, P. R. 1956. (Unpublished data.)

Hinderson, R. 1933. Fortgiftning av nötkreatur genom sötvattensplankton. *Finsk. Veterinartidskrift* 39:179-189 (cited in Fitch et al., 1934).

Hughes, E. O., Gorham, P. R., and Zehnder, A. 1958. Toxicity of a unialgal culture of *Microcystis aeruginosa. Can. J. Microbiol.* 4:225-236.

Ingram, W. M., and Prescott, G. W. 1954. Toxic fresh-water algae. *Amer. Midland Naturalist* 52:75-87.

Jakob, H. 1954. Compatibilities et antagonismes entre algues du sol. *C. R. Acad. Sci.* 238:928-930.

Konst, H. V., et al. 1959. (Unpublished data.)

Lefèvre, M., Jakob, H., and Nisbet, M. 1952. Auto- et hetero-antagonisme chez les algues d'eau douce in vitro et dans les collections d'eau naturelles. *Ann. Sta. Centr. Hydrobiol. Appl.* 4:5-198.

Louw, P. G. J. 1950. The active constituent of the poisonous algae, *Microcystis toxica* Stephens. *So. Afr. Indust. Chemist* 4:62-66.

Mackenthun, K. M., Herman, E. F., and Bartsch, A. F. 1948. A heavy mortality of fishes resulting from the decomposition of algae in the Yahara River, Wisconsin. *Trans. Am. Fisheries Soc.* (for 1945) 75: 175-180.

McLachlan, J., and Gorham, P. R. 1961. Growth of *Microcystis aeruginosa* Kütz. in a precipitate-free medium buffered with Tris. *Can. J. Microbiol.* 7:869-882.

McLachlan, J., and Gorham, P. R. 1962. Effects of pH and nitrogen sources on growth of *Microcystis aeruginosa* Kütz. *Can. J. Microbiol.* 8:1-11.

McLachlan, J., Hammer, U. T., and Gorham, P. R. 1963. Observations on the growth and colony habits of ten strains of *Aphanizomenon flos-aquae* (L.) Ralfs. *Phycologia* 2:157-168.

McLeod, J. A., and Bondar, G. F. 1952. A case of suspected algal poisoning in Manitoba. *Can. J. Public Health* 43:347-350.

Medcof, J. C., et al. 1947. Paralytic shell-fish poisoning on the Canadian Atlantic coast. *Bull. Fish. Res. Bd. Canada*, No. 75. 32 pp.

Mullor, J. B. 1945. Algas toxicas. *Revista de Sanidad, Asistencia Social y Trabajo* 1:95-114.

Olson, T. A. 1951. Toxic plankton. pp. 86-95. *In* Proceedings of Inservice Training Course in Water Works Problems, February 15-16, 1951. University of Michigan School of Public Health, Ann Arbor.

Olson, T. A. 1960a. Water poisoning—A study of poisonous algae blooms in Minnesota. *Am. J. Public Health* 50:883-884.

Olson, T. A. 1960b. (Unpublished data.)

Olson, T. A. 1961. (Private communication.)

Otterstrom, C. V., and Steeman-Nielsen, E. 1939. Two cases of extensive mortality in fishes caused by the flagellate *Prymnesium parvum* Carter. Report of the Danish Biol. Sta., Copenhagen.

Phinney, H. K., and Peek, C. A. 1961. Klamath Lake, an instance of natural enrichment. pp. 22-27. *In* Algae and Metropolitan Wastes. Transactions of the 1960 Seminar. U.S. Dept. Health, Educ. and Welfare. Robt. A. Taft Sanitary Engineering Center, Cincinnati, Ohio.

Prescott, G. W. 1948. Objectionable algae with reference to the killing of fish and other animals. *Hydrobiol.* 1:1-13.

Prescott, G. W. 1961. Algae of the western Great Lakes area (rev. ed.). William C. Brown Co., Dubuque, Iowa.

Proctor, V. W. 1957a. Some controlling factors in the distribution of *Haematococcus pluvialis*. *Ecol.* 38:457-462.

Proctor, V. W. 1957b. Studies of algal antibiosis using *Haematococcus* and *Chlamydomonas*. *Limnol. and Oceanog.* 2:125-139.

Ray, S. M., and Wilson, W. B. 1957. Effects of unialgal and bacteria-free cultures of *Gymnodinium brevis* on fish. *Fishery Bull. of the Fish and Wildlife Service, U.S. Dept. Int.* 57(123):469-496.

Riegel, B., et al. 1949. Paralytic shellfish poison. V. The primary source of the poison, the marine plankton organism, *Gonyaulax catenella*. *J. Biol. Chem.* 177:7-11.

Rose, E. T. 1953. Toxic algae in Iowa lakes. *Proc. Iowa Acad. Sci.* 60:738-745.

Schwimmer, M., and Schwimmer, D. 1955. The role of algae and plankton in medicine. Grune and Stratton, New York and London.

Senior, V. E. 1960. Algal poisoning in Saskatchewan. *Can. J. Comp. Med.* 24:26-31.

Shelubsky, M. 1951. Observations on the properties of a toxin produced by *Microcystis*. *Verh. int. Ver. Limnol.* 11:362-366.

Shilo (Shelubsky), M., and Aschner, M. 1953. Factors governing the toxicity of cultures containing the phytoflagellate *Prymnesium parvum* (Carter). *J. Gen. Microbiol.* 8:333-343.

Simpson, B., and Gorham, P. R. 1958a. Source of the fast-death factor produced by unialgal *Microcystis aeruginosa* NRC-1. *Phycol. Soc. Am. News Bull.* 11:59-60. (abstract).

Simpson, B., and Gorham, P. R. 1958b. (Unpublished data.)

Stephens, E. L. 1945. *Microcystis toxica* sp. n.: A poisonous alga from the Transvaal and Orange Free State. *Trans. Roy. Soc. So. Africa* 32:105-112.

Steyn, D. G. 1945. Poisoning of animals by algae (scum or waterbloom) in dams and pans. Union of So. Africa, Dept. Agr. and Forestry. Govt. Printer, Pretoria.

Thomson, W. K. 1958. Toxic algae. V. Study of toxic bacterial contaminants. DRKL Rept. No. 63, Defence Res. Board of Canada.
Thomson, W. K., Laing, A. C., and Grant, G. A. 1957. Toxic algae. IV. Isolation of toxic bacterial contaminants. DRKL Rept. No. 51, Defence Res. Board of Canada.
Tisdale, E. S. 1931a. Epidemic of intestinal disorders in Charleston, West Virginia, occurring simultaneously with unprecedented water supply conditions. *Am. J. Public Health* 21:198-200.
Tisdale, E. S. 1931b. The 1930-31 drought and its effect upon water supply. *Am. J. Public Health* 21:1203-1215.
Veldee, M. V. 1931. Epidemiological study of suspected water-borne gastroenteritis. *Am. J. Public Health* 21:1227-1235.
Vinberg, G. G. 1954. Toksicheskii fitoplankton. *Uspekhi Sovr. Biologii* 38, 2(5):216-226. [Toxic phytoplankton. N.R.C. Tech. Transl. TT-549. 1955.]
Zehnder, A., and Gorham, P. R. 1960. Factors influencing the growth of *Microcystis aeruginosa* Kütz. *emend.* Elenkin. *Can. J. Microbiol.* 6:645-660.

Extracellular Products of Algae

Marcel Lefèvre

Director, Centre de Recherches Hydrobiologiques
Centre National de la Recherche Scientifique
Gif-sur-Yvette (S. et O.) France

INTRODUCTION

Since ancient times, it has been recognized that fresh-water algae possess particular properties. Filamentous algae contained in natural waters and known as *Conferves* were gathered and used to hasten the healing of sores. On the other hand, numerous observations carried out during the last century showed that certain algae are capable of yielding highly toxic substances, in standing waters or even in the sea, and that these substances can be fatal to man or to the animal who has absorbed them either directly or indirectly. It is only within the last fifteen years, however, that the production of active substances originating from algae, and their effects on living beings, have been studied on a scientific basis.

Unlike most authors who studied essentially the toxic properties of certain algae, Lesage directed his research work differently: he was able to prove that Chlorophyceae and certain Cyanophyceae possess beneficial properties. With the help of his assistant, Feller, he showed that these algae could be used to great advantage in animal and even in human therapeutics. Unfortunately, Lesage was but poorly acquainted with the classification of algae, and even less with their culture. Thus, his work lacked fundamental precision and did not lead to further investigations.

This work was executed in the Laboratories of the Centre de Recherches Hydrobiologiques du C. N. R. S.

This paper was presented at the NATO Advanced Study Institute by Dr. Hedwig Jakob.

Five years ago, other workers took up the question again, this time from a sounder scientific standpoint. Their results on animal and human therapeutics were also most encouraging, and, like Lesage before them, they believed these effects to be due to antibiotic properties of the algae. Quite to the contrary, further studies showed that these algae produce substances which greatly enhance cell division, and that the results obtained are due to a stimulation of the organism's self-defense. This field of study should prove to be most fruitful in medicine, hydrobiology, and in soil research. Only time will tell if such hopes materialize.

HISTORICAL BACKGROUND

The first observation concerning active substances secreted by algae are those of Harder (1917); he showed that *Nostoc punctiforme* produces autotoxic substances. Akehurst (1931) mentions that waters inhabited by algae contain excretion products of unknown chemical nature, capable of acting accessorily as nutrients and of inhibiting or enhancing the growth of other organisms. Lefèvre (1932, 1937) remarks that algae cultivated in clonic colonies excrete agents which tend to limit or even to check their own division (autoinhibition) and which also eliminate most of the bacteria present in the medium. Mast and Pace (1938) likewise report a phenomenon of autoinhibition in cultures of *Chilomonas paramaecium*.

Since 1940, investigations concerning active factors of algal origin have increased. They are undertaken mainly in the U.S.A., Canada, Germany, and France. The works of Pratt *et al.* on *Chlorella vulgaris* have rapidly become classics. They showed that this alga produces active substances which behave as antibiotics for gram-positive as well as for gram-negative bacteria. Certain authors—Flint and Moreland (1946), Lefèvre, Jakob, and Nisbet (1948-1952), Denffer (1948), Jorgensen (1956), and Jakob (1961)—are studying both the antibiotic and the stimulating properties of certain algae on other algae *in vitro.* Others have undertaken to demonstrate the bacteriostatic or bactericidal effects of certain algae *in vitro.* Among these workers, let us mention Klosa (1949), Spoehr *et al.* (1949), Haas (1950), Vischer (1950), Harder and Oppermann (1953), and Emeis (1956). Finally, regarding medical uses,

Lefèvre and Laporte (1957-1962), with the technical help of Chwetzoff, have been working since 1957 on the stimulating properties of agents secreted into the culture medium by certain Cyanophyceae of thermal waters.

It is now generally agreed that natural waters contain active agents secreted by fresh-water or salt-water algae. Of these, the toxic agents have received most attention, because of their dramatic consequences: intestinal disorders often leading to death of human beings and of animals. The following authors have contributed greatly to this field: Tisdale (1931), Fitch *et al.* (1934), Davis (1948), Gunter *et al.* (1948), Stephens (1948), Connell and Gross (1950), and Shilo and Achner (1953). From 1948 to 1952, Lefèvre *et al.*, working with natural waters, have studied the effects of active agents secreted by Cyanophyceae forming waterblooms on other Cyanophyceae and on other algae as well. For the last decade, attempts have been made to obtain mass cultures of algae producing active substances, both toxic and nontoxic, with a view to determining their nature and their properties. Among the most important works are those of Allen (1952), Allen and Arnon (1955), Bishop, Anet, and Gorham (1959), Gorham (1960), Fitzgerald, Gerloff, and Skoog (1952), Hughes, Gorham, and Zehnder (1958), Jakob (1961), Zehnder and Gorham (1960), and Lefèvre, Laporte, and Chwetzoff (1957-1962), Lefèvre *et al.* have mainly investigated those substances favoring cell division rather than those endowed with toxic properties.

EVIDENCE OF ACTIVE SUBSTANCES PRODUCED BY ALGAE

a. In Nature

Large colonies of algae forming "waterblooms" are well known to develop on natural waters. The Cyanophyceae are largely responsible for these proliferations. Under such circumstances, the other species which normally inhabit these waters—*Scenedesmus, Pediastrum, Coelastrum, Ankistrodesmus, Closterium, Euglenia*, etc.—tend to disappear. However, they do not vanish completely, but remain present under a resistant form.

When the algae which have developed intensively finally die and decompose, the latent forms multiply rapidly. The decomposition products of the waterblooms, as produced by bacteria, seem to favor this development. One is led to believe that active substances produced by waterblooms are responsible for the latent state adopted by other species. The waterblooms themselves disappear as a result of autoinhibition; they are unable to tolerate their own excretion products.

A great many active substances are either thermolabile or adsorbed by carbon black. Thus, as a result, natural water containing such substances and which has been heated or treated with carbon black is capable of favoring the development of species it previously inhibited. As we shall see, it is easy to demonstrate, by laboratory experiments, the presence of active substances secreted by algae. When such substances are toxic for man or for animals, the effects may be dramatic; the resulting disorders may be fatal.

Fresh-water algae are not the only ones which produce toxic substances; certain salt-water Dinoflagellata (*Gymnodinum* and *Gonyaulax*) are also known to cause death among fish or, indirectly, among human beings.

b. In the Laboratory

Extensive analyses of algae contained in natural waters or grown in laboratories have proved that these algae contain or excrete into the surrounding medium numerous substances whose characteristics vary according to the species: polysaccharides, amino acids, vitamins, growth factors, steroids, saturated and unsaturated fatty acids, and finally toxic as well as stimulating factors. The latter factors are but little known; their study is the principal aim of this paper.

Harder (1917), Akehurst (1931), Lefèvre (1932, 1937), Mast and Pace (1938), Pratt (1940-1945), Flint and Moreland (1946), and Lefèvre, Jakob, and Nisbet (1948-1952) have shown that algae excrete active substances. Indeed the growth of algae on a synthetic medium ceases long before the exhaustion of the nutrients contained in the medium (as shown by chemical analysis). This may be due to an autointoxication process of the algae. It is striking that up to 50 or 60 mg of organic substances per liter of medium are frequently observed by permanganate titration. If double-distilled water is added to the medium, the concentration of the inhibitive substances decreases; this leads to a revival of algal growth. If the medium is solid, the same result is obtained by washing it with double-distilled water.

Inhibitive or toxic substances can be eliminated by the action of heat (if they are thermolabile) or by adsorption on carbon black; autointoxicated cells will divide again in the medium devoid of these agents. Pratt (1944) reached similar conclusions by continuously eliminating these substances from the culture medium.

All these results have been confirmed: Levring (1945) reported the presence of an autotoxin in cultures of *Sheletonema costatum*. Denffer (1948) obtained the same results with the diatom *Nitzchia palea*. These autotoxic agents prevent mitosis. In his most interesting paper, Jorgensen (1956) investigates the toxic and the stimulating properties of certain species of algae upon other species. The results obtained with gram-positive and gram-negative bacteria have likewise proved that toxic agents are produced by fresh-water algae (Pratt with *Chlorella*) and by soil algae (Harder and Oppermann with *Stichococcus bacillaris* and *Protosiphon botryoïdes*, Jakob with *Nostoc muscorum*). In a masterly way, Bishop, Anet, and Gorham have demonstrated the toxicity of a number of these agents. Starting with large batches of the Cyanophyceae *Microcystis aeruginosa*, they extracted an active substance and showed its highly toxic effect on large and small animals (mice, sheep, and calves).

PRODUCTION OF ACTIVE SUBSTANCES BY ALGAE

a. In Nature

All algae excrete organic substances, soluble in the surrounding medium, but which are not necessarily endowed with antibiotic or stimulating properties. Waterblooms appear mainly on sunny summer days in waters of neutral or weakly alkaline pH (pH varying from 7 to 8.5, sometimes from 7 to 11). They are nearly always composed of blue-green algae, mainly of the genera *Microcystis*, *Oscillatoria*, *Aphanizomenon*, and *Anabaena*. One seldom observes waterblooms in ponds or swamps of acid and fixed pH whose temperatures vary but of a few tenths of a degree during the course of the year or of the day ("humic" waters of forest ponds, for example). In the waters of fluctuating pH values mentioned above, the waterblooms may be caused by *Euglena*, *Peridinium*, *Chlamydomonas*, *Cosmarium*, etc., as well as by filamentous Chlorophyceae—*Spirogyra*, *Cladophora*, etc.—but these algae are harmless to man and animals.

These excretion products are, however, abiotic for other algae and for bacteria, since these are largely eliminated from the surroundings. Due to algal decomposition, the toxicity of waterblooms varies constantly in natural waters; this renders investigations most difficult. During the early stages of decomposition, the algae release active substances and the toxic effect increases immensely ("fast-death factor" or FDF); it then rapidly disappears, giving way to another effect ("slow-death factor" or SDF) which originates from bacterial decomposition. For a certain time, these two effects are additive (Gorham, 1960).

b. In the Laboratory

Under laboratory conditions, the problem is quite different. Unpredictable interactions observed in nature can be eliminated, and the algae grown in well-defined media. Growth conditions, however, are of major importance as regards the production of active agents. This production is conditioned by: 1) Nature of strain; 2) composition of culture medium; 3) nature and size of inoculum; 4) temperature; 5) illumination; 6) agitation of medium; 7) duration of culture; and 8) season of the year.

1) Nature of strain. Active agents are not synthesized by all the strains of a given species. Thus, *muscarin* is produced by Jakob's strain of *Nostoc muscorum*, but not by the Allison strain. Gorham (1960) has likewise reported that the strains of a given species of *Microcystis aeruginosa* do not all produce the FDF. We met with the same problem upon studying the production of stimulating agents by blue-green algae of thermal waters. We noted, for instance, that four strains of *Phormidium uncinatum*, grown under identical conditions, do not give equally active substances.

These laboratory observations are comparable to the ones which can be made in natural waters. It is indeed most fortunate that all the strains of *Microcystis aeruginosa* are not toxic; since this species is widespread in nature, it would lead to many deaths among wild and domestic animals having access to water containing it.

2) Composition of culture medium. The nature and the yield of active agents secreted by algae depend greatly upon the composition of the culture medium: In certain media, an alga will produce active agents, while in others it will produce none, even though it may develop faster and better.

It is impossible to give even an approximate formula of a few media adaptable to many species and which will force them to a maximum production of active agents. Each worker concerned with this problem must therefore make up a medium which will give the best yield not only for a given species but also for a given strain. The media we have used are entirely synthetic; they are composed exclusively of mineral salts.

3) Nature and size of inoculum. The inoculation of a medium is usually done by means of an old inoculum, whose growth capacity has completely or nearly completely disappeared.

The growth curve of newly-transferred cells shows the following phases: a short phase (lag phase) wherein cells divide slowly; a phase of rapid division (logarithmic growth phase), generally lasting about 20 days; and finally, a phase of slow or of no division (stationary phase), which can last very long without leading to death of the cells.

During the lag phase, the young cells are smaller than the parent cells; they also contain less chlorophyll and thus are lighter in color. This becomes even more apparent during the growth phase. As the stationary phase is reached, the cells divide more slowly, become larger, and laden with chlorophyll and food reserve (vibratile mucilaginous bodies of Desmidiaceae, for instance). Everything leads us to believe that the cell assimilates easily and beyond its needs, accumulating reserve material between two divisions.

If by means of a suitable test one can estimate the amount of active substance present in the culture medium during the various growth phases, the following observations can be made: an accumulation of these factors during the 36 or 48 hr following incubation; nearly total disappearance during the logarithmic growth phase; as the rate of division decreases, the active factors reappear, reaching a maximum when the stationary phase is attained. At this point, their production seems stabilized; it neither increases nor decreases. If it is desired to shorten the lag phase so as to obtain large batches of cells rapidly, it is advantageous to start from cells which are still in their phase of rapid division, and thus have not reached the stationary phase. It is also advisable to use an inoculum whose volume is proportional to the volume of the new medium.

4) Temperature. An increase in temperature often leads to an increase in algal growth, but sometimes also to a nonproduction of active substances. Gorham (1960) cultured *Microcystis aeruginosa* between 25°C and 32.5°C. At 25°C the yield in toxic factors was excellent, while at 32.5°C it was extremely low. Lefèvre and Laporte, working on stimulating factors contained in the culture medium of *Phormidium cebennense*, a species found in thermal waters, have operated at 41°C and 32°C. At 41°C growth is astonishing (maximum reached in 8 days), but the filtered medium is totally inactive, whereas at 32°C it is very active.

5) Illumination. Illumination varies from one species to another; it may be continuous or discontinuous. Certain species will stand continuous illumination, while others will not endure more than 15 to 18 hr of illumination per 24 hr. Spoehr, Smith, Strain, Milner, and Hardin claim that poor results are obtained by fluorescent lighting alone. However, fluorescent lighting enables high light intensities without great increase in temperature, and for this reason has often been used (Zehnder and Gorham, 1960).

Light intensity varies within wide ranges: 1000 to 20,000 lux. On an average, the level is 1500 to 2500 lux. Jaag (1945) reports that cultures of blue-green algae are fairly indifferent to light intensity. Likewise, Lefèvre and Laporte have cultured blue-green algae of thermal waters with intensities as low as 500 lux and obtained good yields in active agents. However, for many other algae, the amount of light given must be in relation to the temperature: too great an intensity with too low a temperature gives poor results and vice versa.

6) Agitation of medium. Opinions vary greatly as to the influence of agitation as well as carbon dioxide supply in the production of active agents by algae. This is not surprising, since all species do not require the same physico-chemical conditions to attain maximum growth in minimum time. On the other hand, cultures showing maximum growth do not necessarily give the best yield in active substances.

Many workers have successfully increased the growth of Chlorophyceae by aeration with a 5% carbon dioxide gas mixture. As regards blue-green algae, results disagree. Bishop et al. (1959) obtained good results by aerating cultures of *Microcystis aeruginosa* with air deprived of carbon dioxide (flow rate 2 liters/min for 8 liters of medium). Zehnder and Gorham, working with the same species, report a notable improvement in the yield by constant mechanical shaking, but poor results if the medium is agitated by a 0.3% carbon dioxide gas mixture. Lefèvre, Laporte, and Chwezoff, however, working with blue-green algae of thermal waters (*Oscillatoria* and *Phormidium*), obtained excellent results with a 2% carbon dioxide gas mixture.

7) **Duration of culture.** The results obtained by Lefèvre, Jakob, and Nisbet are in full agreement with those of Pratt et al. concerning the variation with time of the activity within the filtered medium. They report the following (1952): "Thus within a span of nine weeks, a given summer culture of *Pandorina morum* can go through various stages, wherein the medium becomes very active, then less active, inactive and finally active again."

It is as yet difficult to explain what causes the properties of the filtrates to fluctuate, making them antibiotic or stimulating, depending on the age of the culture. Two possibilities must be considered:

1. Many toxic substances such as strychnine are stimulating and beneficial at low concentrations. This would explain why filtrates can appear stimulating during the linear phase of culture growth (usually in the first two or three weeks), at a time when very little active substance is excreted.

2. One may also assume that algae secrete not just one active substance but several, some antibiotic, others stimulating; the amounts secreted would vary with time, and the net result would be determined by the prevalence of one or of a group of substances over the others.

Since 1952, Lefèvre et al. have adopted the latter hypothesis. Recent observations of Lefèvre and Laporte, working on blue-green algae cultures of thermal waters have further confirmed it. Pratt, working with Chlorophyceae, has likewise expressed the same opinion. In view of these facts, Lefèvre et al. have developed a method for rapidly obtaining large quantities of active substances, starting with aged cultures. Indeed, since aged cells transferred into a fresh medium disperse the active substances they have stored, it is only necessary to inoculate a massive dose of old cells into a new medium to recuperate within 24 or 48 hr most of the active substances available. These substances then appear in high concentrations. This is known as *jus de décharge.* It is now generally agreed that the best yield in active substances is obtained after four to six weeks of culture.

8) Season of the year. Most authors have cultivated algae under constant conditions of temperature and light. However, in spite of such precautions, it has generally been observed that cell division is much slower in winter than in summer. Let us quote Lefèvre, Jakob, and Nisbet (1952): "First of all, a seasonal factor seems to intervene: the experiments give poor results from the end of fall to the beginning of spring, even though the cultures are maintained at 24°C and illuminated 15 hr over 24 hr.

"During this period of time, cultures develop slowly and the filtrates are slightly—if at all—active.

"Hence, it appears difficult to force Nature. In the laboratory, algae tend to follow the rhythm imposed upon them in their natural environment and to necessitate a period of rest in winter."

Pratt et al. likewise report a seasonal variation in the composition and activity of marine algae extracts possessing antibiotic properties. Working with *Nostoc muscorum,* Jakob noted that for no apparent reason, certain cultures of a given batch, thus grown in the same conditions of temperature and light, released no active substances; this phenomenon reappeared in one quarter of the cases. Lefèvre and Laporte, studying stimulating substances of certain blue-green algae, have observed that cell division decreases markedly from December to February.

One is thus led to conclude that the season of the year affects both cell division and production of active substances.

A parallel can be drawn between these observations and those of Heatley and Chain (1938), who remarked that certain generations of *Penicillium notatum* did not produce penicillin even though in subsequent cultures it was again synthesized by younger generations.

ACTIVITY TESTS

As noted previously, the production of active substances varies considerably according to the species and strain of alga cultured and the age of the culture. It has thus proved indispensable to elaborate methods enabling the study of the following phenomena: the ability of a strain to synthesize active substances; the time of culture at which production of these substances is at the maximum. Obviously, the tests differ according to whether they aim at showing the antibiotic, toxic, or stimulating properties of the excreted products.

Antibiotic properties. These are determined by methods commonly used in bacteriology to determine the effect on pathological bacteria of substances secreted by fungi and bacteria: paper discs, cup-plate assay, etc. Numerous active substances secreted by algae are antibiotic for both gram-positive and gram-negative bacteria.

Toxic properties. These are determined by oral administration or by injection of the substance into animals. Quantity of substance and time necessary for death to occur constitute the measure of toxicity. White mice and rats as well as domestic animals (sheep and calves) are generally used. Certain animals, e.g., ducks, appear refractory to intoxication by *Microcystis aeruginosa* (Gorham).

Stimulating properties. These have been studied mainly by Lefèvre and Laporte on culture media of blue-green algae of thermal waters. They can be tested on bacterial cultures, plant and animal tissue cultures, and by their mitotic effect on unicellular fresh-water algae: *Cosmarium, Micrasterias, Pediastrum,* for example. The method using *Cosmarium lundelli* has proved both accurate and rapid; it has become our most common working tool. The details of the method will be given further.

ISOLATION AND PROPERTIES OF ACTIVE SUBSTANCES

Since this subject has been extensively and brilliantly reviewed by Gorham, it seems unnecessary for us to dwell on it further. We shall therefore only give a few indications concerning stimulating agents secreted by blue-green algae of thermal waters and cultured by Lefèvre and Laporte.

The research concerning the purification and the chemical properties of these substances has been undertaken by Barbier and Haenni in the laboratories of E. Lederer. Paper chromatography and electrophoresis have yielded the best results to date. These substances appear soluble in water and ethanol, but insoluble in common orgainc solvents—chloroform, petroleum ether, benzene, etc. The filtered medium is concentrated. Preparative paper chromatography leads to a slightly brown substance, oily in appearance, distinctly fluorescent in ultraviolet light.

Further results will be published later by Barbier and Haenni.

Properties of Biologically Active Substances Secreted by Algae

Lefèvre et al. have studied active substances with particular attention to their morphological and cytological effects on the species of alga which produced them or on other species.

The action can be specific: harmful for certain species, yet favorable for others. They can be thermolabile or thermostable, and are often difficult to detach from ion exchangers or carbon black. The properties of a given crude filtrate can be greatly modified by the action of heat or by an adsorbant: from antibiotic, it can become stimulating, and vice versa.

These as well as other observations lead us to believe that a crude filtrate can contain autoantagonistic as well as favorable substances; the total effect measured is then given by the predominance of one substance over the others. Let us likewise stress the fact that active substances can be extremely toxic for animals but not necessarily for all living creatures. Hence FDF extracted from *Microcystis aeruginosa* is very toxic for animals, but not for bacteria; it is without effect on *Bacillus subtilis, Staphylococcus aureus, Escherichia coli,* and *Pseudomonas hydrophila* (Bishop et al., 1959). On the other hand, the strains of *B. subtilis* and *S. aureus* used were sensitive toward penicillin and aureomycin.

It is well known that algae, bacteria, and fungi always release autotoxic agents which slow down their own division. This effect has been called *autoantagonism* by Lefèvre et al.

Autoantagonism *in vitro*. Algae present morphological and cytological reactions toward autoantagonism: hypertrophy and deformation of the cells, anarchic colonial forms in cenobial algae (e.g., *Scenedesmus, Pediastrum*), greasy degeneration, chloroplast modifications, crystallization of the vibratile mucilaginous bodies, excessive mucilage production in *Cosmarium*, massive accumulation of paramylon in Euglenaceae and of carotene in certain Volvocaleae, and so on. These reactions can probably be attributed to factors which inhibit division without affecting assimilation.

Heteroantagonism *in vitro.* This term designates certain phenomena which occur when one adds to a healthy culture a few drops of a filtrate of another species antagonistic to the first one (Lefèvre, Jakob, and Nisbet).

Three situations can be observed:

1. Division is inhibited, whereas assimilation remains undisturbed. The organism accumulates reserve substances, becomes hypertrophic, and finally bursts (Lefèvre et al., 1952; Jorgensen, 1956).

2. Assimilation is inhibited, whereas division remains possible. The organism divides at the expense of its own reserve material, becomes gradually weaker, and finally dies when its supplies are exhausted.

3. The two former actions occur simultaneously: both assimilation and division are disturbed. In this case, the cells lyse more or less rapidly.

FIGS. 1-4. Effects of heteroantagonism in a collection of natural waters: (1) In a pond, waterblooms of slender *Aphanizomenon* have developed; other algae have nearly disappeared and only exist sporadically. (2) As the *Aphanizomenon* disappears (by autoantagonism), other algae divide actively. (3) In spring water supplied to a pond, *Cosmarium lundelli* divides rapidly (whereas in the pond *Oscillatoria planctonica* has divided intensively). (4) *Cosmarium lundelli* inoculated into a filtrate of this pond water then rapidly dies.

FIG. 1.

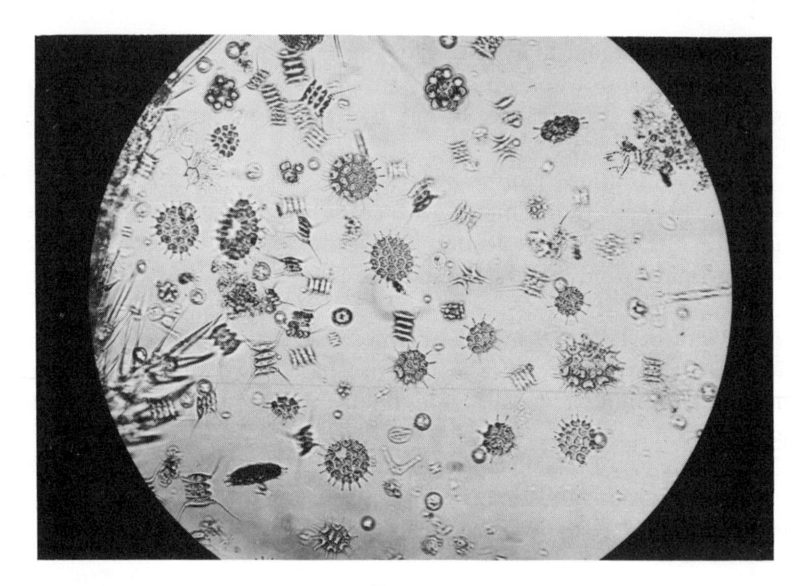

FIG. 2

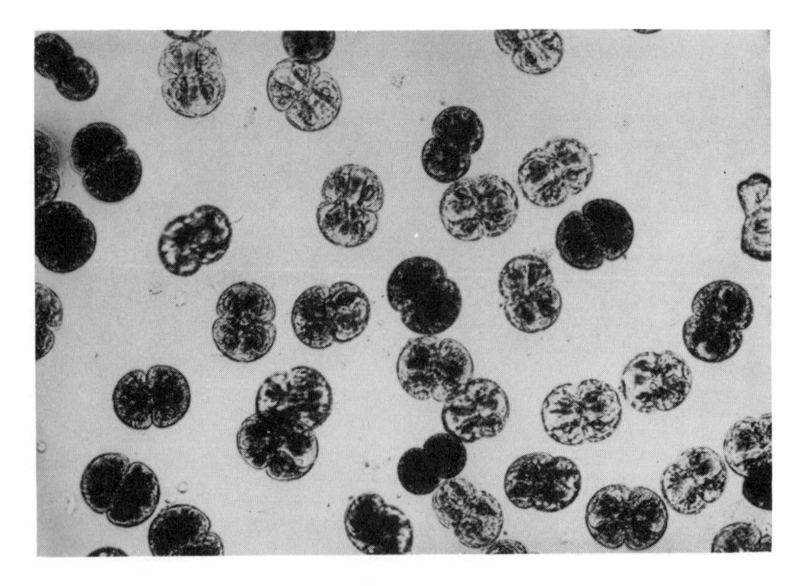

FIG. 3

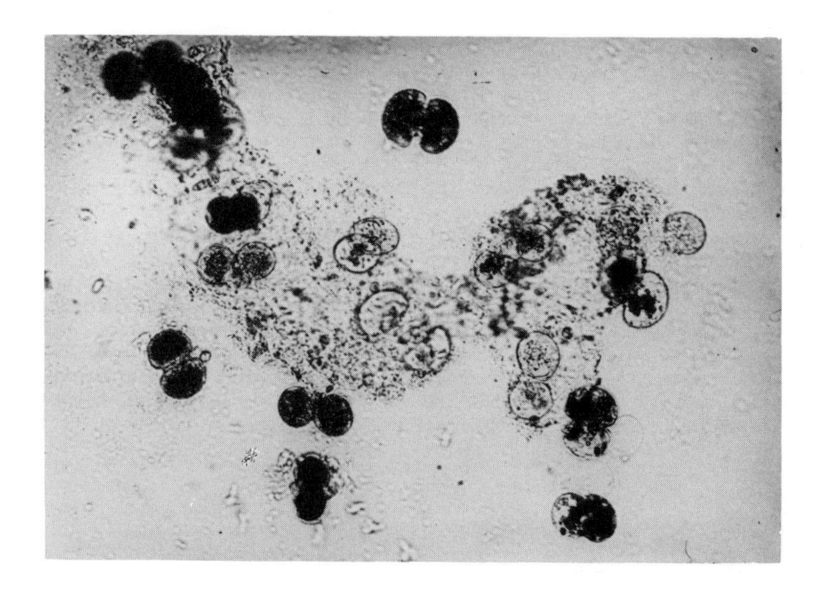

FIG. 4

One can observe the morphological and cytological reactions of an alga toward heteroantagonism either when the active substances abruptly inhibit division, or when one or two divisions still occur before total inhibition results. In the first case, the cells become hypertrophic inasmuch as assimilation is not likewise blocked. If the cell membranes allow it, further morphological changes arise; the cells become deformed, spherical, and present very sharp vacuolation. In *Pediastrum*, for example, the ornamentation of marginal cells disappears, or, as in very slender species, the extremities of the cells become club-like. *Ankistrodesmus* lose their needle form and become spindle-shaped.

But it is mostly when the action of the substance is not immediate, allows one or two divisions, and partly blocks assimilation that one can see the most spectacular reactions. In *Pediastrum*, the cenobia become morular, chloroplasts are broken up, intensive vacuolation occurs, and fatty globules appear. In *Mesotaenium*, which is unicellular and isolated, the new cells now remain linked to one another after division, giving rise to filaments of various lengths. In *Phormidium*, hormogones are no longer formed; the filaments become yellowish, grow excessively in length and twist

FIGS. 5-12. Effects of heteroantagonism in laboratory cultures. *Morphological Reactions*: (5) *Phormidium uncinatum* grown in liquid medium. (6) Same species grown in the same medium to which a culture filtrate of *Pandorina morum* has been added; cells and filaments become longer, are no more fragmented, and become curly. Hormogonies are no longer formed. (7) *Pediastrum clathratum* control cultures in liquid medium. (8) Same species grown in the same medium in the presence of a culture filtrate of *Pandorina morum*. The few cells still capable of dividing only give rise to globules which morphologically do not resemble the normal species. (9) *Pediastrum boryanum* control cultures in liquid medium. (10) Same species grown in the same medium, in the presence of a culture filtrate *Phormidium uncinatum*. Division is blocked, but assimilation remains possible. Cells become much larger; they will finally burst. *Cytological Reactions*: (11) *Pediastrum boryanum* control cultures in liquid medium. (12) Same species composed of same number of cells grown in the same medium, in the presence of a culture filtrate of *Scenedesmus quadricauda*. Cells become larger, pyrenoids disappear, the intracellular structure becomes globular, and enormous vacuoles are formed. The cells die rapidly. (Figures 11 and 12 are at equal magnification.)

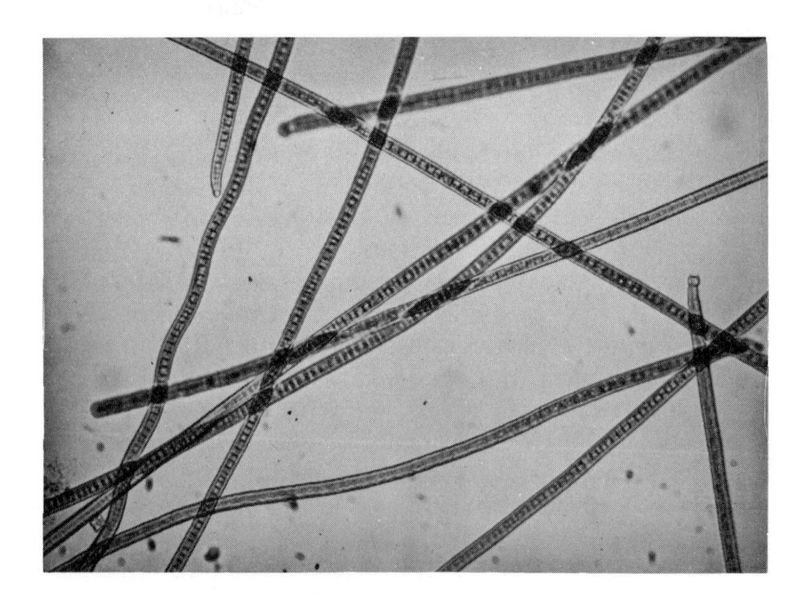

FIG. 5

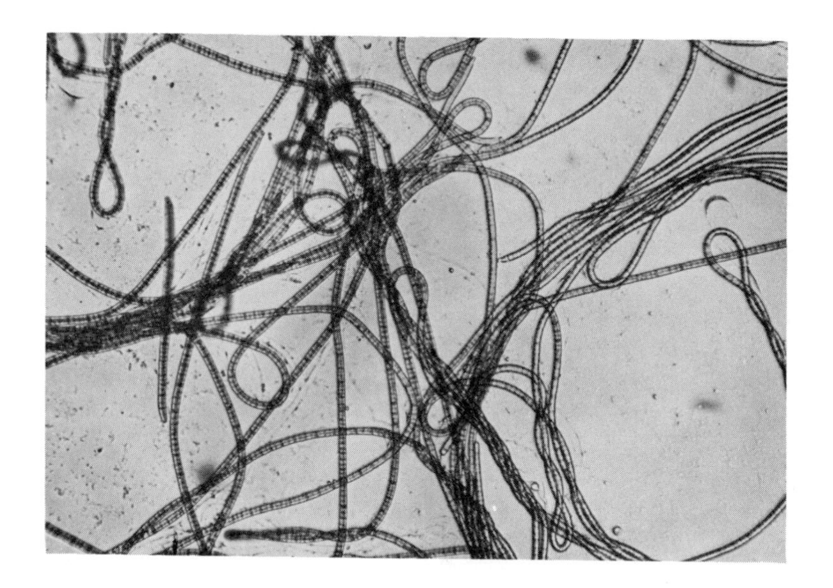

FIG. 6

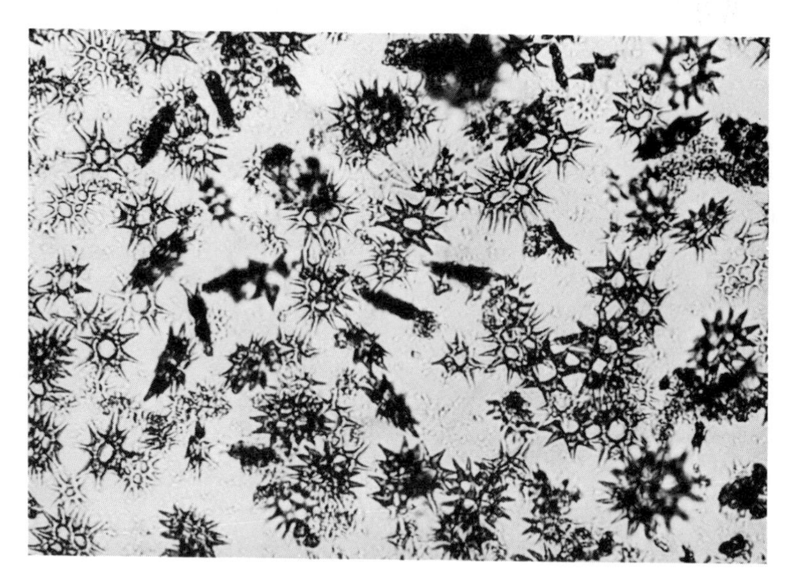

FIG. 7

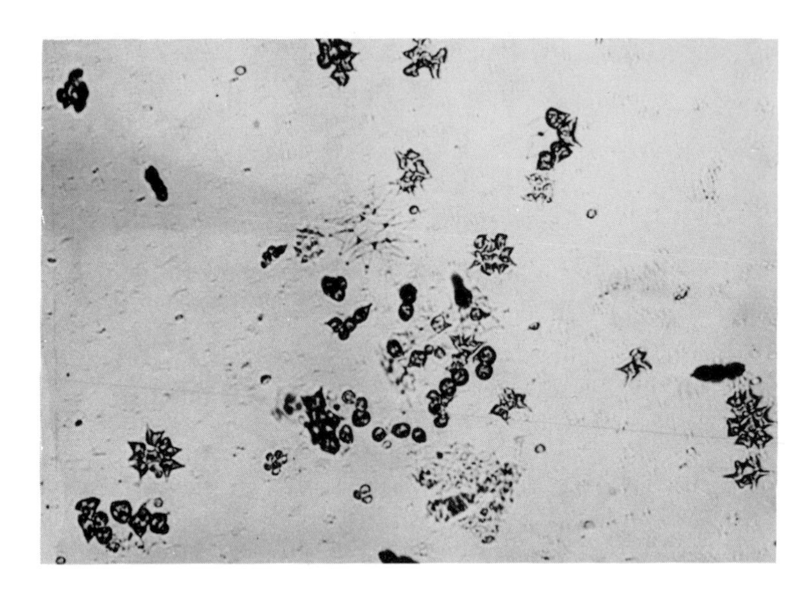

FIG. 8

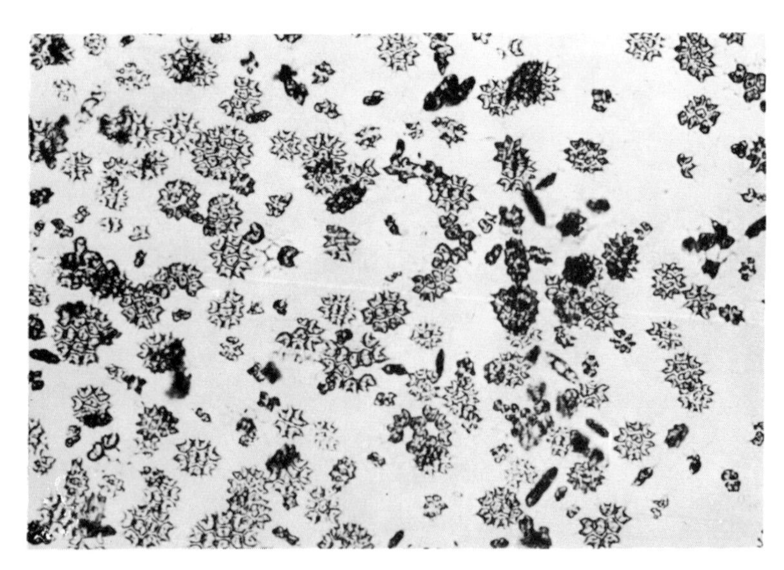

FIG. 9

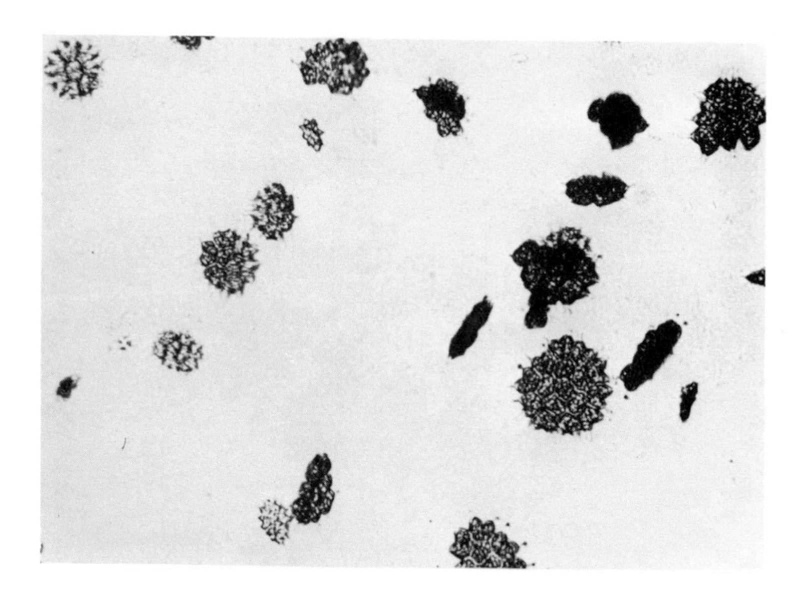

FIG. 10

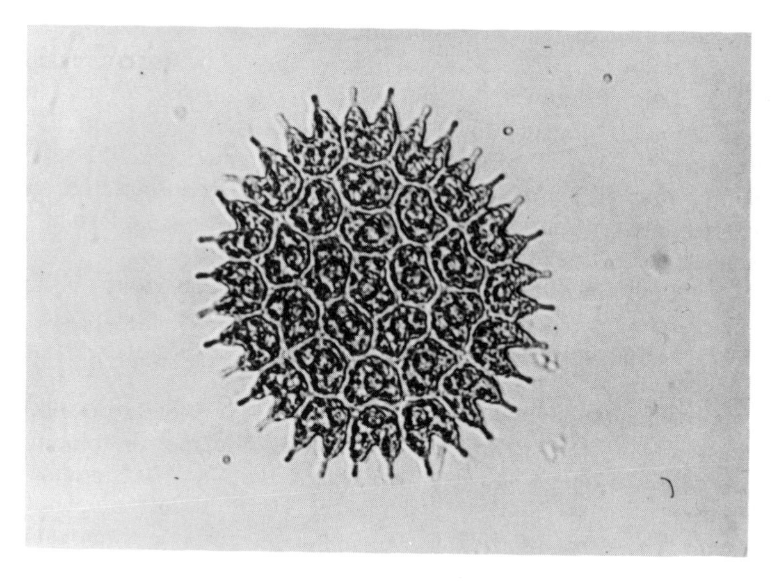

FIG. 11

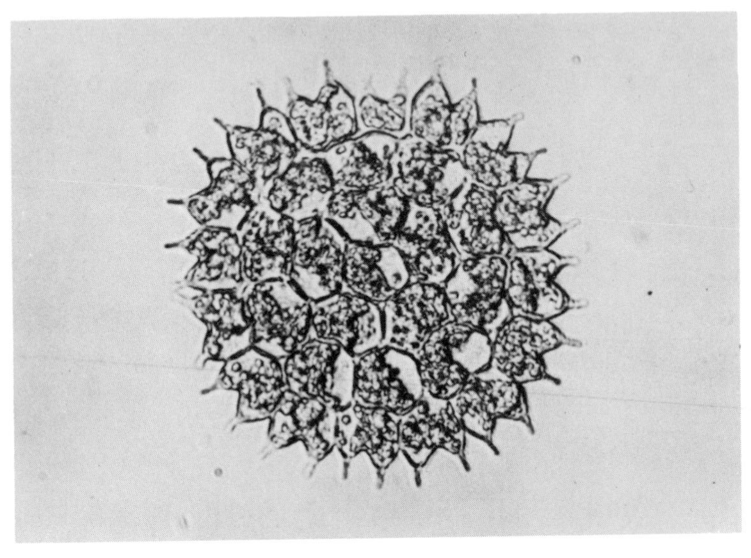

FIG. 12

one upon another. Striae due to transversal cell walls vanish; the filaments appear smooth.

The effect of muscarin on *Cosmarium lundelli* has been extensively studied by Jakob. Five hours after the contact of these cells with an active filtrate, their chloroplasts begin to grow. After 24 hr, vacuolation is considerable, the chloroplasts are affected, and one can see a massive formation of vibratile mucilaginous bodies. Finally, after 48 hr, the chloroplasts are completely retracted, the vacuoles are enormous, the vibratile mucilaginous bodies have disappeared and death ensues.

We shall not refer here to the reactions of animals to toxic substances of algal origin, as this subject will be reviewed by Schwimmer (see also Schwimmer and Schwimmer, 1955).

Autoantagonism in natural waters. When a species develops extensively in standing waters, thus giving rise to waterblooms, there comes a time when it is intoxicated by its own accumulated excretion products and dies. When the water is not completely stagnant, but is renewed slowly, this phenomenon does not occur, since the secretion products are constantly removed and thus cannot accumulate.

Heteroantagonism in natural waters. When one species of algae predominates in standing waters, it can be seen that all other species appear but sporadically, and the number of bacterial species decreases. Lefèvre et al. (1952) suggest that this phenomenon is due to antagonistic substances produced by the predominant species. These observations have been confirmed by those of Rice (1954) and others, who have tried to utilize this massive proliferation in connection with problems of purification of waters containing decaying matter. When the dominant species disappears, the others begin to multiply actively. It even looks as though their proliferation is favored by the decomposition products of the dominant species.

FIGS. 13-17. Effects of autoantagonism in laboratory cultures. *Morphological Reactions*: (13) Young culture of *Pediastrum clathratum*. (14) Old culture of the same species. Division is blocked by autoantagonism, but assimilation is still possible. *Cytological Reactions*: (15) Young culture of *Cosmarium lundelli* (control). (16) Greasy degeneration of the same species in an old culture. Division is blocked. (17) Crystallization of vibratile mucilaginous bodies in a vacuole of an old culture of *Cosmarium botrytis*, solid medium.

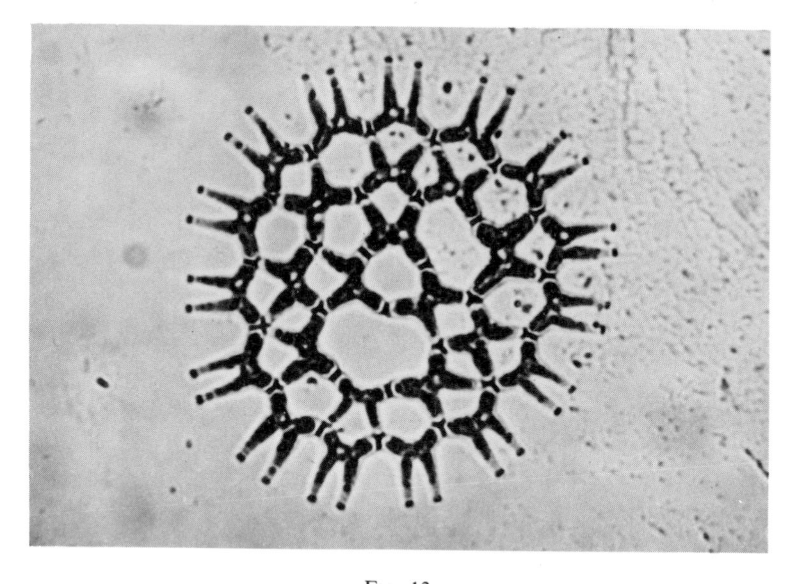

FIG. 13

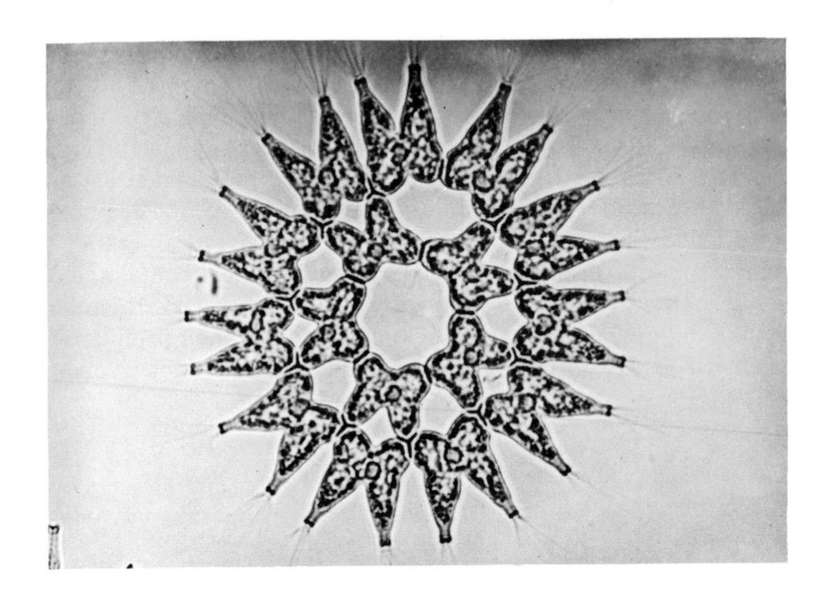

Fig. 14

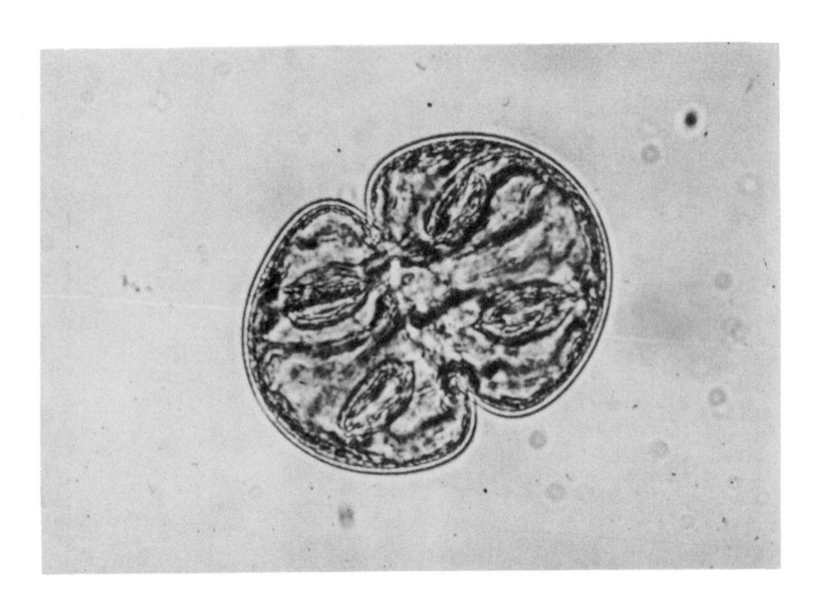

Fig. 15

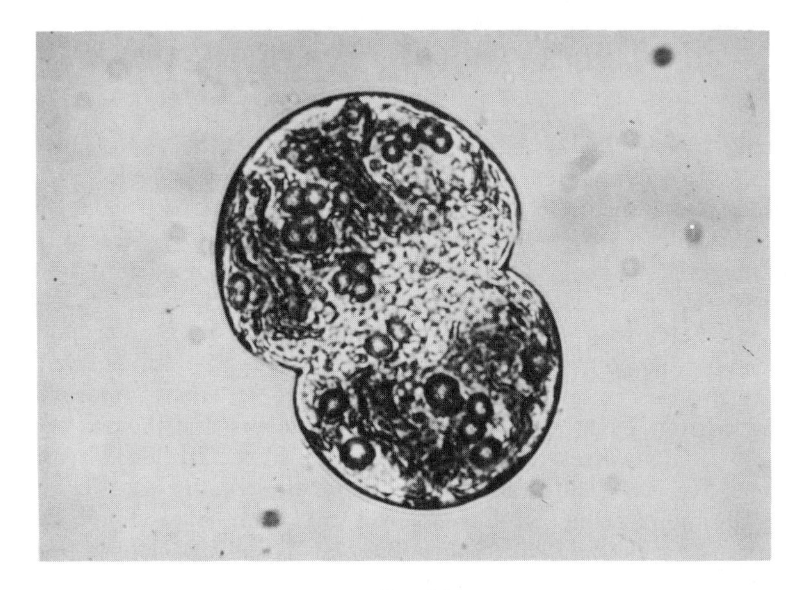

FIG. 16

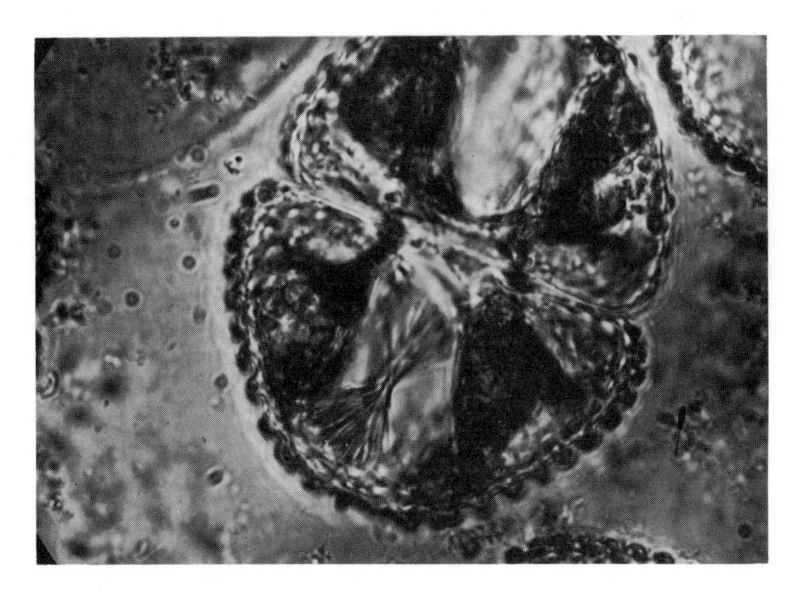

FIG. 17

SPECIAL RESEARCHES CONCERNING STIMULATING SUBSTANCES PRODUCED BY CERTAIN ALGAE

We have already referred to the fact that certain algae produce active substances (Akehurst, 1931; Lefèvre et al., 1952; and Jorgensen, 1956). Lefèvre et al. concluded that fresh-water algae can be used in therapeutics. Feller (1948) had already worked along the same line as Lefèvre. He obtained positive results in the treatment of infected wounds and in the healing of scars by algal cultures. These researches were taken up by Lefèvre and Laporte in 1957.

Patients with ulcers and with atonic wounds were successfully treated with cultures of blue-green algae. The same effects were always observed: infection disappeared, atonic wounds improved and began to bud, with normal healing. It was first thought that these results were due to an antibiotic substance produced by the algae. However, all efforts to prove that these cultures contained an antibiotic remained fruitless. On the contrary, a stimulating action on bacterial division was observed. These experiments were confirmed by others done on anaerobic bacteria in a laboratory of the Pasteur Institute, in Paris.

Experiments done on animal tissue cultures likewise showed a stimulation of cell division. Tissue cultures of chick embryo fibroblasts were used in this work. Growth is twice as fast as in the controls; outside the region of growth, one observes an intensive formation of macrophages (Laboratoire de therapeutique, Biological Institute in Montpellier). Experiments on plant tissue cultures (tubers of Jerusalem artichokes) done in the laboratories of the Phytotron Institute in Gif-sur-Yvette showed an appreciable increase in weight in the presence of filtrates of blue-green algae cultures: up to 50% as compared with controls. Both division and growth of cells are enhanced. We then tested the action of these filtrates on isolated cells of certain unicellular fresh-water algae: *Micrasterias papillifera, Pediastrum boryanum,* and *Cosmarium lundelli.* In every case a stimulation of growth rate resulted.

The test on *Cosmarium lundelli* was used to determine the activity of various filtrates and thus to select the species of algae which yield the most interesting substances. The procedure of this test is as follows:

The test organism is *Cosmarium lundelli* var. *ellipticum*, strain of Lefèvre 1928, a unicellular Desmidiacea which reproduces by simple division.

The culture medium (known as "LH") used to culture the blue-green algae of thermal waters is composed of about 2 g of mineral salts per liter of double-distilled water. Fresh-water algae such as *Cosmarium* can only live and divide if this medium is very much diluted. In a few hours, the cells of *Cosmarium* die in the LH medium diluted by one half, whereas if a few drops of an active filtrate of blue-green algae are added the lethal action is avoided, division occurs, and the colony grows normally.

Practically, the titration is done in the following manner:

Ten ml of LH medium diluted to one half is pipetted into a series of 6-cm-diameter Petri dishes. One series receives one, another two, another three drops, and so on, of the filtrate subjected to the test. Each one of the dishes is inoculated with one drop of a suspension of *Cosmarium lundelli*, containing about 50 cells. A control series without filtrate is inoculated in the same manner.

The dishes are maintained at 22 to 24°C and illuminated with 1200 lux during 15 hr out of 24 hr. A low-power microscope suffices to observe *in situ* the evolution of the *Cosmarium* cells.

The results are as follows:

After 4 to 5 hr, the cells of the controls are altered, and interior disturbances appear. If the filtrate is but slightly active, the cells in the series having received but one drop behave like the controls: they lyse within the first 24 hr following incubation. The titer is given by the first concentration in which the cells do not lyse but begin to divide normally and in a reasonable time. If the first concentration is six drops, or if division only begins 36 or 48 hr after inoculation, the filtrate is considered only slightly active.

Generally, one or two drops of an active filtrate suffice to observe normal division of the cells in less than 24 hr. The number of divisions then increases with the number of drops of active filtrate added. In this case, the multiplication rate of the *Cosmarium* cells is the limiting factor: it is one division per 24 hr. Thus, even if all the cells divide, one cannot observe faster divisions than one per 20 to 24 hr. The first division can at times be observed approximately 15 hr after incubation.

To summarize, the activity titer is measured by the number of drops and time elapsed since inoculation required for the first division to appear. This test gives very reproducible results.

FIGS. 18 and 19. Cosmarium test: (18) *Cosmarium lundelli* in "LH" medium diluted to one half, 24 hr after inoculation; cells lyse and die. (19) Same species grown in the same medium to which a small amount of stimulating culture medium of blue-green algae from thermal waters has been added; after 15 hr, nearly all the cells are dividing and they appear quite normal.

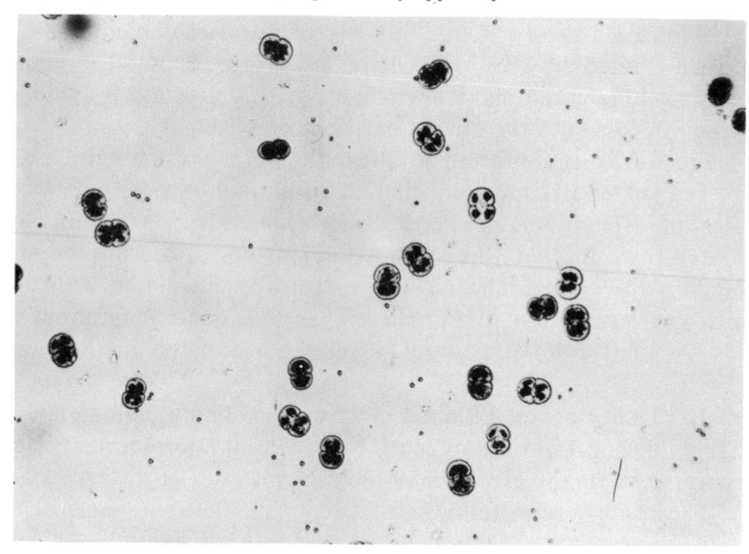

FIG. 18

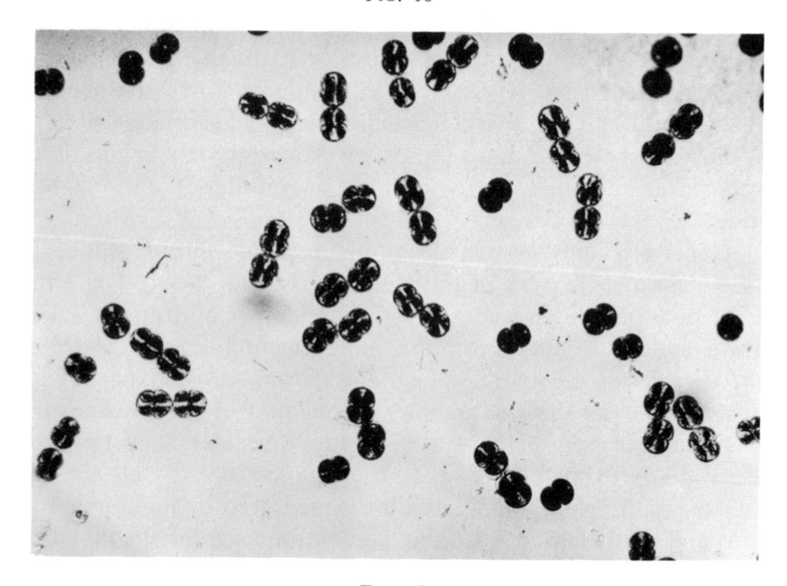

FIG. 19

Origin and Culture of the Strains of Blue-Green Algae

The strains of blue-green algae giving the best results to date have been isolated in unialgal cultures from thermal waters of Dax and Bagnères de Bigorre (southern France). The temperature of these springs is about 50°C. Despite slight morphological differences, they are all related to the species *Phormidium* Gom.

The cultures can be effected either on solid medium (in Petri dishes) or in liquid culture (in penicillin flasks). The liquid cultures are made up of 1.5 liters of LH per flask; a constant flow of a mixture of air and carbon dioxide (2%) bubbles through the culture. The solid medium consists of the liquid medium to which 8-10% of agar-agar—previously washed and dissolved by heat—has been added.

Cultures are maintained at 30 to 32°C and are illuminated by 1200 to 1500 lux from incandescent lamps for 15 hr out of 24 hr. Under these conditions, the cultures can be used after 12 to 15 days. The titration is effected by the previously described *Cosmarium* test.

Properties of the Filtrates

a. Physico-chemical and chemical properties. Crude filtrates are clear or slightly amber in color, becoming darker as they are concentrated. After bacteriological filtration, they are stable for several weeks at room temperature and for over a year at 5°C. Boiling for 60 min partly destroys their biological activity.

As noted earlier, E. Lederer and his co-workers, Barbier and Haenni, are attempting to determine the chemical structure of the active and inactive substances contained in the filtrates.

These stimulating substances have so far proved to be soluble in water and ethanol, insoluble in common organic solvents, fluorescent in ultraviolet light; they do not, however, give the classical tests indicating the presence of indole-3-acetic acid.

b. Biological properties. 1. *In the laboratory.* As we have already pointed out, tests effected on various organisms—bacteria, plant tissue cultures, animal tissue culture, unicellular algae—all seem to give the same result, namely, stimulation of growth. It is highly unlikely that this action is due to a common substance known to be required for growth, since cells of such different origin as animal and algal cells, for example, most certainly have very different requirements.

2. *Clinical assays.* The assays are done by means of local application of algal cultures on agar. The first clinical assays were done in 1957 at the "Station Thermale" of Bagnères de Bigorre (specializing in the treatment of rheumatism) on patients suffering from unhealthy wounds, from scars due to prolonged immobility and which had not healed despite classical treatments. Nine persons have been treated, all of them successfully. The infected wounds begin to bud and heal completely. The healing is at times so rapid that a silver nitrate treatment is necessary to restrain it. Seven other patients have been treated in a hospital in Paris, all with success. Patients with fistula who have undergone telecobalt therapy have likewise been treated; pain disappears rapidly and the wounds heal without relapse.

Treatments applied to various animals, mainly cats and dogs, have resulted in the healing of badly infected wounds which would otherwise have been fatal (Veterinarian Institute of Alfort).

Over 40 clinical assays either on human beings or on animals have been carried out and have given results often characterized as "spectacular" by the operators. These results as well as those of Lesage and Feller clearly show the therapeutic interest of such substances and are most encouraging.

CONCLUSIONS

The studies concerning active substances secreted by algae are still too recent and too few for us to estimate their importance.

However, from a purely scientific point of view, they are likely to clarify a great many questions in hydrobiology which have hitherto remained unsolved.

In soil research, little has been done to show the part played by these substances, but certain authors do not doubt it to be most important.

The field of research of phycotherapy is just at its beginning, but the first results obtained seem very encouraging. In human therapeutics, a certain tendency to return to old methods of phytotherapy is arising and it is certain that phycotherapy can contribute greatly to relieve human suffering in the future.

We sincerely hope that there are many research teams working with this frame of mind.

ACKNOWLEDGMENTS

We are greatly indebted to Miss A. L. Haenni and Miss H. Jakob for their help in the translation of this report.

LITERATURE CITED

Akehurst, S. C. 1931. Observations on pond life. *J. Roy. Mic. Soc.* 51:237-265.
Allen, M. B. 1952. The cultivation of Myxophyceae. *Arch. Microbiol.* 17:34-53.
Allen, M. B., and Arnon, D. 1955. Studies of nitrogen-fixing blue-green algae. *Plant Physiol.* 30:366-372.
Bishop, T., Anet, J., and Gorham, R. 1959. Isolation and identification of the fast-death factor in *Microcystis aeruginosa* NRC-1. *Canad. J. Biochem. Physiol.* 37:453-471.
Connell, C., and Gross, B. 1950. Mass mortality of fish associated with the protozoan Gonyaulax in the Gulf of Mexico. *Science* 112:359-363.
Davis, C. 1948. *Gymnodinium brevis:* A cause of discolored water and animal mortality in the Gulf of Mexico. *Bot. Gaz.* 109:358-360.
Denffer, D. 1948. Über einen Wachstum-hemmstoff in alternden Diatomeenkulturen. *Biol. Zentr.* 67, No. 7.
Feller, B. 1948. Contribution à l'étude des plaies traitées par un antibiotique dérivé des algues. Thèse vétérinaire. Paris (Alfort.).
Fitch, C. P., et al. 1934. Waterbloom as a cause of poisoning in domestic animals. *Cornell Veterinarian* 24:1-40.
Fitzgerald, G., Gerloff, G., and Skoog, F. 1952. Studies on chemicals with selective toxicity to blue-green algae. *Sewage and Ind. Wastes* 24:888-896.
Flint, L. H., and Moreland, C. F. 1946. Antibiosis in the blue-green algae. *Am. J. Botany* 33:218.
Gorham, P. R. 1960. Toxic waterblooms of blue-green algae. *Can. Vet. J.* 1(6): 235-245.
Gunter, G., et al. 1948. Catastrophic mass mortality of marine animals and coincident phytoplankton bloom on the west coast of Florida. *Ecol. Monogr.* 18:309-327.
Harder, R. 1917. Ernährungsphysiologische Untersuchungen an Cyanophyceen, hauptsächlich dem endophytischen *Nostoc punctiforme. Z. Bot.* 9:145.
Harder, R., and Oppermann, A. 1953. Über antibiotische Stoffe bei den Grünalgen *Stichococcus bacillaris* und *Protosiphon botryoïdes. Arch. f. Microbiol.* 19:398-401.
Hughes, O., Gorham, R., and Zehnder, A. 1958. Toxicity of unialgal culture of *Microcystis aeruginosa. Can. J. Microbiol.* 4:225-236.
Ingram, W., and Prescott, W. 1954. Toxic fresh-water algae. *Am. Midland Naturalis* 52(1):75-87.
Jaag, O. 1945. Experimentelle Untersuchungen über die Variabilität einer Blaualge unter dem Einfluss verschieden starker Belichtungen. *Verh. Naturfors. Basel.* 56:28-40.
Jakob, H. 1961. Compatibilités, antagonismes et antibioses entre quelques algues du sol. *Rev. Gen. Bot.* 1-74.

Jorgensen, G. 1956. Growth-inhibiting substances formed by algae. *Physiol. Plantar.* 9:712-726.
Ketchum, B., and Redfield, A. 1949. Some physical and chemical characteristics of algae growth in mass culture. *J. Cell. Comp. Physiol.* 33:(3)281-300.
Klosa, J. 1949. Ein neues Antibiotikum aus Grünalgen. *Z. f. Naturf.* 4b:187.
Lefèvre, M. 1932. Recherches sur la biologie et la systématique de quelques algues obtenues en cultures. *Rev. Algol.* 6, fac. 3-4:313-338.
Lefèvre, M. 1937. Technique des cultures cloniques de Desmidiées. *Ann. Sc. Nat., Bot.*, Ser. 10, 19.
Lefèvre, M., and Nisbet, M. 1948. Sur la sécrétion par certaines espèces d'algues, de substances inhibitrices d'autres espèces d'algues. *Compt. Rend. Acad. Sc.* 226:107-109.
Lefèvre, M., and Jakob, H. 1949. Sur quelques propriétés des substances actives tirées des cultures d'algues d'eau douce. *Compt. Rend. Acad. Sc.* 229:234-236.
Lefèvre, M., Jakob, H., and Nisbet, M. 1950. Sur la sécrétion par certaines Cyanophytes, de substances algostatiques dans les collections d'eau naturelles. *Compt. Rend. Acad. Sc.* 230:2226-2227.
Lefèvre, M., Jakob, H., and Nisbet, M. 1952. Auto et hétéroantagonisme chez les algues d'eau douce in vitro et dans les collections d'eau naturelles. *Ann. de la Stat. Centr. d'Hydrob. appl.* 4:5-198.
Lesage, A. 1945. Phycothérapie et Phycéine. *Bull. Acad. Vet.* 18:272.
Levring, T. 1945. Some culture experiments with marine plankton diatoms. *Med. Oceanogr. Inst. Göteborg.* 3, No. 12.
Lewin, R. A. 1956. Extracellular polysaccharides of green algae. *Canad. J. Microbiol.* 2:665-672.
Lund, W. 1957. Chemical analysis in ecology illustrated from lake district tarns and lakes. Algal differences. *Proc. of the Limnol. Soc. of London, Sess.* 167, Pt. 2, 165-175.
Mast, S. O., and Pace, D. M. 1938. The effect of substances produced by *Chilomonas paramecium* on the rate of reproduction. *Physiol. Zool.* 11:359-382.
McLachlan, J., and Gorham, R. 1961. Growth of *Microcystis aeruginosa* Kütz. in a precipitate-free medium buffered with tris. *Can. J. Microbiol.* 7:869-882.
Pratt, R. 1942. Some properties of the growth inhibitor formed by *Chlorella* cells. *Am. J. Botany* 29(2):142-148.
Pratt, R. 1943. Retardation of photosynthesis by a growth-inhibiting substance from *Chlorella vulgaris. Am. J. Botany* 30(1):32-33.
Pratt, R. 1944. Influence on growth of *Chlorella* of continuous removal of chlorellin from the culture solution. *Am. J. Botany* 31(7):418-421.
Pratt, R. 1945. Influence of the age of the culture on the accumulation of chlorellin. *Am. J. Botany* 32(7):405-408.
Pratt, R., and Fong, J. 1940. Further evidence that *Chlorella* cells form a growth-inhibiting substance. *Am. J. Botany* 27(6):431-436.
Pringsheim, E. 1949. Pure cultures of algae. Cambridge. I Vol.
Rice, T. R. 1954. Biotic influences affecting population growth of planktonic algae. *Fish. Bull. U. S.*, No. 87, p. 227.
Saunders, W. 1957. Interrelations of dissolved organic matter and phytoplankton. *Bot. Rev.* 23(6):389-410.
Schwimmer, M., and Schwimmer, D. 1955. The role of algae and plankton in medicine. Vol. I, pp. 1-85. New York and London.
Shilo, M., and Aschner, M. 1953. Factors governing the toxicity of cultures containing the phytoflagellate *Prymnesium parvum. J. Gen. Microbiol.* 8:333-343.
Spoehr, H. 1944-1945. Chlorellin and similar antibiotic substances. *Carnegie Inst. Washington Yearbook* 44:65-71.

Spoehr, H., and Milner, H. 1949. Composition chimique de *Chlorella.* Influence des conditions de milieu. *Plant Physiol. U.S.A.* 24:120-149.

Spoehr, H. A., et al. 1949. Fatty acid antibacterials from plants. *Carnegie Inst. Wash. Publ.* No. 586.

Stephens, E. 1948. *Microcystis toxica.* sp. nov., a poisonous alga. *Hydrobiologia* 1:14.

Tisdale, E. S. 1931. Epidemic of intestinal disorders in Charleston, West Virginia, occurring simultaneously with unprecedented water supply conditions. *Am. J. Public Health* 21:198-200.

Vischer, W. 1950. Symposium über die Biologie des Bodens im Schweizerischen Nationalpark. Botanische Untersuchungen. *Ber. d. Schweiz. Naturf. Ges.* 130:86-92.

Zehnder, A., and Gorham, R. 1960. Factors influencing the growth of *Microcystis aeruginosa* Kütz. *emend.* Elenkin. *Canad. J. Microbiol.* 6:645-660.

Algae and Medicine

David Schwimmer and Morton Schwimmer

New York Medical College
Metropolitan Medical Center, New York

I. Animal Intoxications Associated with Algae

(By Morton Schwimmer*)

This discussion will be concentrated almost exclusively on the toxic effects of microscopic algae, i.e., phytoplankton. Table I (1) shows a simplified diagramatic scheme of the interrelationship of algae and plankton. The form our contribution will take is not that of a report of an original laboratory study. Rather it will be a digestion and distillation of what has gone before, as seen from a medical perspective.

Much data on toxic algae have been obscured in diverse esoteric journals (2-5)—and I suspect few of our medical colleagues are constant readers of the *American Midland Naturalist* (6), the *Journal of the American Water Works Association* (7), or the various veterinary journals published in a dozen different languages. Some algal classics have further been masked by such deceptive titles as *The*

TABLE I

Interrelationships of Algae and Plankton

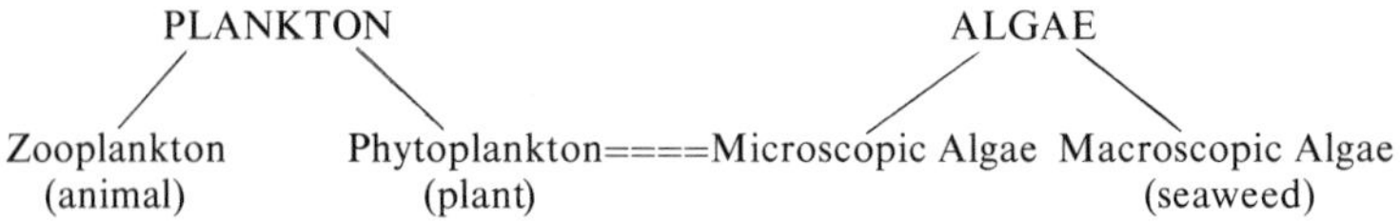

PLANKTON			ALGAE	
Zooplankton (animal)	Phytoplankton (plant)	====	Microscopic Algae	Macroscopic Algae (seaweed)

*Clinical Instructor in Medicine, New York Medical College, Metropolitan Medical Center, New York.

Importance of Upwelling Water to Vertebrate Paleontology and Oil Geology (8). We have been most fortunate in having alert colleagues in key libraries who have been able to ferret out even the most obscure articles with algal implications (9).

ALGAE ISOLATED FROM NORMAL ANIMAL ALIMENTARY TRACTS

The best reviews on the presence of algae in the normal animal alimentary tract are those of Simons (10) and Langeron (11, 12). The latter published his classic review, *The Parasitic Oscillatoriae of the Digestive Canal of Man and Animals,* in the French journal, *The Annals of Parasitology.* Since 1836, when Valentin (13) first reported the isolation of algae in normal animal alimentary tract (*Hygrocrocis intestinalis* in the intestine of a cockroach), 47 separate isolations of algae in normal animal alimentary tracts have been achieved (14-25). Hosts included myriapods, toads, guinea pigs, pigs, goats, sheep, horses, hens, deer, boar, agouti, rats, viscacha, mice, frogs, ducks, euphasids, penguins, and various water birds. The predominant algal genera were *Arthromitus, Oscillospira, Simonsiella, Alysiella, Anabaeneolum, Blastocystis, Phaeocystis*—and many diatoms. Algae have been found in every part of the digestive system.

ANIMAL INTOXICATIONS FROM ALGAL BLOOMS
(See Table II)

Since Francis' report in 1878, over 60 separate episodes of fresh-water algal toxicity in animals have been reported in the literature (26-101). However, as Olson observed (65-67), there are undoubtedly many small outbreaks of authentic algal poisoning which remain unreported for the reason that they are small or because there is a tendency on the part of both veterinarian and farmer to attribute peculiar animal losses to causes more widely known and better understood than toxic algal poisoning.

The genera of blue-green algae incriminated have been *Nodularia, Rivularia, Aphanizomenon, Oscillaria, Anabaena, Microcystis,*

TABLE II

Animal Intoxications from Natural Algal Blooms

System	Algae Involved	Victims	Manifestations	Post-Mortem Findings
Gastro-intestinal	*Nodularia* *Rivularia* *Anabaena* *Microcystis* *Nostoc* *Coelosphaerium* *Aphanizomenon* *Oscillatoria*	sheep, cattle, hogs, horses, dogs, ducks, gulls, monkey	Weakness, nausea, severe thirst, retching, diarrhea, sudden purgation, blood-covered, hard-feces.	Hemorrhage of palate, sloughing of stomach mucous membrane, gastroenteritis, bloody patches on mucous membrane of stomach and intestines, nonhemorrhagic inflammation of mucous membrane of intestinal tract, serous and serosanguineous ascites, atrophic colitis.
Hepatic	*Microcystis* *Anabaena* *Coelosphaerium* *Aphanizomenon*	sheep, cattle, dogs, sow, ducks, gulls, geese, fish	Jaundice, photosensitivity of skin.	Liver congested, spotty, dark red to black, enlarged, brittle or flabby; spleen enlarged, congested, dark, mottled; bloody ascites.
Neuro-muscular	*Nodularia* *Anabaena* *Coelosphaerium* *Aphanizomenon* *Gloeotrichia* *Nostoc* *Microcystis*	horses, dogs, cats, turkeys, geese, fish, shore birds, wildlife	Convulsions, staggering gait, partial or general paralysis, opisthotonus, muscular twitching, thrashing of legs, tonic spasms, blinking of eyes, craning of neck, extreme weakness, falling, lethargy, subnormal temperature, stupor, unconsciousness, death.	Congestion of cerebral and spinal vessels, and dura mater.
Respiratory	*Anabaena* *Microcystis* *Aphanizomenon* *Coelosphaerium*	hogs, cattle, sheep, horses, dogs, pigs, fish, geese, wildlife, water birds	Labored and hurried respiration, slimy discharge from nostrils, foamy discharge from mouth, choking, cyanosis.	Patchy congestion of lungs, pulmonary hyperemia, foamy slime in lower bronchi; lungs filled with blood; pleural effusion; serous or serosanguineous pulmonary edema.
Cardio-vascular	*Nodularia* *Anabaena* *Microcystis*	sheep, hogs, cattle, horses, mules, hares, gulls, ducks, birds	Pulse rapid, thin or weak.	Heart flaccid, dilated, injected; subpericardial and endocardial petechial hemorrhage; sanguineous, serous, or serosanguineous pericardial effusions.

Coelosphaerium, and *Nostoc.*

The outbreaks of toxicity have been in such diverse locales as Australia, U.S.A., Germany, Union of South Africa, Canada, Hungary, Finland, Sweden, Argentina, Bermuda, Israel, and Russia. Among the victims were livestock (including sheep, cattle, hogs, chickens, and ducks), domestic animals (as horses, dogs, and cats); also shore-birds, land birds, and wildlife.

Gastrointestinal: The commonest clinical manifestations have been those involving the gastrointestinal tract. Most often mentioned have been nausea, vomiting, diarrhea, and thirst. A composite picture of the post-mortem findings in the gastrointestinal tract ranges from simple inflammations to petechiae and gross hemorrhage. Necrosis and atrophy have also been described, as have various kinds of ascites (7, 26, 28, 34-38, 41, 48, 77, 81-84).

Hepatic: The chief clinical manifestations have been jaundice and photosensitivity of the skin. Hepato-splenomegaly has been prominent on autopsy, with varying degrees of congestion and necrosis. Ascites, occasionally frankly bloody, has also been noted (34-40, 58, 72, 77, 81-84).

Neuromuscular: Some of the neuromuscular symptoms have been rather severe and dramatic. Outstanding have been spasms, twitchings, and convulsions; weakness, incoordination, and paralysis; and lethargy verging into stupor and death. Post-mortem findings have paradoxically been reported to show only congestion of the cerebro-spinal blood vessels and meninges (7, 26-29, 31, 34-40, 41, 43, 48, 54, 55-56, 60-61, 65-67, 72-75, 77, 81-84).

Respiratory: Animals exposed to toxic algae have manifested striking respiratory symptoms, including mild to severe dyspnea and cyanosis. Choking has often been present, with wheezing or even frank foamy discharges from the nostrils.

Pathologic examinations have shown hyperemia, pleural effusions, and pulmonary edema (34-40, 43, 48, 58, 60-61, 64, 77, 81-84).

Cardiovascular: The circulatory manifestations—in contrast to the neuromuscular ones—have been less prominent clinically than on autopsy. Weak and rapid pulse have been described, as well as the probably associated respiratory symptoms noted in the previous section. Post-mortem examinations have indicated striking damage, the heart often being flaccid and dilated. Hemorrhages and pericardial effusions have also been prominent (26, 34-38, 43, 48, 77).

EXPERIMENTAL INTOXICATIONS WITH NATURAL ALGAL BLOOMS (See Table III)

To corroborate the role of algae as causative agents in the etiology of animal intoxications, diverse experiments have been performed with the algal blooms collected from areas of toxic incidents.

There are no fewer than 32 such scientific studies reported in the literature to date, encompassing 431 individual experiments, utilizing literally thousands of test animals.

TABLE III

Experimental Intoxications with Natural Algal Blooms

System	Algae Involved	Test Animals	Experiments	Manifestations	Post-Mortem Findings
Gastro-intestinal	*Nodularia* *Microcystis* *Anabaena* *Nostoc* *Lyngbya* *Aphanizomenon* *Coelosphaerium* *Gloeotrichia*	sheep, cattle, rabbits, guinea pigs, mice, pigs, rats, ducks, pigeons, fish	Fed, injected subcutaneously and intra-peritoneally.	Poor appetite, emaciation, refusal of food and water for 24 hr, repeated swallowing, extreme salivation, abdominal swelling, constipation, defecation, hard feces covered with bloody slime.	Bloody patches on mucous membrane of stomach wall, inflammation of mucous membrane of stomach and intestines, intestinal mucosa partially desquamated and hemorrhagic, congealed blood in intestines, intestines contracted or dilated, anal region moist with sticky slime, serous or serosanguineous ascites.
Hepatic	*Microcystis* *Anabaena* *Nostoc*	mice, cats, rabbits, cattle, sheep, ducks, chickens, frogs, fish, rats, pigs	Fed, injected subcutaneously and intra-peritoneally.	Listlessness, weight loss, jaundice, ascites, cirrhosis.	Liver slightly swollen, mottled, or markedly congested, enormously dilated; dark red or black, very brittle or even flabby; months or even a year later, brownish yellow, hard and very tough livers. Gall bladder distended. Spleen enlarged and hard. *Microscopic*: Liver cells swollen, cytoplasm granular and vacuolated, albuminous degeneration, general cellular damage with particularly severe injury to liver parenchyma; hepatic injury followed through successive stages of acute parenchymatous, hydropic, and fatty degeneration to necrosis in the centers of the lobule; marked engorgement of centers of lobules.

Neuro-muscular	*Nodularia* *Anabaena* *Aphanizomenon* *Microcystis* *Nostoc* *Lyngbya* *Coelosphaerium* *Gloeotrichia*	sheep, cattle, rabbits, guinea pigs, pigeons, mice, cats, ducks, chickens, frogs, rats, fish, Daphnia	Fed, injected, and immersed in water.	Restlessness, weakness, rolling on sides, colonic spasm, opisthotonus, blinking, convulsions, trembling, tremors, ataxic gait, partial and complete paralysis, muscular contractions of hind extremities and throat, apathy, staggering, prostration, coma, high jumps, piloerection, nervousness, blindness, flaccid paralysis, pupils markedly constricted and later dilated, involvement of sense of equilibrium and spatial orientation, decline in body temperature.	Muscular hyperemia, congested dura mater.
Respiratory	*Anabaena* *Aphanizomenon* *Microcystis* *Nostoc* *Lyngbya* *Coelosphaerium* *Gloeotrichia*	rabbits, guinea pigs, pigeons, mice, cats, rats, cattle, sheep, fish	Drenched, fed, injected intramuscularly, intraperitoneally and intravenously.	Dyspnea, tachypnea, sneezing, coughing, salivation, gasping, cyanosis, death.	Lungs full of blood, petechial hemorrhages, hemorrhages in alveoli and bronchi, serous or serosanguineous pleural effusions, pulmonary edema.
Cardio-vascular	*Nodularia* *Anabaena* *Aphanizomenon* *Microcystis* *Nostoc* *Lyngbya* *Coelosphaerium*	rats, cattle, mice, fish, sheep, cats, frogs, ducks, chickens	Fed, injected subcutaneously and intraperitoneally; tissues perfused.	Tachycardia, pallor, pulse intermittently unobtainable, drop in blood pressure, vasospasm of ears, tails, and conjunctivae, arrhythmias, bradycardia, shock, circulatory collapse, increase in heart beat and blood pressure just before death.	Venous congestion: heart flaccid, dilated three to four times normal volume; pericardial effusion, petechial hemorrhages; edema. *Microscopic*: early degeneration of the myocardium; acute parenchymatous and hydropic degeneration, focal necrosis as well as hyperemia.
Renal	*Microcystis* *Anabaena* *Nostoc* *Lyngbya* *Aphanizomenon* *Coelosphaerium*	rabbits, guinea pigs, rats, cattle, mice, fish, pigs	Fed, injected intraperitoneally and subcutaneously.	Uncontrolled or difficult urination.	Acute parenchymatous and hydropic degeneration and focal necrosis as well as hyperemia, albuminous degeneration, slight nephrosis, cloudy swelling, hemorrhage.

Not only were the investigators able to reproduce similar clinical manifestations and post-mortem findings, but also, because they were able to give selective dosages both orally and parenterally, more extensive system disorders could be induced.

The genera of blue-green algae tested were *Nodularia, Microcystis, Anabaena, Nostoc, Lyngbya, Aphanizomenon, Coelosphaerium,* and *Gloeotrichia.*

Test animals included sheep, cattle, pigs, cats, rabbits, guinea pigs, mice, rats, chickens, ducks, pigeons, fish, frogs, and the cladoceran *Daphnia.*

As indicated in Table III, the manifestations can best be divided according to the systems involved.

Gastrointestinal symptoms included anorexia, extreme salivation, abnormal swallowing motions, abdominal distention and disturbance of intestinal motility. As in the spontaneous toxic episodes, there was evidence of varying degrees of inflammation or hemorrhagic congestion in the whole tract, as well as serous and bloody ascites (7, 31, 34-40, 54, 62, 65-67, 72-74, 77, 81-84, 90, 92-93).

Hepatic damage following administration of suspected algae was indicated by anorexia, weight loss, jaundice, and ascites. Far more striking were the post-mortem findings, which varied from minor to extreme damage, depending upon the size of the administered dose. In the acute cases the liver was markedly congested, dark, and either brittle or flabby; on microscopic examination, hepatic cells were swollen, the cytoplasm granular and vacuolated with evident albuminous degeneration. In other animals, hepatic damage could be followed through successive stages of acute parenchymatous, hydropic and fatty degeneration, to centro-lobular necrosis. In animals receiving repeated dosages, and surviving a number of months, the *livers* and *spleens* showed characteristic gross and microscopic findings of cirrhosis (7, 34-40, 58, 62, 73, 74, 77, 78, 79, 81-84, 89, 90, 91).

As detailed in Table III, the experimental administration of toxic algae produced the complete gamut of *neuromuscular disturbances,* from minor restlessness through spasms, convulsions, paralyses, coma, and death. Quite disappointing, as in the spontaneous toxicities, was the great sparsity of post-mortem findings. These were limited to muscular hyperemia and congestion of the dura mater (7, 26, 31, 34-38, 48, 49-53, 54, 58, 62, 64, 65-68, 72, 73, 74, 78, 81-84, 89, 90, 92, 93).

The *respiratory disorders* induced in the experiments included chiefly, dyspnea, tachypnea, sneezing, coughing, salivation, wheezing, cyanosis, and death. Pathological lesions included pulmonary edema, petechial and gross hemorrhages, in the parenchyma and alveoli, and serous or serosanguineous pleural effusions (7, 30-40, 48, 54, 58, 65-68, 74, 78, 89-90, 91, 93, 94).

The *cardiovascular manifestations* could be better observed in the experimental animals than in the natural toxic incidents. Clinical findings included pallor and vasospasm of ears, tails, and conjunctivae. There were also tachycardia, variable strength of pulse, arrhythmias, fall in blood pressure, and death.

As in the natural incidents, autopsies showed the *hearts* flaccid and dilated, especially on the right side. There were also petechial and generalized hemorrhages as well as pericardial effusions.

Microscopically there were noted acute parenchymatous and hydropic degeneration, and focal necrosis (26, 39, 40, 48, 62, 73, 74, 78, 81-84, 89-91).

Similar microscopic changes were noted in the *kidney,* with evidence also of moderate nephrosis and albuminous degeneration (7, 31, 54, 58, 65-68, 73, 74, 77, 78, 91, 92, 93).

Some interesting *laboratory findings* were noted:

Mason and Wheeler (89), after injecting *Microcystis aeruginosa,* reported an initial hyperglycemia, and terminally a marked reduction in hematocrit, hemoglobin, red blood cells, and total serum proteins. Using the same organism in rabbits and rats, Louw and Smit (58) noted the development of anemia.

Shelubsky (78) [Shilo], again with *Microcystis,* found the algal ashes to be nontoxic. He could not inactivate the algal toxin with human blood rich in esterase. Antisera produced by Shelubsky did not form any visible precipitation with pure toxic fractions, nor did they neutralize their toxicity. Since purified toxic fractions failed to elicit any antibody formation, he concluded the toxin was neither an antigen nor a hapten.

Davidson (7) tested *Nostoc rivulare,* injecting unautoclaved crude extract intraperitoneally. He reported hypothermia, hyperglycemia, tachycardia, and other physiologic changes quantitatively for the first time.

During the first 10 hr after an injection, body temperature dropped 20.8°F. The heart rate, measured by electrocardiograph, was reduced by 180 beats/min. The respiration increased by 60/min. Although Mason and Wheeler (89) had indicated an increase in blood sugar in many animals, Davidson found a decrease in blood sugar of 19 mg/100 ml blood after the 10-hr interval. The blood coagulation time was reduced from 5 to 3 min. The red blood cell count at the end of 6 hr was reduced by 2,500,000/cm^3.

EXPERIMENTAL STUDIES WITH ALGAL CULTURES (See Table IV)

Having observed apparent toxicity of naturally occurring algal blooms, on both spontaneous and experimental administration, there remained only the development of unialgal cultures to study more intensively their true toxic potential.

Since 1935, 22 separate investigators have conducted such studies and 95 distinct toxicity studies have been performed on laboratory cultured algae (21, 98, 99, 100, 102-108).

The algae tested have included greens as well as blue-greens—*Chlorella* and *Scenedesmus* among the former; and *Prototheca, Anabaena, Oscillatoria, Coelosphaerium, Microcystis,* and *Aphanizomenon* among the others.

The animals tested were a monkey, kittens, guinea pigs, rabbits, chickens, sheep, calves, and mice.

Modes of administration included introduction into rectum and colon, oral feeding, and intraperitoneal injection.

In practically all respects, the clinical and autopsy findings paralleled those seen on administration of naturally occurring toxic blooms.

TABLE IV
Animal Toxicity Studies with Laboratory-Cultured Algae

System	Algae Tested	Test Animals	Test	Manifestations	Post-Mortem Findings
Gastro-intestinal	*Prototheca* *Anabaena* *Oscillatoria* *Coelosphaerium* *Microcystis* *Aphanizomenon* *Chlorella*	kittens, monkey, guinea pigs, rabbits, chickens, sheep, calves	Introduced into rectum and colon, fed, and injected intraperitoneally.	Diarrhea, colitis, anorexia, weight loss, constipation.	Pronounced inflammation of the intestinal tract and peritoneum.
Hepatic	*Anabaena* *Microcystis* *Aphanizomenon* *Coelosphaerium* *Chlorella* *Scenedesmus*	mice, guinea pigs, rabbits, sheep, calves	Fed and injected intraperitoneally.	Steady weight loss, roughness of fur, scales on snout, tail and skin.	Liver pale, often blotchy, dark, engorged; pronounced hyperemia with congestion and hemorrhage; moderate to severe hepatic necrosis.
Neuro-muscular	*Microcystis* *Aphanizomenon* *Coelosphaerium* *Anabaena* *Chlorella*	mice, guinea pigs, rabbits, chickens, sheep, calves	Fed and injected intraperitoneally.	Pallor, piloerection, convulsions, partial paralysis of hind legs, spasmodic movements, twitching of legs, convulsive jumps, dragging of legs, alternating periods of restlessness and apathy, loss of equilibrium.	
Respiratory	*Microcystis* *Aphanizomenon* *Coelosphaerium* *Anabaena* *Chlorella*	mice, guinea pigs, rabbits, chickens, sheep, calves	Fed and injected intraperitoneally.	Difficulty in breathing, nose became reddish-purple.	Lung hemorrhage.
Cardio-vascular	*Microcystis* *Coelosphaerium* *Aphanizomenon* *Anabaena*	mice, guinea pigs, rabbits, sheep, calves, chickens	Fed and injected intraperitoneally.	Pallor of tail, ears and eyes.	Greatly reduced blood supply in both peripheral and visceral circulation.

Of added particular interest are some reports on *green* algae, generally thought to be nontoxic and widely considered as potential large-scale food sources. Herold and Fink (104-106), in 1958, in Germany, fed *Scenedesmus* to mice, and the animals developed hepatic necrosis. In Japan, Arakawa and his colleagues (107-108), in 1960, observed diarrhea, weight loss, and decreased egg-laying in hens fed either decolorized or undecolorized *Chlorella*. When they fed *Chlorella* to mice, many developed diarrhea and died.

TUMOR FORMATIONS ASSOCIATED WITH ALGAE
(See Table V)

Some interesting observations are available on tumor formations associated with algae.

In 1923, Langeron (11-12) reported that *Anabaena cycadea* produced tuberization of cycadacea roots. Ciferri and Redaelli (19, 109-112), in Italy, produced transitory granulomata in guinea pigs with subcutaneous and intraperitoneal injections of *Prototheca portoricensis*. Mariani (21) was able to reproduce their experiment with even greater success; he produced a caseous granuloma with caseous centers and with lymph node metastases. In Russia, Newiadomski (20) reported local neoplastic reactions following subcutaneous injection of *Blastocystis enterocola* in rats.

More recently, Davidson (7, 113-114), injecting various types of *Nostoc rivulare* filtrates and extracts in mice, was able to produce tumors on shoulders, backs, and abdominal areas with subcutaneous or intraperitoneal injections.

Another potential tumor former is carageenin. This algal derivative (alginate), a sulfated polygalactose extract of Irish Moss, the red alga *Chondrus crispus*, is widely used as a food stabilizer and also as an important ingredient in medical lubricating jellies. Since 1953, when Robertson and Schwartz (115) first reported it in the *Journal of Biological Chemistry* over a dozen other reports (116-124) in the literature have attested to carrageenin's ability when injected intraperitoneally to form granulomatous tumors composed primarily of collagen fibers.

My brother, Dr. David Schwimmer, will discuss this further with relation to human beings.

TABLE V
Tumor Formations Associated with Algae

<table>
<tr><th>Reference</th><th>Organism</th><th>Host</th><th colspan="3">Findings</th></tr>
<tr><td>1923 Langeron (France)</td><td>Anabaena cycadea</td><td>roots of Cycadaceae</td><td colspan="3">In association with two nitrifying bacteria, produced tuberization of certain roots of Cyadaceae.</td></tr>
<tr><td>1935 Ciferri and Redaelli (Italy)</td><td>Prototheca portoricensis</td><td>guinea pig</td><td colspan="3">Subcutaneous and intraperitoneal injections in the guinea pig were able to cause a transitory granulomatous lesion.</td></tr>
<tr><td>1937 Newiadomski (Russia)</td><td>Blastocystis enterocola</td><td>rats</td><td colspan="3">Algal culture extract introduced under the skin brought about a local reaction considered neoplastic in nature.</td></tr>
<tr><td>1942 Mariani (Italy)</td><td>Prototheca portoricensis</td><td>guinea pig</td><td colspan="3">When guinea pigs were injected intraperitoneally, a granuloma with semi-liquid purulent, caseous centers was produced locally; metastases resulted following diffusion via the lymphatics to the lymph nodes.</td></tr>
<tr><td>1959 Davidson (U.S.A.)</td><td>Nostoc rivulare Kütz.</td><td>mice</td><td>Injected Subcutaneously</td><td>Injected Intraperitoneally</td><td>External Application</td></tr>
<tr><td></td><td></td><td>Sublethal amount of unautoclaved aqueous filtrate</td><td></td><td>Produced tumors on shoulder at 6 months or longer</td><td>No effect</td></tr>
<tr><td></td><td></td><td>Sublethal amount of unautoclaved crude extract</td><td>Animals that survived the immediate effects of injections developed large tumors on either shoulders or backs within 3 months and died within 9 months.</td><td>Algae Concentration Dosage
0.08 mg/ml water [0.0231 mg/g body weight] }
0.40 mg/ml water [0.0534 mg/g body weight] } all survived: all with tumors at 3 months.
1.20 mg/ml water [0.0742 mg/g body weight] }
1.60 mg/ml water [0.0933 mg/g body weight] death at 60 hr.
Tumors appeared on the shoulders and abdominal regions of the mice within 6 months.</td><td>External application produced heavy scales on the shaved skin. Scales were shed in 4 days and the hair grew again within a few weeks.</td></tr>
<tr><td></td><td></td><td>Sublethal amount of unautoclaved aqueous extract</td><td></td><td>Animals that survived the immediate effects of injections . . . developed large tumors on their shoulders and backs within 3 months and died within 9 months.</td><td></td></tr>
</table>

MISCELLANEOUS

It has been reported repeatedly in the literature that *fish kills* quite often are caused by toxic algae (52, 78). Nevertheless, even in New York City, *fish kills* are practically always credited to more obscure causes (125-127), e.g., clogging of the gills with sand, botulism, anoxia, rather than checking the algae present, for inherent toxicity. The same can be said for "heat stroke" in fish (128-129).

Another disorder that we feel may well have an algal etiology is Brisket disease in cattle. Hecht et al. (130-135), in Utah, have looked upon this as a form of pulmonary hypertension in animals due to high altitude, similar to such a condition in humans.

However, the high altitude (above 8,000 feet), the geography, the seasonal distribution, presence of water pot holes, together with symptoms of diarrhea, dyspnea, cardiac, pulmonary and hepatic involvement, in young or new animals, and with subsidence of disorder upon departure from the region, all suggest strongly the incrimination of algae.

The same should be considered in the cattle disorder called the *borrachera* in the plains (*llanos*) of Venezuela.

All the above manifestations, individually and collectively, have been produced experimentally by algal toxins.

The same thought should be entertained when animals such as dogs and cats develop distemper-like behavior during so-called summer "dog days."

II. Human Intoxications Associated with Algae

(By David Schwimmer*)

There is no logical reason to suppose that man is less susceptible than animals to noxious algae, and the impressive data on animal intoxications suggest that algae might also adversely affect human beings.

This does indeed happen. The chief modes of algal entry into

* Associate Clinical Professor of Medicine, New York Medical College, Metropolitan Medical Center, New York.

the human body are by ingestion, inhalation, contact, or parenteral injection (136-145). Ingestion embraces food and drink, including wine, fruits, milk, fish, and animals with algae on or in them. Contact includes not only ordinary touching, but also swimming or immersion in infested waters. The effects on man are more likely to be chronic, therefore more cryptogenic, and the resultant disorders may well end up — if not wrongly attributed to other causes—as respectably "idiopathic," "primary," or "essential" diseases. That humans are less subject than animals to acute (and lethal) intoxication is perhaps due to man's olfactory delicacy—a reluctance to ingest enough stagnant water malodorous from decaying algae. (Even in animals, as indicated by Dr. Morton Schwimmer, acute intoxications are often difficult to study and document, because the episodes are unscheduled, and the toxicity of a given batch of algae may rapidly diminish.)

The delayed chronic human effects are generally inapparent simply because of physicians' meager awareness of the toxic potential of algae. Thus the medical historian may with consummate ease overlook indirect exposure to (or ingestion of) toxic algae via fish, animals, fruits, and vegetables.

Before discussing *toxic* algae, it ought to be mentioned that there are some algal residents in normal animals and human beings (See Table VI) (10-12, 20, 23, 146, 147). Notable among these are *Simonsiella, Oscillospira, Anabaenolium, Phaeocystis*, and some species of *Chlorella*, isolated from varying portions of the digestive tract. As Tiffany (22) has said, the entozoic forms may even be "necessary to comfort and gastronomic happiness." Recent work has indicated that algae parasitic or saprophytic in animals may become holophytic on exposure to light. Even more important may be the possibility that algae could imitate the fungi, which are having a field day in the human body since we began upsetting nature's balances by the widespread use of antibiotics, antimetabolites, steroids, and pesticides; and algae with toxic potential *do* exist in many municipal water supplies (9, 148).

We have arbitrarily chosen to discuss human algal intoxications under the following headings:

1. Gastrointestinal
2. Respiratory
3. Dermatologic
4. Ichthyosarcotoxic
5. Miscellaneous

TABLE VI

Algae Isolated from Normal Humans

Reference	Year	Locale	Algal Organism	Subject	Site
Rabenhorst (cited in Weber)	1887	England	*Pleurococcus Beigeli* (Protococcus)	human	hair of the nape of the neck
Müller	1906	Germany	Oscillaria ("Scheiben-bakterien") *Simonsiella Mulleri* Schmid	student	mouth
			Oscillaria ("Scheiben-bakterien") *Simonsiella Mulleri* Schmid	humans	coating of teeth and in saliva
Simons	1922	Germany	*Simonsiella Mulleri* Schmid	humans	the saliva and in buccal cavity
			Simonsiella crassa Schmid	2 men	buccal cavity
Langeron	1923	France	*Simonsiella Mulleri* Schmid	man	salivary sediment—chiefly on waking—and in scrapings from the buccal cavity
			Simonsiella crassa Schmid	man	buccal cavity
			Anabaeniolum Brumpti	man	in stools or in cecal contents
			Anabaeniolum minus	man	intestine
Newiadomski	1937	Russia	*Blastocystis enterocola*	human	feces
Mariani	1942	Italy	*Blastocystis enterocola*	human	feces

TABLE VII

Human Gastrointestinal Disorders Associated with Algae

Year	Locale and Author	Victims	Algae Involved	Manifestations of Toxicity
1842	London, England (Farre, 1844; Küchenmeister, 1857)	35-year-old married female	*Oscillatoria intestini* Küchenmeister	Dyspepsia, griping, bowel obstruction
1930	Puerto Rico (Ashford, Ciferri, and Dalmau, 1930)	woman	*Prototheca portoricensis*	"Atypical sprue"
1930	Puerto Rico (Ashford, Ciferri, and Dalmau, 1930)	woman	*Prototheca portoricensis* var. *trispora*	"Suspicious of sprue"
1930	Suburbs of Washington, D.C., U.S.A. (Tarbett and Frank, cited in Tisdale, 1931)	many families	Unidentified algae	Sudden onset of nausea, vomiting, epigastric pain, diarrhea with cramps of 1-4 days duration
1930	Charleston, West Virginia, U.S.A. (Tisdale, 1931; Veldee, 1931; Tarbett, cited in Tisdale, 1931)	8,000 to 10,000 people	Blue-green algae	Sudden onset of nausea, vomiting, epigastric pain, diarrhea with cramps of 1-4 days duration
1931	Ironton and Portsmouth, Ohio, U.S.A. (Waring, cited in Tisdale, 1931; Veldee, 1931)	many people	Unidentified algae	"Intestinal influenza"
1930–1931	Louisville, Kentucky, U.S.A. (Tisdale, 1931; Veldee, 1931)	many people	Unidentified algae	"Intestinal disorders"
1930–1931	Weston, West Virginia, U.S.A. (Tisdale, 1931)	many people	Unidentified algae	"Intestinal disorders"
1930–1931	Sisterville, Ohio, U.S.A. (Tisdale, 1931)	many people	Unidentified algae	"Intestinal disorders"

1925, 1929, 1930	Yellowstone National Park, Wyoming, U.S.A. (Spencer, 1930)	500 people		Nausea, vomiting, diarrhea, cramp, pains of short duration (6-48 hr). Frontal headaches
1931	Huntington, West Virginia; Ashland, Kentucky; Cincinnati, Ohio, U.S.A. (Veldee, 1931)	thousands of people	"Algae"	Abdominal pain, nausea, vomiting and diarrhea
1940	New Jersey, U.S.A. (Nelson, in Monie. 1940)	humans	*Anabaena*	"Gastrointestinal disorders"
1959	Gull Lake, Saskatchewan, Canada (Dillenberg, 1959; Dillenberg and Dehnel, 1960; Senior, 1960)	Oregon tourist	*Microcystis*	Headache, nausea, and gastro-intestinal upset
1950	Govan, Long Lake, Saskatchewan, Canada (Dillenberg, 1959; Senior, 1960)	10 children at a camp	*Anabaena*	Diarrhea and vomiting
1959	Fort Qu'Appelle, Echo Lake, Saskatchewan, Canada (Dillenberg, 1959; Dillenberg and Dehnel, 1960; Senior, 1960)	Dr. M., a physician, practicing part-time	1. *Microcystis* 2. *Anabaena circinalis*	Crampy stomach pains, nausea, vomiting, painful diarrhea, fever, headache, weakness, pains in muscles and joints
1960	Regina, Saskatchewan, Canada (Dillenberg, 1962)	physician's 4-year-old son	*Aphanizomenon*	Abdominal pain, nausea, vomiting, diarrhea, wooziness, headache, thirst
1961	Saskatchewan, Canada (Dillenberg, 1962)	4 students	1. *Microcystis* 2. *Anabaena*	Headaches, general malaise. loose stools

1. GASTROINTESTINAL (See Table VII)

The digestive tract understandably has harbored more algae than any other bodily system, because most algae are ingested. The earliest disorder documented was that by Dr. Edward Farre, an English physician, in 1842, in a paper before the Microscopical Society in London (149-152). He described a 35-year-old female who "... suffered lately from slight dyspepsia. Six days ago, after suffering considerable griping pains in the bowels, which continued for twelve hours, she passed *per anum* a number of shreds, which being discharged with some difficulty, and causing an obstruction of the bowel, her attention was thereby attracted, and some of the shreds were pulled away by herself, so that there can be no question as to the source whence they were derived." Farre identified these shreds as *Oscillatoria*, and he stated their origin probably was the drinking water "supplied by the ordinary service-pipes of the metropolis."

After Farre, we have two women with sprue-like syndrome reported from Puerto Rico by Ashford and his colleagues (153). The isolated alga was *Prototheca portoricensis*.

Next we have an interesting series of episodes occurring in 1930-1931 in Washington and in various communities in the Ohio River Valley, including Louisville (154-156). These outbreaks were of acute epidemic proportions, with hundreds to thousands of individuals involved.

That these episodes were related is indicated by their common water supply, as well as by their strikingly similar symptomatology. They were "characterized by a sudden onset, pain in the region of the stomach, usually nausea or vomiting or both, and followed by diarrhea of varying severity. Those ill had essentially no fever ... duration ... varied with severity—usually from less than one day to upward of four days." Medical observers considered the manifestations similar to those produced by chemical irritants or purgatives; they felt that "... the character of the onset and the ensuing symptoms do not suggest a disease caused by a bacterial infection. ..." Indeed, in Tisdale's report (154,155) it is stated that "... the bacteriological records continuously indicated a drinking water of safe quality." In some instances the diagnosis was a little more facile—"intestinal flu"—but fortunately without invoking the ubiquitous "virus."

Although no exact species identifications were offered, the authors were positive on the presence of unusually heavy algal growth (some of the blue-green variety) following a great drought, and that these imparted a bad taste and smell to the water. These blooms were notably unresponsive to any water purification methods. Even though the epidemiological data point to algae, and the symptoms resemble those of proven algal intoxication, there are some who feel these Ohio Valley outbreaks are "unconvincing" and that more proof is required (156). These comments apply also to the digestive disturbances occurring in crops at Yellowstone National Park in 1925, 1929, and 1930 (157), and to the New Jersey episode in 1940 mentioned by Nelson (158).

We are greatly indebted to Dillenberg and his colleagues (81-88) in Saskatchewan for a series of extremely well-documented cases, noted after many instances of animal fatalities. The first concerned a tourist from Oregon who fell ill with headache, nausea, and gastrointestinal upset the night after swimming in a lake. The clinical diagnosis had been enteritis or amebic dysentery, but the stool showed only *Microcystis.* Reported recovery some 24 hr later with chloramphenicol was probably a spontaneous phenomenon, rather than being due to the antibiotic.

Anabaena was the responsible organism in ten children who developed diarrhea and vomiting a day after bathing in algae-covered Long Lake. Another observation by Dillenberg concerned a physician who slipped and fell into Echo Lake and accidentally swallowed a half pint of lake water. Three hours later he experienced crampy stomach pains and nausea, then vomited several times. Two hours later he had painful diarrhea. The next morning he had a fever of 102°, a splitting headache, pains in muscles and joints, and weakness. Examination of the slimy green stools revealed innumerable spheres of *Microcystis* and some *Anabaena* chains, but no other pathogens. The physician recovered in several days (81-85).

In a recent letter we learned that Dr. Dillenberg's own four-year-old son fell into the lake, swallowed an undetermined amount of lake water, then developed abdominal pain, nausea, vomiting, and diarrhea; next day he had wooziness, headache, and great thirst. *Aphanizomenon* was found in both vomitus and stool. In 1961, Dillenberg also had occasion to see four students who developed headache, malaise, and diarrhea after swimming in water heavily infested with *Microcystis* and *Anabaena.*

TABLE VIII

Human Respiratory Disorders Associated with Algae

Year	Locale and Author	Victims	Algae Involved	Manifestations of Toxicity
1916	West Coast of Florida, U.S.A. (Taylor, 1917)	many people	Dinoflagellates	Sneezing, coughing, chest tightness, dyspnea, sore throat, stuffed nose
1934–1935	Texas Coast, U.S.A. (Lund, 1935)	humans	"Heavy inshore plankton growth"	Irritation
1934	Muskego Lake, Waukesha County, Wisconsin, U.S.A. (Heise, 1949)	42-year-old man	*Oscillatoriaceae*	Itching of eyes, complete blockage of nose
1935	Muskego Lake, Waukesha County, Wisconsin, U.S.A. (Heise, (1949)	same man, one year later	*Oscillatoriaceae*	Itching of eyes, complete blockage of nose, plus mild asthma
1936–1946	North Lake, Waukesha County, Wisconsin, U.S.A. (Heise, 1949)	same patient	*Oscillatoriaceae*	Nasal discharge and blockage, asthma
1945	Lake Keesus, Waukesha County, Wisconsin, U.S.A. (Heise, 1949)	39-year-old woman	*Oscillatoriaceae*	Swollen eyelids, blocked nares, generalized urticaria
1946	Lake Keesus, Waukesha County, Wisconsin, U.S.A. (Heise, 1949)	same patient	*Oscillatoriaceae*	Swollen eyelids, blocked nares, generalized urticaria

1946–1947	Captiva Island, Florida, U.S.A. (Gunter, Williams, Davis, and Smith, 1948)	humans	*Gymnodinium brevis*	Burning of eyes, stinging of nostrils, hard cough
1946–1947	Captiva Island, and other islands off the west coast of Florida, U.S.A. (Galtsoff, 1948)	humans	*Gymnodinium brevis*	Burning in throat, nostrils and eyes; sneezing and coughing
1946–1947	West Coast of Florida, U.S.A. (Hunter and McLaughlin, 1958)	humans	*Gymnodinium brevis*	Irritation of respiratory tract
1947	Venice, Florida, U.S.A. (Thompson cited in Woodcock, 1948)	humans	*Gymnodinium* sp.	Hard cough, burning in respiratory tract
1947	Venice, Florida, U.S.A. (Woodcock, 1948)	author and two companions	*Gymnodinium* sp.	Hard cough, burning in respiratory tract
1947	Venice Beach, Florida, U.S.A. (Woodcock, 1948)	author and two companions	*Gymnodinium* sp.	Throat irritation
	Lower west coast of Florida, U.S.A. (Ingle, 1954)	people near shoreline	*Gymnodinium brevis*	Irritation of eyes, nose and throat

Dillenberg has mentioned another case which was interesting, although not gastrointestinal. It concerned a 12-year-old boy who, after swimming in a swimming hole, fell acutely ill with high fever, unconsciousness for six hours, labored breathing, and later with generalized pains, especially of joints and muscles. The stool contained *Aeromonas, Spirogyra,* and *Mougeotia,* while the swimming hole water also had abundant *Microcystis aeruginosa.* Dillenberg felt the *Aeromonas* probably caused the boy's patchy pneumonia, but that the sudden onset and the coma tie in much better with an algal toxin.

2. RESPIRATORY (See Table VIII)

In the respiratory group of diseases associated with algae, there are 14 separate episodes reported from 1916 to date (161-169). They can be roughly divided into two groups. The first group includes those associated with the *Oscillatoraceae,* with contact achieved by swimming in infested waters. Most of these were documented by Heise (162,163). The manifestations included itching of the eyes, blockage of the nose, and sore throat. Asthma and generalized urticaria were also present in some cases, and the afflictions were clearly demonstrated by Heise to be allergic in nature.

The second group comprises those disturbances generally produced by *Gymnodinium brevis,* with the organism wafted to the eyes and nose of the victims near the shore as "gas," vapor, or droplets (161, 164-169). Symptoms produced were burning of the eyes, sneezing, hard cough, chest tightening, and dyspnea. Most of these manifestations were due to direct chemical irritation, but allergy may have played a part in a few patients who also developed asthma.

3. DERMATOLOGIC (See Table IX)

Some of the cases in this category are controversial. The 67 ocean bathers reported by Sams (170) had itchy erythematous wheals, chiefly in the areas covered by bathing suits—possibly related to pressure or to retention of the irritating organism against the skin. Showering immediately after ocean exposure prevented

TABLE IX

Human Skin Disorders Associated with Algae

Year	Locale and Author	Victims	Algae Involved	Manifestations of Toxicity
1937–1949	Lower east coast of Florida, U.S.A. (Sams, 1949)	67 ocean bathers	"Plankton"	Erythematous wheals (in areas covered by bathing suit), itching, fever
1950	Lake Carey, Pennsylvania, U.S.A. (Cohen and Reif, 1953)	4-year-old girl	*Anabaena*	Erythematous papulo-vesicular dermatitis
1951	Lake Carey, Pennsylvania, and Canada (Cohen and Reif, 1953)	same patient, one year later	*Anabaena*	Erythematous papulo-vesicular dermatitis
1952	Lake Carey, Pennsylvania, U.S.A. (Cohen and Reif, 1953)	same patient, two years later	*Anabaena*	Erythematous papulo-vesicular dermatitis
1953	Pennsylvania, U.S.A. (Cohen and Reif, 1953)	swimmers	Blue-green algae	Itching, swelling and redness of conjunctivae
1958	Oahu, Hawaii, U.S.A. (Grauer, 1959; Banner, 1959; Grauer and Arnold, 1962)	125 cases received treatment; hundreds of mild unreported cases	*Lyngbya majuscula* Gomont	Itching and burning of skin, erythema, blisters, desquamation in areas covered by bathing suit
1959	Oahu, Hawaii, U.S.A. (Grauer and Arnold, 1962)	31-year-old medical officer	*Lyngbya majuscula* Gomont	Burning, stinging and itching of skin—then erythematous papules and vesicles in areas under bathing suit
1959	Oahu, Hawaii, U.S.A. (Grauer and Arnold, 1962)	9-year-old niece	*Lyngbya majuscula* Gomont	Burning, stinging and itching of skin—then erythematous papules and vesicles in areas under bathing suit
1959	Oahu, Hawaii, U.S.A. (Grauer and Arnold, 1962)	2 other adults	*Lyngbya majuscula* Gomont	Burning, stinging and itching of skin—then erythematous papules and vesicles in areas under bathing suit
1961	Georgia, U.S.A. (Hardin, 1961)	people who swam in sea water off Florida coast	"Ocean organism"	Burning, stinging and itching of skin—then erythematous papules and vesicles in areas under bathing suit

the dermatitis. Younger children also often had fever and malaise. Similar episodes of dermatitis in other parts of the world were mentioned by various discussers of Sams' paper, and some were stated to be associated with "red tide," which often is caused by *Gymnodinium*. Other authors have felt that "bather's itch" is due to some type of nematocyst (171,172).

Cohen and Reif (173) made a careful study of repeated episodes of a definitely allergic case of papulo-vesicular dermatitis occurring in a youngster whenever she swam in a lake with *Anabaena*. Special sensitivity was demonstrated to phycocyanin derived from the *Anabaena*. The same authors also reported an acute allergic conjunctivitis in many other individuals swimming in lakes containing blue-green algae.

The 125 Hawaiian cases reported by Grauer and Arnold (174,175), and by Banner (176), had a dermatitis resembling in its bathing suit distribution those described by Sams (170). These cases, however, were much more severe. The redness was followed in a few hours by blistering and deep desquamation, leaving a moist, bright red, tender and painful area, often appearing like a burn, even going on to scarring in severe cases. The authors considered the eruptions to be due to primary irritation—not allergy—by *Lyngbya majuscula*, after controlled studies in many human volunteers. They also stressed the prophylactic value of cleansing the exposed area with soap and water.

An interesting dermatologic sidelight is that offered by de Almeida and his colleagues in Brazil in 1946 (see Table X). They reported the isolation of *Chlorella* or *Chlorococcum* from lesions of three patients with mycoses—one with renal actinomycosis, one with pulmonary and lingual blastomycosis, and one with lymphatic-integumentary blastomycosis. In a fourth patient with actinomycosis, only vegetations suspicious of algae were found. We are unable to say what role the algae may have played here—whether they were symbiotic, whether they grew in a patient weakened by another chronic disease, or whether they were able to grow because of therapy administered for the mycoses.

In addition to these various spontaneous episodes of illness associated with algae—and bear in mind Dr. Gorham's superb demonstrations of toxins—there have been various clinical experiments with algae.

Szendy (19) in Brazil, in 1940, thought he had isolated algae

TABLE X

Human Mycoses Associated with Algae

Year	Locale and Author	Victims	Algae Involved	Disorder
1913	Bahai, Brazil (Torres, cited in de Almeida, Lacaz, and Forattini, 1946)	patient	"Green vegetation"	Fistula in iliac fossa and lumbar region due to actinomycosis
1946	Sao Paulo, Brazil (de Almeida, Lacaz, and Forattini, 1946)	white woman	*Chlorella vulgaris* or *Chlorococcum*	Renal actinomycosis
1946	Sao Paulo, Brazil (de Almeida, Lacaz, and Forattini, 1946)	white farmer	*Chlorella vulgaris* or *Chlorococcum*	Pulmonary blastomycosis and an ulcerous lesion of the tongue
1946	Sao Paulo, Brazil (de Almeida, Lacaz, and Forattini, 1946)	man	*Chlorella vulgaris* or *Chlorococcum*	Blastomycosis of the lymphatic-integumentary type. Material was taken from ulcerous lesions of the left inter-labial sulcus

from patients with metapneumothorax and pleural effusions; Mariani (21) in Italy, two years later, could not duplicate these results. A creditable series of studies was made by Woodcock (168), who, in 1947, had reported the severe respiratory irritations in Florida patients near the shore. Using human volunteers, Woodcock first sprayed into the nose and throat an unconcentrated reddish sea water (containing very many *Gymnodinium brevis*), and many times reproduced the cough and respiratory tract burning experienced along local beaches during onshore winds. The same test made with *clear* sea water produced *no* irritation. Next, Woodcock and two companions took to sea. During 5 hr with no wind or white caps, there was no respiratory trouble; however, during a 30-min squall, coughing and respiratory burning occurred, only to subside rapidly after the wind and spray had died down. Woodcock then repeatedly produced respiratory irritation with vapor rising from a pan of *red* sea water heated to boiling; the irritation regularly ceased when the boiling stopped. It could not be reproduced when relatively clear sea water, *without Gymnodinium*, was used. It was also determined that the irritant vapor could be filtered out when the volunteers held absorbent cotton over the nose and mouth. Woodcock also found that simple storage of *red* water for several weeks did not impair its respiratory irritant qualities.

Heise, who had reported clinical episodes of ocular and nasal irritations in swimmers exposed to *Oscillatoraceae* (162, 163), ran a series of allergic studies in 1949 to further his understanding. His first subject was a 57-year-old male whose hay fever was aggravated by swimming in a lake. The patient gave an immediate positive skin reaction to a 1:100,000 dilution of extract of *Oscillatoraceae* scum collected at the lake shore. Passive transfer tests were also positive. When 0.03 cc of 1:1000 dilution was given intracutaneously, asthma occurred within 20 min, and a 4-cm skin wheal at the injection site presented in the next 4 hr. Final allergic proof was achieved when Heise (163) was able to desensitize the patient with potent algal extract, to the point where the symptoms disappeared and the man could swim happily ever after!

Heise further tested the same algal antigen in two large groups of human volunteers. Those who had had no symptoms failed to react. Those in whom swimming produced "hay fever" reacted to high dilutions; in fact, one 39-year-old woman required epinephrine for the very severe local reaction.

In 1951, Heise also demonstrated that 10 patients, sensitive to *Oscillatoraceae*, were also highly allergic to *Microcystis*. At the same time, skin testing of 50 "normal" people with both *Oscillatoraceae* and *Microcystis* antigens failed to elicit any reaction.

An experiment of a different sort was tried on five young volunteers by McDowell and four associates (178-180). In order to determine the efficacy of algae as a human food source, they fed preparations of *Chlorella ellipsoides* and *Scenedesmus* in gradually increasing amounts in an isocaloric diet, from 10 g daily to 500 g. The results favored the Malthusians, for the subjects developed early abdominal distension, eructation, flatulence, and bulky dry bowel movements, all progressively more severe. Later they also suffered from nausea, vomiting, abdominal and rectal pain, headache, malaise, and weight loss.

4. ICHTHYOSARCOTOXICOSIS

In addition to the direct toxicities of algae hereto recorded—and the effects demonstrated yesterday by Dr. Gorham—there are the indirect algal effects seen after the ingestion of certain fish. The earliest reported episode of fish poisoning is by Meyer (181) in the Virgin Islands in 1530, and since then many incidents have been reported all over the world, with many thousands of victims, with over 500 documented deaths.

The best known are the periodic acute paralytic shellfish poisonings from eating mussels and clams (181-189). Symptoms are mainly neurological, with prickling and numbness of the mouth and fingers, ataxia and incoordination, ascending paralysis, and respiratory failure. Death may result in one to twelve hours. Outbreaks have generally occurred during the summer months. Linder (189), in 1928, first suggested that the toxicity might derive from the food eaten by the fish. Meyer (181), Sommer (181, 182), and their coworkers later showed that toxicity was present only when the shellfish ingested large quantities of the dinoflagellate *Gonyaulax*. Warm weather and increased nutrients in the water may make the algae grow so well that the sea may appear rusty red by day and phosphorescent at night. These changes, incidentally, have been used by knowing natives as warning signs.

The toxin is believed to be a highly potent heat-stable alkaloid with effects similar to those of strychnine, muscarine, and aconitine. The poison is concentrated in the livers or digestive organs; thus the "dark meat" is more toxic. The fish themselves are not affected, and they will continue to be toxic to predators for several weeks after the *Gonyaulax* have disappeared from their diet.

Autopsies have showed no structural changes in the victims of *acute* shellfish poisoning, probably because the toxin acted so fast. In the *chronically* poisoned experimental animals, however, definite changes have been found in the medullary ganglion cells, and in the large cells of the anterior horn of the spinal cord. Damage is also seen in the Golgi apparatus of some of the spinal ganglia. Neural mitochondria are normal, but those in the renal convoluted tubules show damage. The latter ties in with the clinical renal disturbances reported by some in human shellfish poisoning (187, 188).

Besides mussels and clams, other *periodically* poisonous fish are found all over the world, but in *localized* sectors. Their toxicity is unrelated to spoilage or bacterial contamination. These include chiefly surgeonfish, triggerfish, pompano, porcupine fish, wrasse, snapper, filefish, surmullet or goatfish, moray eel, sea bass or grouper, and barracuda.

John Randall (190) has altered the classification of Halstead and Lively (191, 192) to include *Gymnothorax* (Moray eel) poisoning as a severe *Ciguatera* type, and to exclude *Scombroid* poisoning as perhaps bacterial. The manifestations are chiefly neuromuscular, with numbness, nausea, headache, myalgia, weakness, motor incoordination, convulsions, coma, and respiratory paralysis—all in appropriate progression, depending upon severity. Mortality is probably under 5%.

Randall considers a blue-green alga as the source of the toxin. The localization of poisonous fish thus would depend on the availability of nutrients for algal growth, and algae as fish food.

Another interesting disease which needs mention here is *Haffkrankheit*, first described in 1924 (193) as an epidemic sickness involving chiefly sailors in Königsberghaff in East Prussia. It recurred several times there in later years, and also in Sweden near Lake Ymsen (57). Epidemics have also been recorded in Russia, where it has gone under the names of *Yuksov* and *Sartlan Disease* (194-196), after the locations of the outbreaks.

The disease has a sudden onset 15-20 hr after eating of fish, and is characterized by weakness, muscular pain and tenderness, producing spasms and immobility, and some digestive upset. The urine soon thereafter becomes brownish-black. The picture may clear within a few hours or days, or go on to death from uremia in about 1% of cases.

The basic pathophysiology is a myoglobinuria.

Children rarely have had the disease, but it has been quite extensive and virulent in the epidemic areas in cats, foxes, and shore birds. Many causes have been invoked, including various chemicals (arsines, selenium, sulphites), bacteria and viruses.

It is now generally conceded that it is due to periodic ingestion of toxic algae by some fish in the basic food cycle. The periodicity of epidemics is related to the concurrence of favorable parameters for algal nutrition and growth, and for wind and current localizations in a given area. That certain individual patients are recurrently stricken may represent some individual metabolic susceptibility.

5. MISCELLANEOUS

In line with the granulomata and neoplasms produced in animals by various algal substances [including the recent work of Davidson (7, 113, 114) in Texas] is the interesting article by Reed, Smith, and Sternberg from Tulane (197). They reported five cases of foreign body granuloma resulting from lubricating jelly which utilized carrageenin. The lesions were apparently associated with traumatic urethral instrumentation and introduction of abundant amounts of jelly under pressure. Subcutaneous injection in rats of the two types of lubricating jelly used in the humans produced identical granulomata (115-124).

Because of an almost romanticized trend to predict the use of algae in closed ecological systems—space ships and atomic subs—for O_2–CO_2 exchange (1 kg algae in 100 liter suspension will respire one 70-kg man) and even for complete recycling of biological elements (198), *a warning must be issued.* It is that green plants, including algae, produce small quantities of *carbon monoxide* (199-201). Syrrel S. Wilks noted an unusually high level of carbon monoxide in human muscle into which much green vegetation had been imbedded during a fatal plane accident. Wilks then measured the carbon monoxide formation of algae and various other green plants, and found that carbon monoxide formation did occur in normal photosynthesis in light, and also with injury to the organisms. Clearly, then, to prevent asphyxiation in a closed system, a means must be developed for disposing of the carbon monoxide.

Since most algae are aquatic and since this is an era of central water supply (even if it is only a village pond or tank), basic principles of epidemiology lead us to a number of intriguing speculations concerning the causative role of algae in certain disease syndromes.

Let us first take cholera. It has a long, respectable history dating back to the 7th century A.D. in India, and is estimated to have disposed of some 66,000,000 natives since then. It is an acute specific infection involving primarily the lower portion of the ileum, and it is spread by contaminated water or food. And the causative organism is clearly known to be the *Vibrio comma*, despite previous suppositions that it was due to "miasmas, effluvia, or malaria." McGuire (202), however, feels that high temperature and high relative humidity, accompanied by intermittent rains, form the most favorable atmosphere for development of cholera. A more specific evaluation of environmental factors (geogens) was given by Cockburn and Cassanos (204) in 1960, in a study of cholera in Bengal, where the disease is endemic. The peak periods are in the hot dry months. Cholera vibrio thrives in highly alkaline media, too alkaline for other enteric organisms. Cockburn and Cassanos believed that in hot dry weather, the heavy algal growths in the village water tanks are a *key* factor and that they raise pH so high that cholera vibrio is favored over other organisms and epidemics occur. The infections traditionally end with the onset of the monsoon.

The interrelationships of algae and bacteria were touched on by Dr. Gorham, who feels the "slow-death factor" is probably bacterial. In this connection, there is also great interest in Floyd Davidson's studies (101) of the effects of blue-green algal extracts upon the growth of bacteria. In some instances there was inhibition of growth of *Salmonella enteritidis, Salmonella typhosa, Shigella dysenteriae*, and *Staphylococcus aureus*. In other instances, one species of bacteria was *stimulated* by low concentration of algal extract and *inhibited* by high levels. In still other cases an algal extract stimulated one bacterial species and inhibited another. Ryther (205) expressed similar thoughts on inhibitory effects of phytoplankton.

And, of course, you have heard Dr. Jakob's good work today. And Dillenberg (84, 85) also wrote us: "I am inclined to believe that symbiotic relations exist between algal toxins and bacterial or viral manifestations. I firmly believe this in the case of Botulism C poisoning of ducks, but also that it may hold for many other conditions, as yet not clarified in their epidemiological pattern. Polio? Very worthwhile to investigate! The seasonal pattern would fit in most cases. . . ."

Now, if we consider the question of polio, there are certain epidemiological facts:

1. Although polio virus is always around, there are not always clinical cases, even in unvaccinated individuals.
2. People with digestive and neurological symptoms during the polio season are often automatically diagnosed as polio—although similar manifestations may occur from ECHO and Coxsackie viruses, and from algal intoxications.
3. Polio outbreaks are almost invariably associated with water—for drinking or swimming.
4. The polio season coincides with the time of greatest growth and development of toxic algae.
5. The meteorologist Landers (206) correlates the 1954 Tallahassee epidemic with low humidity and drought.
6. The meteorologist Father Gherzi (207) associates Montreal polio epidemics with *maritime* air currents (as opposed to polar air currents).

Nigrelli has said (208): "There can be no doubt that antibiosis is a wide-spread limiting factor even among such organisms as fishes and invertebrates as well as the phyto—and zoo—plankton." It is thus fascinating to read the subsequent report of a group from Bethesda that: "Antibacterial and antiviral substances have been isolated from the abalone and the oyster. Antibacterial activity against *Streptococcus pyogenes,* as well as antiviral activity against influenza virus and polio-virus, was demonstrated *in vitro* and *in vivo.*"

And, if we consider Davidson's (101) concept of algal alteration of pathogen virulence; Dr. Jakob's thoughts on phycologic secretions; Cockburn's (203) notion of algal alteration of pH; Rusk's (209) report that a conch diet in Bimini has long prevented polio (possibly via Hardy's inhibition theory); and Seller's (210) idea that high CO_2 tissue concentration produces paralysis in polio (algae *can* alter CO_2)—then . . .

Are we not entitled to look more closely at the relationship between algae and polio or polio-like syndromes?

In the same spirit, we must have another good look at hepatitis—

both acute and chronic forms, especially in view of the tremendous increase of the former in recent years.

In 1946 Ashworth and Mason (91), in Texas, using *Microcystis aeruginosa* in an exhaustive study on albino rats, were able to produce acute hepatitis and jaundice. The rapid acute changes included acute parenchymatous, hydropic, and fatty degeneration to necrosis and congestion at the center of the lobule. Regeneration began in 3-5 days, with complete recovery in 30 days, from *single* administration of toxin (incidentally, similar acute necrosis was seen in the kidneys).

This work corroborated that of Steyn (34-38) in 1944 and 1945. He reported acute hepatitis and jaundice in a variety of animals, and also stressed "chronic hardening of the liver" (or cirrhosis) due to *Microcystis*. Steyn also thought human hepatic involvement was likely. Stephens (39,40) in South Africa, Mullor (62) in Argentina, O'Donoghue and Wilton (77) in the U.S.A., and Dillenberg (81-84) in Canada are only a few others who have reported hepatic involvement.

The three main algae implicated in these disorders have been *Microcystis, Anabaena*, and *Aphanizomenon*. As my brother has indicated, these species are very often present in domestic water supplies declared quite fit for human consumption. Grant and Hughes (92,93), among others, have indicated that ordinary standard water purification procedures have no effect on algae (see Table XI). (As a matter of fact, this is why in 1950 the Canadian Defense Research Board began considering algal toxins as possible water-denying agents for potential chemical warfare.)

My brother has visited numerous distilleries and wineries—strictly as an avocation!—here and abroad, and has been especially impressed by the repeated presence of blue-green algae in water supplies, pipes, and vats (9).

Reverting to acute hepatitis, there is a fascinating report in the Medical Journal of Australia (211) in January 1958 of a water-borne outbreak in New South Wales:

> *Twenty* students (ages 17-19) attended a picnic. *Nineteen* drank raw river water. *Fifteen* developed vomiting, diarrhea, anorexia. *Six* developed frank acute hepatitis 4-5 weeks later, while *five* had subclinical hepatitis. Now—only those attending the picnic got hepatitis, with *no secondary cases*. The season was hot and dry, the water polluted. Although no algae were identified, the implications are clear enough.

TABLE XI

Algae with Known Toxic Potential Identified in Municipal Water Supplies

U.S.A.		Abroad	
Illinois	*Anabaena*	Brisbane, Australia	*Microcystis* *Anabaena*
Minnesota	*Anabaena*		
Summit, New Jersey	1) *Aphanizomenon* 2) *Anabaena*	River Ouse Reservoir, England	*Oscillatoria*
Lexington, Kentucky	*Aphanizomenon*	Scotland	*Oscillatoria*
Tiffin, Ohio	*Anabaena*	Regina and Moose Jaw, Saskatchewan, Canada	*Anabaena* *Microcystis*
Eau Claire, Wisconsin (swimming)	1) *Microcystis* 2) *Nostoc*	Weyburn, Saskatchewan, Canada	*Anabaena* *Aphanizomenon*
New York City	1) *Anabaena* 2) *Aphanizomenon*	Switzerland	*Oscillatoria*
	3) *Microcystis* 4) *Coelosphaerium*	Pinsk Province, U.S.S.R.	*Microcystis*
Storm Lake, Iowa	*Anabaena*	Isle Margarita, Venezuela	*Oscillatoria* *Anabaena*
Missouri River water	*Anabaena*		
Mississippi River water	*Oscillatoria*		

There are also recent reports of cockroaches and a chimpanzee acting as vectors for hepatitis (212, 213)—and you will recall that animals do harbor algae.

Also, even in those hepatitis outbreaks (in New Jersey and Mississippi) traced to "infected" shellfish (clams and oysters), no check was made on the presence of toxic algae (214-216). This, despite reports by Ryther (205), Davis (217), Chanley (218), Gurevich (219), and Goryunova (220) that algae can and do have toxic effects on shellfish.

And finally, it is common knowledge that detergents, so popular in the past 15 years, do have the capacity to promote algal growth, probably via phosphate compounds (221, 222). It may be no accident that the increased incidence of hepatitis parallels the increased use of detergents.

Again, it must be mentioned that hepatitis can be caused by toxic algae alone, or by some interaction with the elusive hepatitis virus, should it actually be isolated.

These involvements we have described are of tremendous interest, but we physicians are even more concerned with the *chronic* algal intoxications resulting from long-standing exposure to repeated small doses of toxin. This applies not only to cirrhosis of the liver, but also to some of the neuropathies whose origins are mysterious and obscure:

1. The various scleroses, especially the amyotrophic lateral sclerosis of the Chamorro tribe in the Marianas, whose incidence is 100 times that of the rest of the world (420/100,000) and produces 8-10% of adult deaths. An extensive 1955 survey (223) guessed it might be genetic. We have been perturbed that no consideration was given to toxic algae in a maritime and fish-eating population. It is at least gratifying to hear a recent opinion that the illness might well be toxic in origin, even if the blame was laid on certain nitrile-containing nuts which are ground into flour by the natives (224).
2. The peculiar paralytic "Jamaican Sickness" described in the British West Indies by Dr. E. K. Cruickshank (University College of the West Indies, Kingston, Jamaica, B.W.I.) (225) and attributed by some to the native "bush tea."
3. The unsolved Filipino "nightmare sickness" or "Bangotgot," fatal to over 100 healthy men in recent years following an acute hallucinatory episode, with disturbingly negative autopsy findings (176, 226).

Another wide-open field for speculation is that of dental cavitation. It is the consensus of dentists that cavities are secondary to something which can penetrate enamel. Although infrequent, algae have been isolated from the normal human mouth (10-12), especially *Oscillatoria*. Considering that *Oscillatoria* are often present in drinking water, we might expect them in the mouth more frequently (9,148).

Now consider certain facts:

1. *Oscillatoria* have often been reported to have the capacity to dissolve concrete in dams and pools (227-231).

2. Peyer (232) documented repeatedly that algae and fungi could bore holes in the teeth of fish and other animals.

3. Zipkin (233) indicated that copper sulfate (an algicide) inhibits caries in rats and hamsters when added to diet or drinking water.

4. The efficacy of fluoridation in preventing caries may be related to the capacity of its sister halide, chlorine, to act as an algicide (234).

These data certainly suggest that algae may contribute to the problem of dental decay.

I should like to insert a quotation here:

> "As soon as people understand that something more than a mere analysis by a chemist or an inoculation of guinea pigs is necessary before the purity of a water supply can be ascertained, there will be a greater demand for those who can make accurate microscopical examination, and we may hope for an increase in our knowledge of this most important aspect of a most important subject."

It is interesting that G. T. Moore (235) said this in 1899!

In closing, I would like to remind you of Santayana's observation that those who cannot remember the past are condemned to repeat it!

LITERATURE CITED

1. Schwimmer, M., and Schwimmer, D. 1955. The role of algae and plankton in medicine. Grune & Stratton, New York. 85 pp.
2. Nova Scotia Research Foundation. 1953. Utilization of seaweed. Bibliography No. 2. Utilization of algae. Halifax. [47] leaves. (Supersedes Bibliography No. 1, 1952.)
3. Nova Scotia Research Foundation. 1955. Selected bibliography on algae. No. 3. Halifax, [73] leaves.
4. Nova Scotia Research Foundation. 1958. Selected bibliography on algae. No. 4. Halifax, 109 pp.
5. Nova Scotia Research Foundation. 1960. Selected bibliography on algae. No. 5. Halifax, 220 pp.
6. Ingram, W. M., and Prescott, G. W. 1954. Toxic fresh-water algae. *Am. Midland Naturalist* 52:75.
7. Davidson, F. F. 1959. Poisoning of wild and domestic animals by a toxic water-bloom of *Nostoc rivulare* Kütz. *J. Am. Water Works Assoc.* 51:1277.
8. Brongersma-Sanders, M. 1948. The importance of upwelling water to vertebrate paleontology and oil geology. *Verhandel. K. Nederl. Akad. Wetensch. Afd. Natuurkunde,* Sect. 2, Deel 45, No. 4, p. 1.
9. Chamberlain, W. J. 1948. Effects of algae on water supply. University of Queensland, Brisbane, Department of Chemistry. Papers, Vol. 1, No. 29. 80 pp., 60 plates.
10. Simons, H. 1922. Saprophytische *Oscillarien* des Menschen und der Tiere [Saprophytic *Oscillaria* in man and animals]. *Centralbl. f. Bakt. I. Abt. Originale.* 88:501.
11. Langeron, M. 1923. Les *Oscillariees* parasites du tube digestif de l'homme et des animaux [The parasitic *Oscillatoriae* in the digestive canal of man and animals]. *Ann. de Parasit.* 1(1):75.

12. Langeron, M. 1923. Les *Oscillariees* parasites du tube digestif de l'homme et des animaux [The parasitic *Oscillatoriae* in the digestive canal of man and animals]. *Ann. de Parasit.* 1(2):113.
13. Valentin, G. 1836. *Hygrocrocis intestinalis,* eine auf der lebendigen und ungestört functionierenden Schleimhaut des Darmkanales vegetierende Conferve [*Hygrocrocis intestinalis,* a Conferva vegetating on the living, normally functioning mucosa of the intestinal canal]. *In* Repertorium für Anatomie und Physiologie, Berlin. Vol. 1, p. 110.
14. Leidy, J. 1904. Observations. *Proc. Acad. Nat. Sci.* 4:225 (1849). *Reprint: Smithsonian Misc. Collections.* 46, No. 1477, p. 12.
15. Robin, C. P. 1853. Histoire naturelle des végétaux parasites qui croissent sur l'homme et sur les animaux vivants [Natural history of plant parasites growing in man and live animals]. J. B. Baillière, Paris. pp. 359-360; 404-405.
16. Müller, R. 1911. Zur Stellung der Krankheitserreger im Natursystem [On the role of pathogens in nature]. *München. Med. Wochschr.* 58:2246.
17. Collin, B. 1912-1913. Sur un ensemble de protistes parasites des batraciens (note préliminaire). d. *Arthromitus batrachorum* n. sp. [Preliminary note on the association of plant parasites and batraciens]. *Arch. de zoologie exp. et gen. Notes et revue* 51(3):63-64.
18. Chatton, E., and Pérard, C. 1913. Schizophytes du caecum du cobaye. I. *Oscillospira Guilliermondi* n.g., n. sp. [Schizophytes in the guinea pig cecum]. *Compt. rend. soc. de biol.* 74:1159.
19. Redaelli, P., and Ciferri, R. 1935. La patogenicita per gli animali di alghe acloriche coprofite del genere *Prototheca* [Pathogenicity of coprophytic achloric algae of the genus *Prototheca* in animals]. *Boll. soc. ital. biol. sper.* 10:809. *Also in French:* Pouvoir pathogène pour les animaux des algues coprophytes achloriques du genre *Prototheca.* Observations sur les protothecaceae [Pathogenicity of coprophytic achloric algae of the genus *Prototheca* in animals. Observations on protothecaceae]. *Boll. sez. ital. soc. internaz. microbiol.* 7:316.
20. Newiadomski, M. M. 1937. Blastozystentumoren [Blastocyte tumors]. *Zentralbl. f. Bakt. I. Abt. Originale* 138:244.
21. Mariani, P. L. 1942. Ricerche sperimentali intorno ad alcune alghe parassite dell'uomo [Experimental researches on some parasitic algae of man]. *Boll. soc. ital. di microbiol.* 14:113.
22. Tiffany, L. H. 1935. Algae of bizarre abodes. *Sci. Month.* 40:541.
23. Tiffany, L. H. 1958. Algae, the grass of many waters. Second ed. C. C. Thomas, Springfield, Illinois. 199 pp.
24. Messikommer, E. 1948. Algennachweis in Entenexkrementen [Evidence of algae in duck feces]. *Hydrobiologia, Acta Hydrobiologica* 1:22.
25. Sieburth, J. M. 1959. Gastrointestinal microflora of antarctic birds. *J. Bact.* 77:521.
26. Francis, G. 1878. Poisonous Australian lake. *Nature* 18:11.
27. Porter, E. D. Investigations of supposed poisonous vegetation in the waters of some of the lakes of Minnesota. Dept. of Agric. Report 1881/86. Biennial Report of the Board of Regents, No. 4, Suppl. No. 1, p. 95.
28. Stalker, M. On the Waterville cattle disease. University of Minnesota, Dept. of Agric. Report 1881/86. Biennial Report of the Board of Regents, No. 4, Suppl. No. 1, p. 105.
29. Arthur, J. C. Second report on some algae of Minnesota supposed to be poisonous. University of Minnesota. Dept. of Agric. Report 1881/86. Biennial Report of the Board of Regents, No. 4, Suppl. No. 1, p. 109. *Idem. Bull. Minnesota Acad. Nat. Sci.* 3:97, 1885.

30. Nelson, N. P. B. 1903-1904. Observations upon some algae which cause "water bloom." *Minnesota Bot. Studies* 3:51. (Minnesota Geolog. and Nat. History Survey. Reports of Survey. Bot. Ser. VI.)
31. Fitch, C. P., et al. 1934. "Water bloom" as a cause of poisoning in domestic animals. *Cornell Veterinarian* 24:30. *Idem.* Paper No. 1248 Journal Series, Minnesota Agricultural Experiment Station.
32. Gortner, R. A. [On toxic water bloom]. *Cited in* J. E. Tilden 1935. The algae and their life relations. Univ. of Minnesota Press, Minneapolis. p. 473.
33. Seydel, E. 1913. Fischsterben durch Wasserblüte [Fish deaths due to water-bloom]. *Mitt. Fischerei-Ver. Brandenburg n.s.* 5(9):87.
34. Steyn, D. G. 1943. Poisoning of animals by algae on dams and pans. *Farming in South Africa* 18:489.
35. Steyn, D. G. 1944. Poisonous and non-poisonous algae (waterbloom, scum) in dams and pans. *Farming in South Africa* 19:465.
36. Steyn, D. G. 1944. Vergifting deur slyk (algae) op damme en panne [Poisoning by algae on dams and pans]. *South African M. J.* 18:378.
37. Steyn, D. G. 1945. Poisoning of animals by algae (scum or waterbloom) in dams and pans. Pretoria, Union of South Africa. Dept. of Agric. and Forestry. 9 pp.
38. Steyn, D. G. 1945. Poisoning of animals and human beings by algae. *South Africa J. Sc.* 41:243.
39. Stephens, E. L. 1948. *Microcystis toxica* sp. n.: a poisonous alga from the Transvaal and Orange Free State. (Summary of paper read May 16, 1945, Roy. Soc. of South Africa.) *Hydrobiologia, Acta Hydrobiologica (The Hague)* 1:14.
40. Stephens, E. L. 1949. *Microcystis toxica* sp. n.: a poisonous alga from the Transvaal and Orange Free State. (Read May 16, 1945.) *Tr. Roy. Soc. South Africa* 32(pt. 1):105.
41. Cotton, H. L. 1914. Algae poisoning. *Am. J. Vet. Med.* 9:903.
42. Náday, L. 1914. Az állóvizék virágzása [Water-bloom of stagnant water]. *Természettud. Közlöny (K. Magyar természettudomanyi társulat. Budapest)* 46:432.
43. Gillam, W. G. 1925. The effect on live stock, of water contaminated with fresh water algae. *J. Am. Vet. Med. Assoc.* 67 (n.s. 20):780.
44. Howard, N. J., and Berry, A. E. 1933. Algal nuisances in surface waters. *Canad. Pub. Health J.* 24:377.
45. One hundred twenty seven hogs, 4 cows die after drinking water from [Big Stone] lake, stock stricken, last Saturday, all die in short time, lake water sent in for analysis. *Wilmot Enterprise,* Wilmot, S. Dak., Sept. 24, 1925.
46. Farmer tells some news [on stock poisoning in Big Stone Lake]. *Wilmot Enterprise*, Wilmot, S. Dak., Oct. 1, 1925.
47. Woodcock, E. F. 1927. Plants of Michigan poisonous to livestock. *J. Am. Vet. Med. Assoc.* 25:475.
48. Hindersson, R. 1933. Forgiftning av nötkreatur genon sötvattensplankton [Poisoning of cattle by fresh-water plankton]. *Finsk. Vet. Tidskrift* 39:179.
49. Prescott, G. W. 1933. Some effects of the blue-green algae, *Aphanizomenon flos-aquae*, on lake fish. *Collecting Net, Woods Hole, Mass.* 8:77.
50. Prescott, G. W. 1938. Objectionable algae and their control in lakes and reservoirs. *Reprint: Louisiana Municipal Review, Shreveport,* Vol. 1, Nos. 2 and 3 (July/August, Sept./Oct.).
51. Prescott, G. W. 1939. Some relationships of phytoplankton to limnology and aquatic biology. *In* Problems of lake biology. *Publ. Am. Assoc. Adv. Sc.* 10:65.
52. Prescott, G. W. 1948. Objectionable algae with reference to the killing of fish and other animals. *Hydrobiologia, Acta Hydrobiologica (The Hague)* 1:1.

53. Prescott, G. W. 1951. Algae of the western Great Lakes area, exclusive of desmids and diatoms. Bloomfield Hills, Mich. Cranbrook Inst. of Science, 1951. (Cranbrook Inst. of Science. Bull. No. 31.) p. 44.
54. McLeod, J. A., and Bondar, G. F. 1952. A case of suspected algal poisoning in Manitoba. *Canad. J. Pub. Health* 43:347.
55. Deem, A. W., and Thorp, F. 1939. Toxic algae in Colorado. *J. Am. Vet. Med. Assoc.* 95:542.
56. Durrell, L. W., and Deem, A. W. 1940. Toxic algae in Colorado. *J. Colorado-Wyoming Acad. Sc.* 2(6):18.
57. Berlin, R. 1948. Haff disease in Sweden. *Acta Med. Scand.* 129:560.
58. Louw, P. G. J. 1950. The active constituent of the poisonous algae, *Microcystis toxica* Stephens. With a note on experimental cases of algae poisoning in small animals, by J. D. Smit. *South African Industrial Chemist* 4:62.
59. Remer, F. *Cited in* Quin, A. H. 1943. Sheep poisoned by algae. *J. Am. Vet. Med. Assoc.* 102:299.
60. Rose, E. T. 1953. Toxic algae in Iowa lakes. *Proc. Iowa Acad. Sc.* 60:738.
61. Firkins, G. S. 1953. Toxic algae poisoning. *Iowa State College Veterinarian* 15:151.
62. Mullor, J. B. 1945. Algas tóxicas [Toxic algae]. *Rev. Sanidad, Asistencia Social y Trabajo (Santa Fé, Argentina)* 1:95.
63. Mullor, J. B., and Wachs, A. M. 1948. Algunas caracteristicas del alga toxica "Anabaena venenosa" [Some characteristics of toxic alga "Anabaena venenosa"]. Congreso Sudamericano de Quimica, 4th. Trabajos presentados, 1948, 1:326-327.
64. Mackenthun, K. M., Herman, E. F., and Bartsch, A. F. 1948. A heavy mortality of fishes resulting from the decomposition of algae in the Yahara River, Wisconsin. *Tr. Am. Fisheries Soc.* 75:175 (1945).
65. Olson, T. A. 1949. History of toxic plankton and associated phenomena. Algae-laden water causes death of domestic animals; nature of poison. *Sewage Works Engineering* 20(2):71.
66. Olson, T. A. 1951. Toxic plankton. Proc. Inservice Training Course in Water Works Problems, Univ. of Michigan, School of Public Health, Ann Arbor, Michigan. p. 86.
67. Olson, T. A. 1952. Toxic plankton. *Water and Sewage Works* 99:75.
68. Olson, T. A. 1960. Water poisoning—a study of poisonous algae blooms in Minnesota. (Laboratory section abstracts of papers presented at 87th Annual Meeting, 1959.) *Am. J. Pub. Health* 50:883.
69. Oliveiro, L. de, Nascimento, R. do, Krau, L., and Miranda, R. 1956. Diagnóstico biológico das mortandades de peixes na Lagoa Ridrogo de Freitas; nota prévia [Biological diagnosis of the mortality of fish in the Rodrigo de Freitas Lagoon; preliminary notes]. *Brasil Med.* 70:125.
70. Oliveiro, L. de, Nascimento, R. do, Krau, L., and Miranda, R. 1957. Observações hidrobiologicas e mortandade de peixas na Lagoa de Freitas [Hydrobiologic observations and the mortality of fish in the Rodrigo de Freitas Lagoon]. *Mem. Inst. Oswaldo Cruz* 55:211.
71. Bailey, J. W. 1959. Algae can poison cattle. *Jersey J.* 6:5.
72. Brandenburg, T. O., and Shigley, F. M. 1947. "Water bloom" as a cause of poisoning in livestock in North Dakota. *J. Am. Vet. Med. Assoc.* 110:384.
73. Barnum, D. A., Henderson, J. A., and Steward, A. G., 1950. Algae poisoning in Ontario. *Milk Producer* 25:312.
74. Stewart, A. G., Barnum, D. A., and Henderson, J. A. 1950. Algal poisoning in Ontario. *Canad. J. Comp. Med.* 14:197.
75. MacKinnon, A. F. 1950. Report on algae poisoning. *Canada J. Comp. Med.* 14:208.

76. Scott, R. M. 1952. Algal toxins. *Public Works* 83:54.
77. O'Donoghue, J. G., and Wilton, G. S. 1951. Algal poisoning in Alberta. *Canad. J. Comp. Med.* 15:193.
78. Shelubsky (Shilo), M. 1950. Observations on the properties of a toxin produced by *Microcystis*. Internat. Assoc. Theoret. and Applied Limnology. (11th International Limnological Congress, Brussels, 1950.) Proceedings 11:362.
79. Shilo (Shelubsky), M., and Aschner, M. 1953. Factors governing the toxicity of cultures containing phytoflagellate *Prymnesium parvum* Carter. *J. Gen. Microbiol.* 8:333.
80. Vinberg, G. G. 1955. Toxic phytoplankton. *Translated from: Uspekhi Sovr. Biologii* 38, 2(5):216. (National Research Council of Canada. Technical Translation TT-549.) Ottawa. 25 leaves.
81. Dillenberg, H. O. 1959. Toxic waterbloom in Saskatchewan. Presented before the 14th Annual Meeting INCDNCM, August 26-29, 1959, at Washington State College, Pullman, Washington.
82. Dillenberg, H. O., and Dehnel, M. K. 1960. Toxic waterbloom in Saskatchewan, 1959. *Canad. M. A. J.* 83:1151.
83. Dillenberg, H. O., and Dehnel, M. K. 1961. "Waterbloom poisoning." Fast and "slow death" factors isolated from blue-green algae at Canadian N R C Laboratories. *World-Wide Abstr. Gen. Med.* 4 (No. 4):20.
84. Dillenberg, H. O. 1961. Case reports of algae poisoning. Personal communication.
85. Dillenberg, H. O. 1962. Case reports of algae poisoning. Personal communication.
86. Water made children ill. 1959. *Saskatoon Star-Phoenix*, Saskatoon, Canada, July 22, 1959.
87. There is no ground for an algae scare. 1959. Qu'Appelle Lake tourist trade falls off. *The Leader-Post*, Regina, Canada, July 20, 1959 and July 21, 1959.
88. Senior, V. E. 1960. Algal poisoning in Saskatchewan. *Canad. J. Comp. Med.* 24:26.
89. Mason, M. F., and Wheeler, R. E. 1942. Observations upon the toxicity of blue-green algae (Am. Soc. Biol. Chemists. 36th Ann. Meeting, Boston). *Fed. Proc.* 1(1^2):124.
90. Wheeler, R. E., Lackey, J. B., and Schott, S. 1943. A contribution on the toxicity of algae. Pub. Health Rep. 57:1695 (Nov. 6, 1942). *Abstract in: J. Am. Vet. Med. Assoc.* 102:230.
91. Ashworth, C. T., and Mason, M. F. 1946. Observations on the pathological changes produced by a toxic substance present in blue-green algae *(Microcystis aeruginosa). Am. J. Path.* 22:369.
92. Grant, G. A. 1953. Toxic algae. I. Development of toxicity in blue-green algae. Canada. Defence Research Board. Defence Research Chemical Laboratories. Report No. 124. Project No. D52-20-20-18. Ottawa. 11 leaves.
93. Grant, G. A., and Hughes, E. O. 1953. Development of toxicity in blue-green algae. *Canad. J. Pub. Health* 44:334.
94. McIvor, R. A., and Grant, G. A. 1955. Toxic algae. II. Preliminary attempts to prepare toxic extracts. Canada. Defence Research Board. Defence Research Chemical Laboratories. Rep. No. 190. Project No. D52-20-20-18. Ottawa. 7 leaves.
95. Chaput, M., and Grant, G. A. 1958. Toxic algae. III. Screening of a number of species. Canada. Defence Research Board. Defence Research Laboratories. Report No. 279, Project No. D52-20-20-18. Ottawa. 6 leaves.
96. Thomson, W. K., Laing, A. C., and Grant, G. A. 1957. Toxic algae. IV. Isolation of toxic bacterial contaminants. Canada. Defence Research Board.

Defence Research Kingston Laboratory. Report No. 51. Project No. D52-20-20-18. Ottawa. 9 leaves.

97. Thomson, W. K. Toxic Algae. V. Study of toxic bacterial contaminants. Canada. Defence Research Board. Defence Research Kingston Laboratory. Report No. 63. Project No. 63. Project No. D52-20-20-18.
98. Hughes, E. O., Gorham, P. R., and Zehnder, A. 1958. Toxicity of unialgal culture of *Microcystis aeruginosa*. *Canad. J. Microb.* 4:225.
99. Zehnder, A., Hughes, E. O., and Gorham, P. R. 1956. Giftige Blaualgen. [Toxic blue algae]. Abstract. *Verh. Schweiz. Naturforsch. Ges.* 136:126.
100. Bishop, C. T., Anet, E. F. L. J., and Gorham, P. R. 1959. Isolation and identification of the fast-death factor in *Microcystis aeruginosa* NRC-1. *Canad. J. Biochem. Physiol.* 37:453.
101. Davidson, F. F. 1959. Effects of extracts of blue-green algae on pigment production by *Serratia marcescens*. *J. Gen. Microbiol.* 20:605.
102. Gorham, P. R. 1960. Toxic waterblooms of blue-green algae. *Canad. Vet. Jour.* 1:235.
103. Gorham, P. R. Laboratory studies on the toxins produced by waterblooms of blue-green algae. Read to the Laboratory Sect. Am. Pub. Health Assoc., Annual Meeting, Detroit, Mich., Nov. 14, 1961. Accepted for publication by *Am. J. Pub. Health.*
104. Fink, H., Schlie, I., and Herold, E. 1954. Über die Eiweissqualität einzelliger Grünalgen und ihre Beziehung zur alimentären Lebernekrose der Ratte. XI [On the protein quality of unicellular green algae and its relation to alimentary liver necrosis of the rat. XI]. *Naturwissenschaften* 41:169.
105. Fink, H., and Herold, E. 1957. Über die Eiweissqualität einzelliger Grünalgen und ihre Lebernekrose verhütende Wirkung. II [On the protein quality of unicellular green algae and its preventive action in necrosis of the liver]. *Hoppe Seyler's Z. f. Physiol. Chem.* 307:202.
106. Fink, H., and Herold, E. 1958. Über die Eiweissqualität einzelliger Grünalgen und ihre Lebernekrose verhütende Wirkung. III. Über den Einfluss des Trocknens auf das diätetische Verhalten der einzelligen Zuchtalge *Scenedesmus obliquus*. [On the protein quality of unicellular green algae and its preventive action in necrosis of the liver. III. On the effect of drying on the dietetic behavior of the unicellular cultured alga *Scenedesmus obliquus*]. *Hoppe Seyler's Z. f. Physiol. Chem.* 311:13.
107. Arakawa, S., et al. 1960. Experimental breeding of white leghorn with *Chlorella*-added combined feed. *Japan. J. Exp. Med.* 30(3):185.
108. Arakawa, S., et al. 1960. Experimental breeding of mice with *Chlorella*-added combined feed and their resisting power against dysentery bacilli. *Yokohama Med. Bull.* 11:186.
109. Redaelli, P., and Ciferri, R. 1935. Une nouvelle hypothese sur la nature du *Blastocystis hominis* [A new hypothesis concerning the nature of *Blastocystis hominis*]. *Boll. sez. ital. soc. internaz. microbiol.* 7:321.
110. Redaelli, P., and Ciferri, R. 1936. Argomenti a favore di una sistemazione del genere "Blastocystis" nelle Algae [Arguments for classification of genus "Blastocystis" with algae]. *Boll. ist. sieroterap. milanese* 15:154.
111. Redaelli, P., and Cifferi, R. 1940. Ulteriori osservazioni su *Blastocystis* della rana e conferma della natura e posizione sistematica dell'alga [Further observations on *Blastocystis* of the frog and confirmation of the nature and systematic position of the alga]. *Mycopathologia* 2:239.
112. Ciferri, R. 1961. Algheosı [Algoses]. Enciclopedia Medica Italiana. Aggiornamento. Venezia, Roma, Istituto per la Collaborazione Culturale, Vol. 1, p. 140.

113. Davidson, F. F. 1961. Personal communication *re* tumor formation.
114. Davidson, F. F. 1962. Personal communication *re* tumor formation.
115. Robertson, W. v. B., and Schwartz, B. 1953. Ascorbic acid and the formation of collagen. *J. Biol. Chem.* 201:689.
116. Robertson, W. v. B., and Hinds, H. 1956. Polysaccharide formation in repair tissue during ascorbic acid deficiency. *J. Biol. Chem.* 221:791.
117. Williams, G. 1957. A histological study of the connective tissue reaction to carrageenin. *J. Path. Bact.* 73:557.
118. Nante, L. 1957. Azione dell' ACTH e del prednisolone sul granuloma sperimentale da caragenina [Action of ACTH and of prednisolone on experimental carrageenin granuloma]. *Pathologica* 49:547.
119. Jackson, D. S. 1957. Connective tissue growth stimulated by carrageenin. 1. The formation and removal of collagen. *Biochem. J.* 65:277.
120. Slack, H. G. B. 1958. Connective tissue growth stimulated by carrageenin. 3. The nature and amount of polysaccharide produced in normal and scorbutic guinea pigs and the metabolism of a chondroitin sulphuric acid fraction. *Biochem. J.* 69:125.
121. Robertson, W. B., Hiwett, J., and Herman, C. 1959. The relation of ascorbic acid to the conversion of proline to hydroxyproline in the synthesis of collagen in the carrageenin granuloma. *J. Biol. Chem.* 234:105.
122. Benitz, K. F., and Hall, L. M. 1959. Local morphological response following a single subcutaneous injection of carrageenin in the rat. *Proc. Soc. Exptl. Biol. Med.* 102:442.
123. Fisher, E. R., and Paar, J. 1960. Carrageenin granuloma in the guinea pig and rat. I. Effect of hydrocortisone, estradiol and mast cell depletion on its histological and histochemical features. *Arch. Path.* 70:565.
124. Paronetto, F., et al. 1962. Experimental focal fibrosis of liver produced by carrageenin or quartz; evolution of the lesion and influence of cortisone. *Lab. Invest.* 11:70.
125. Devlin, J. C. 1961. Mystery ailment kills Bronx fish. *New York Times,* May 8, p. 43.
126. Death of fish here puzzles park men. 1962. *New York Times,* July 29, p. 49.
127. Haff, J. O. 1961. Jersey wildlife hit by botulism. Waterfowl and marsh birds dying by thousands in Hackensack area. *New York Times,* Sept. 20, p. 31.
128. Huntsman, A. G. 1946. Heat stroke in Canadian maritime stream fishes. *J. Fisheries Res. Board Canada* 6:476.
129. Bailey, R. M. 1955. Differential mortality from high temperature in a mixed population of fishes in southern Michigan. *Ecology* 36:526.
130. Hecht, H. H., et al. 1959. Brisket disease. *Tr. Assoc. Am. Physicians* 72:157.
131. Newson, I. E. 1915. Cardiac insufficiency at high altitude. *Am. J. Vet. Med.* 10:837.
132. Glover, G. H., and Newson, L. E. 1918. Further studies on brisket disease. *J. Agric. Res.* 15:409.
133. Cuba Caparo, A., and Copaira y Elmo de la Vega, M. 1955. Mal de montana cronico en vacunos (Brisket disease) [Chronic mountain sickness in cattle]. *Ann. Fac. de med. Lima* 38:222.
134. Alexander, A. F., and Jensen, R. 1959. Gross cardiac changes in cattle with high mountain (brisket) disease in experimental cattle maintained at high altitudes. *Am. J. Veterinary Res.* 20:680.
135. Hecht, H. H., et al. 1962. Brisket disease. II. Clinical features and hemodynamic observations in altitude-dependent right heart failure of cattle. *Am. J. Med.* 32:171.
136. Suit, R. F. 1949. Parasitic diseases of citrus in Florida. *Florida, Univ., Agr. Expt. Stas. (Gainesville), Bull.* No. 463. 112 p.

137. Insalata, N. F. 1952. Balking algae in beverage water. *Food Engineering* 24:72.
138. Lamcke, K., and Kühn, R. 1936. Discoloration of wall tile by plant organisms. *Chem. Abstr.* 30:1194.
139. Kasparek, B. 1936. Beiträge zur Diagnose des Ertrinkungstodes durch den Nachweis von Planktonorganismen in Lunge und Duodenum [Contributions to diagnosis of death by drowning through evidence of plankton organisms in the lung and duodenum]. *Deutsche Ztschr. ges. gerichtl. Med.* 27:132.
140. Buhtz, G., and Burkhardt, W. 1938. Die Feststellung des Ertränkungsortes aus dem Diatomeenbefund der Lungen [Determination of place of drowning by evidence of diatoms in the lung]. *Deutsche Z. ges. gerichtl. Med. Originalien* 29:469.
141. Incze, G. 1942. Fremdkörper in Blutkreislauf Ertrunkener [Foreign bodies in the blood stream of drowned persons]. *Zentr. allg. Path.* 79:176.
142. Tamaska, L. 1950. Vizihullák esontvelöjének diatom-tartalmáról [On the diatom content of the bone marrow of drowned persons]. *Orvosi Hetilap* 90:509. *English abstract in: Excerpta Med. (Sect. V)* 3:153.
143. Naeve, W. 1956. Zur praktischen gerichtsmedizinischen Anwendung des Diatomeennachweises im "grossen Kreislauf" [Practical medico-legal aspects of diatoms in the vascular system]. *Deutsche Z. ges. gerichtl. Med.* 45:364.
144. Einbrodt, H. J. 1957. Der phasenmikroskopische Nachweis von Diatomeen in Lungen [Phase microscopic evidence of diatomeae in lungs]. *Deutsche Z. ges. gerichtl. Med.* 46:235.
145. Tamaska, L. 1961. Über den Diatomeennachweis im Knochenmark der Wasserleichen [Demonstration of diatoms in the bone marrow following death by drowning]. *Deutsche Z. ges. gerichtl. Med.* 51:398-403. *Also* Über den Diatomeennachweis im Knochenmark der Wasserleichen [Demonstration of diatoms in the bone marrow following death by drowning]. *Zacchia, Ser.* 2, 24:263.
146. Weber, A. Etude sur les algues parasites des paresseux [Study on parasitic algae in sloths]. *Natuurk. Verhandel., Holland. Maatsch. Wetensch., Haarlem, Ser. 3,* 5, part 1, p. 1.
147. Harper, R. M. 1950. Algae on animals: a bibliographical note. *Ecology* 31:303.
148. U. S. Public Health Service. Robert A. Taft Sanitary Engineering Center, Cincinnati. Algae in water supplies, by C. M. Palmer. Wash., U. S. Govt. Print. Off., 1959. 88 p. (U.S. Public Health Service. Publ. 657.)
149. Farre, A. 1844. On the minute structure of certain substances expelled from the human intestine, having the ordinary appearance of shreds of lymph, but consisting entirely of filaments of a Confervoid type, probably belonging to the genus *Oscillatoria. Tr. Roy. Microscop. Soc., London* 1:92.
Idem. In Kuchenmeister, G. F. H. 1857. On animal and vegetable parasites of the human body. Transl. from the 2nd German edit. by E. Lankaster. Sydenham Soc., London. Vol. 2, p. 264.
150. Küchenmeister, G. F. H. 1855. *Oscillaria intestini. In his* Die in und an dem Körper des lebenden Menschen vorkommenden Parasiten. 2. Abt. Die pflanzlichen Parasiten. B. G. Teubner, Leipzig, p. 26.
151. Küchenmeister, G. F. H. 1857. *Oscillaria intestini. In his* On animal and vegetable parasites of the human body. Transl. from the 2nd German edit. by E. Lankaster. Sydenham Soc., London. Vol. 2, p. 136.
152. Hallier, E. 1886. *Oscillaria intestini. In his* Die pflanzlichen Parasiten des menschlichen Körpers. W. Engelmann, Leipzig. p. 100.
153. Ashford, B. K., Ciferri, R., and Dalmau, L. M. 1930. A new species of *Prototheca* and a variety of the same isolated from the human intestine. *Arch. f. Protistenkunde* 70:619.

154. Tisdale, E. S. 1931. Epidemic of intestinal disorders in Charleston, W. Va., occurring simultaneously with unprecedented water supply conditions. *Am. J. Pub. Health* 21:198.
155. Tisdale, E. S. 1931. The 1930-1931 drought and its effect upon public water supply. *Am. J. Pub. Health* 21:1203.
156. Veldee, M. V. 1931. An epidemiological study of suspected water-borne gastroenteritis. *Am. J. Pub. Health* 21:1227.
157. Spencer, R. R. 1930. Unusually mild recurring epidemic simulating food infection. *Pub. Health Rep.* 45:2867.
158. Nelson, T. C. 1941. Discussion of "Algae control" paper presented by W. D. Monie, 1941, at the New Jersey Section Meeting. *J. Am. Water Works Assoc.* 33:716.
159. Taylor, H. F. 1917. A mortality of fishes on the west coast of Florida. *Science* 48:367.
160. Taylor, H. F. 1917. Mortality of fishes on the west coast of Florida. Rep. U. S. Comm. Fish. 1917, App. III. 24 pp.
161. Lund, E. J. 1934-1935. Some facts relating to the occurrences of dead and dying fish on the Texas coast during June, July and August, 1935. Ann. Rep. Texas Game, Fish, Oyster Comm. 1934-35, p. 47.
162. Heise, H. A. 1949. Symptoms of hay fever caused by algae. *J. Allergy* 20:383.
163. Heise, H. A. 1951. Symptoms of hay fever caused by algae. II. Mycrocystis. *Ann. Allergy* 9:100.
164. Gunter, G., Smith, F. G. W., and Williams, R. H. 1947. Mass mortality of marine animals on the lower west coast of Florida, November 1946-January 1947. *Science* 105:256.
165. Gunter, G., et al. 1948. Catastrophic mass mortality of marine animals and coincident phytoplankton bloom on the west coast of Florida, November 1946-August 1947. *Ecol. Monogr.* 18:309.
166. Galtsoff, P. S. 1949. The mystery of the red tide. *Sci. Month.* 68:109.
167. Hutner, S. H., and McLaughlin, J. J. A. 1958. Poisonous tides. *Sci. Amer.* 199(2):92.
168. Woodcock, A. H. 1948. Note concerning human respiratory irritation associated with high concentrations of plankton and mass mortality of marine organisms. *J. Marine Res.* 7(1):56.
169. Ingle, R. M. 1954. Irritant gases associated with red tide. University of Miami, Coral Gables, Fla. Marine Laboratory. Special Service Bull. No. 9. 4 pp.
170. Sams, W. M. 1949. Seabather's eruption. *Arch. Dermat. Syph.* 60:227.
171. Ayres, S. 1949. Discussion of W. M. Sams' *Seabather's eruption. Arch. Dermat. Syph.* 60:236.
172. Hardin, F. F. 1961. Seabather's eruption. *J.M.A. Georgia* 50:450.
173. Cohen, S. G., and Reif, C. B. 1953. Cutaneous sensitization to blue-green algae. *J. Allergy* 24:452.
174. Grauer, F. H. 1959. Dermatitis escharotica caused by a marine alga. *Hawaii Med. J.* 19:32.
175. Grauer, F. H., and Arnold, H. L. 1961-1962. Seaweed dermatitis; first report of a dermatitis-producing marine alga. *Arch. Dermat.* 84:720. *Abstracts in J. A. M. A.* 178:194; *Modern Med.* 30:138.
176. Banner, A. H. 1959. A dermatitis-producing alga in Hawaii; preliminary report. *Hawaii Med. J.* 19:35.
177. De Almeida, F., Forattini, O., and da Silva Lacaz, C. 1946. Consideracoes sobre tres casos de micoses humanas, de cujas lesoes foram isoladas ao lado

dos cogumelos responsaveis, algas provavalmente do genero *Chlorella* [Observations on three cases of human mycosis in which algae of the genus *Chlorella*, in addition to the causative mycotic organisms, were isolated from the lesions]. *Ann. Fac. de med. da Univ. de Sao Paulo* 22:295.
178. McDowell, M. E., et al. 1960. Algae feeding in humans: acceptability, digestibility and toxicity. (44th Ann. Meeting Fed. Amer. Soc. Exp. Biol., Chicago. 1960. Abstracts) *Fed. Proc.* 19(1[1]):319.
179. Powell, R. C., Nevels, E. M., and McDowell, M. E. 1961. Algae feeding in humans. *J. Nutrit.* 75(1):7.
180. Algae, fish flour, chemical diets tested for therapy, space travel. *Scope Weekly*, Kalamazoo, Mich., May 4, 1960.
181. Meyer, K. F., Sommer, H., and Schoenholz, P. 1928. Mussel poisoning. *J. Prev. Med.* 2:365.
182. Sommer, H., and Meyer, K. F. 1935. Mussel poisoning. *California and West. Med.* 42:423.
183. Sommer, H., and Meyer, K. F. 1937. Paralytic shell-fish poisoning. *Arch. Path.* 24:560.
184. McFarren, E. F., et al. 1956. Public health significance of paralytic shell-fish poison: a review of literature and unpublished research. *Proc. Natl. Shellfisheries Assoc.* 47:114.
185. Covell, W. P., and Whedon, W. F. 1937. Effects of the paralytic shell-fish poison on nerve cells. *Arch. Path.* 24:411.
186. Sapeika, N. 1953. Actions of mussel poison. *Arch. internat. de pharmacodyn. et de therap.* 93:135.
187. Porge, J. F. 1952. Les nephropathies d'origine alimentaire [Nephropathies of alimentary origin]. *Bull. et mém. soc. méd. Paris* 156:39.
188. Dérot, M. 1952. La nephrite apres ingestion des moules, role de l'allergie [Nephritis after ingestion of mussels, role of allergy]. *Presse med.* 60:316.
189. Sommer, H., et al. 1937. Relation of paralytic shell-fish poison to certain plankton organisms of the genus *Gonyaulax*. *Arch. Path.* 24:537.
190. Randall, J. E. 1958. A review of *Ciguatera*, Tropical Fish Poisoning, with a tentative explanation of its cause. *Bull. Marine Science of the Gulf and Carribean* 8(3):236.
191. Halstead, B. W., and Lively, W. M. 1954. Poisonous fishes and ichthyosarcotoxism. *U. S. Armed Forces Med. J.* 5:157.
192. Habekost, R. C., Fraser, I. M., and Halstead, B. W. 1955. Observations on toxic marine algae. *J. Washington Acad. Sc.* 45:101.
193. Jeddeloh, B. Z. 1939. Haffkrankheit [Haff disease]. *Erg. inn. Med.* 57:138.
194. Laskin, V. E. 1948. Kistorii vozniknoveniya i izucheniya yuksovskoy (gaffskoy) bolezni [Study of the history and origin of Yuksov (Haff) disease]. *Gigiena i Sanitariya*, No. 10, p. 44.
195. Basharina, A. A., and Kurochkina, Z. V. 1949. Gaffskaya (yuksovskaya) bolezn v Karelo-Finskoy SSR [Haff (Yuksov) disease in the Karelian-Finnish Soviet Socialist Republic]. *Gigiena i Sanitariya*, No. 2, p. 31.
196. Laskin, V. E. 1939. Yuksovskaya bolezn [Yuksov (Haff) disease]. *Sovetskii Vrach. Jurnal*, No. 9, p. 501.
197. Reed, R. J., Smith, J. L., and Sternberg, W. H. 1961. Granulomas induced by surgical lubricating jelly. *Am. J. Clin. Pathol.* 36:41.
198. Gafford, R. D., and Craft, C. E. 1959. A photosynthetic gas exchanger capable of providing for the respiratory requirement of small animals. School of Aviation Medicine, USAF, Randolph AFB, Texas. Report 58-124.
199. Wilks, S. S. 1957. Algae hazard found facing space travel. *Aviation Week* 67:65.

200. Wilks, S. S. 1959. Carbon monoxide in green plants. *Science* 129:964.
201. Bowman, R. O., and Thomae, F. W. 1961. Long-term nontoxic support of animal life with algae. *Science* 134:55.
202. [McGuire, in] The Geogens of Cholera. *Sciences (N. Y. Acad. Sc.)*, Vol. 1, No. 17, Feb. 1, 1962.
203. Reimann, H. A. 1962. Infectious diseases; annual review of significant publications. *Arch. Intern. Med.* 109:76.
204. Cockburn, T. A., and Cassanos, J. G. 1960. Epidemiology of endemic cholera. *Pub. Health Rep.* 75:791.
205. Ryther, J. H. 1954. Inhibitory effects of phytoplankton upon the feeding of *Daphnia magna* with reference to growth, reproduction, and survival. *Ecology* 35:522.
206. Landers, H. 1956. Weather aspects of Tallahassee polio-like outbreak. *J. Florida Med. Assoc.* 43:355.
207. Montreal priest links polio with prevailing air currents. *Med. Tribune*, 1960.
208. Nigrelli, R. F. 1962. Antimicrobial substances from marine organisms. (Sect. of Biol. and Med. Sc., New York Acad. Sc., Feb. 12, 1962.) *Tr. New York Acad. Sc.*, Ser. 2, 24(5):496.
209. Rusk, H. 1961. Research in the seas. II. Informal inquiry suggests possibility conch diet may lower the polio rate. *New York Times*, Feb. 5, p. 57.
210. CO_2 tension found polio-damage factor. *Med. Tribune*, March 12, 1962.
211. Wallace, E. C. 1958. Infectious hepatitis: report of an outbreak, apparently water-borne. *Med. J. Australia* 45:101.
212. Roth, L. M., and Willis, E. R. 1960. The biotic associations of cockroaches. Smithsonian Institution. Miscellaneous Collections. 141.
213. Hillis, W. D. 1961. An outbreak of infectious hepatitis among chimpanzee handlers at a United States Air Force Base. *Am. J. Hyg.* 73:316.
214. Mason, J. O., and McLean, W. R. 1962. Infectious hepatitis traced to the consumption of raw oysters. *Am. J. Hyg.* 75:90.
215. Sullivan, W. 1961. Hepatitis traced to an oysterman. Many recent cases tracked to river in Mississippi. *New York Times*, Nov. 19.
216. Dougherty, W. J., and Altman, R. 1962. Viral hepatitis in New Jersey. *Am. J. Med.* 32:704.
217. Davis, H. C. 1953. On food and feeding of larvae of the American oyster, *C. Virginica*. *Biol. Bull.* 104:334.
218. Davis, H. C., and Chanley, P. E. 1955. Effects of some dissolved substances on bivalve larvae. *Proc. Natl. Shellfisheries Assoc.* 46:59.
219. Gurevich, F. A. 1949. Sinezelenye vodorosli i embriony presnovodnykh jivotnykh [Blue-green algae in embryos of limnetic animals]. *Doklady Akad. Nauk S.S.S.R. N. S.* 68:939.
220. Goryunova, S. V. 1955. Yavlenie khishchnichestva u sinezelenykh vodorosley [Occurrence of carnivorous blue-green algae]. *Mikrobiologiya* 24:271.
221. Weaver, P. J. 1960. A report of the latest research work investigating the effects of household detergents in water and sewage. Reprint: *Water & Sewage Works*, Dec. 1960.
222. Indestructible detergents. *Sci. Amer.* 201:89, Dec. 1959.
223. Amyotrophic lateral sclerosis (progressive motor neurone disease), an epidemiological study; an exhibit at the Meeting of the American Medical Association, Atlantic City, June 6-10, 1955, by U. S. Department of Health, Education, and Welfare, Public Health Service, National Institutes of Health, National Institute of Neurological Diseases and Blindness, 1955. Govt. Print. Off., Wash. 4 pp.

224. Guam Revisited—Neurological Aspects. 1962. (Based on a paper presented by R. S. Schwab at the 630th meeting, Boston Society of Psychiatry and Neurology, May 18, 1961.) *New England J. Med.* 266:262.
225. Cruickshank, E. K. 1960. (Univ. College, Kingston, Jamaica.) Personal communication.
226. One-hundred Seven "Nightmare Deaths" Still Mystery in Hawaii. *New York Herald Tribune,* Feb. 19, 1956.
227. Kriss, A. E., et al. O vliyanii microorganizmov na beton gidrotekhnicheskikh soorujeniy [The effect of microorganisms on the concrete of hydrotechnical constructions].
228. Myers, H. C. 1947. The role of algae in corrosion. *J. Am. Water Works Assoc.* 38:322.
229. Doell, B. C. 1954. Effect of algae-infested water on the strength of concrete. *J. Am. Concrete Inst.* 26:333.
230. Isachenko, V. 1936. The corrosion of concrete. *Dokl. Acad. Sci. U.S.S.R.,* Moscow, No. 7, pp. 287-289.
231. Oborn, E. T., and Higginson, E. C. 1954. Biological corrosion of concrete. Joint report by the Field Crops Research Branch, Agricultural Research Service, Beltsville, Maryland, U. S. Dept. of Agric.; and Bureau of Reclamation, U. S. Dept. of Interior. (Laboratory Report no. C-735, Engineering Laboratories, Office of Assistant Commissioner and Chief Engineer, Bureau of Reclamation, Denver, Colo.) January 15, 1954. 8 pp., 5 plates.
232. Peyer, B. 1945. Über Algen und Pilze in tierischen Hartsubstanzen [On algae and fungi in hard animal substances]. *Arch. Julius Klaus Stift. Vererbungsforsch.* 20(Suppl.):496.
233. Zipkin, I. 1961. Chemical agents affecting experimental caries. *J.A.M.A.* 177:310.
234. Ringer, W. C., and Campbell, S. J. 1955. Use of chlorine dioxide for algae control at Philadelphia. *J. Am. Water Works. Assoc.* 740.
235. Moore, G. T. 1900. Algae as a cause of the contamination of drinking water. *Am. J. Pharm.* 72:25.

Mass Culture of Microalgae for Photosynthetic Gas Exchange

Richard J. Benoit

Chemical Engineering Section, Research and Development
General Dynamics Electric Boat, Groton, Connecticut

The history of technology relating to the subject of this paper is as old as modern chemistry. The first closed-gas cycle experiments, in which the physiology of a green plant and an animal were combined, were performed in the eighteenth century by Joseph Priestley, the discoverer of oxygen. The first suggestion that photosynthesis be used to solve the respiratory gas supply problem in space vehicles is attributed to J. B. S. Haldane and to others connected with the early years of the British Interplanetary Society. In the United States, Kraft Ehricke at the Air Force School of Aviation Medicine shortly after the war suggested to Professor Jack Myers that he extend his well-known studies on algal physiology into the new realm of air revitalization. The Carnegie Institution of Washington had a long-standing interest in photosynthesis, and in 1942, the possibility of culturing *Chlorella* on a large scale for valuable materials, especially antibiotics, was first investigated. The work of the Institution and its fellows and guests on mass culture of microalgae was reported in 1953 in the Institution Publication 600 (1), which has been reprinted twice, in 1955 and 1961.

In 1956, the U.S. Navy Office of Naval Research convened a symposium (2) at which Myers, Bassham, and others gave estimates of the size and energy limits that might eventually be achieved based on known growth and gas exchange rates for *Chlorella*. In 1958, the first International Symposium on Submarine and Space Medicine (3) was held at the Navy Submarine Base, New London, Connecticut. The second symposium was held in Stockholm in 1960. Its proceedings had not been published at the time of this writing. The first quantitatively acceptable closed-cycle experiments were carried out in Professor Myers' laboratory (4); since then, closed-cycle tests involving algae cultures and animals (including man) have been carried out in a number of laboratories, but no practical application of photosynthetic gas exchange has been made. The reasons why applications are still unrealized can best be given in terms of the general problem of life support in enclosed or confined spaces.

Man's minimum physiological needs include air (oxygen), water, food, and sanitation (disposal of his body wastes). Those needs can be met by completely unrelated systems, but over an indefinitely long time period, the closed ecosystem involving a continuously managed culture of microalgae will be most economical in a logistic sense.

Short-range solutions to a man's physiological needs consist of store-and-use or carry-and-consume systems. Oxygen is carried in high-pressure or supercritical storage tanks; carbon dioxide is absorbed on lithium hydroxide or other alkali, food is condensed but otherwise conventional; the complete ration of water is either carried along or a partial stored ration of water is augmented with water condensed on air conditioning units; wastes are either stored in isolated tanks or are jettisoned. For prolonged voyages or isolations, those approaches obviously become logistically suboptimal. Partially regenerative or conservative systems are required for isolations of intermediate duration. Such systems include, for example, liquid-amine carbon dioxide scrubbers or concentrators (molecular sieve beds), electrolytic oxygen generators, catalytic reactors for CO_2 reduction, distillation or electrodialysis apparatus for recovery of water from urine. In the catalytic reaction of CO_2 and hydrogen from electrolysis, water formed would be recycled to electrolysis; other end-products can be hydrocarbons or even solid carbon depending on reaction conditions and catalysts. The last system is still under development but seems achievable in the very near future. For more complete recovery of the oxygen contained in body wastes, a waste incinerator is envisioned, which would add only less than five percent of the oxygen required per cycle. It is theoretically possible to make carbohydrate from the catalytic reaction of carbon dioxide and water, but it is difficult to conceive of a high yield of edible carbohydrate from the process.

An alternative to the inorganic catalytic process is a fermentation culture of hydrogen-fixing bacteria, but so far only low growth rates have been reported. Hydrogen-fixing bacteria would be used as a complement to electrolysis, of course. Whichever combination were used, food would be stored for the whole mission.

The main purpose of this paper, however, is not to discuss electrochemical alternatives to the photosynthetic life support system, so let us proceed with the discussion of that system.

The synoptic equation for photosynthesis is commonly written:

$$\underset{\text{gas}}{CO_2} + H_2O \longrightarrow (CH_2O) + \underset{\text{gas}}{O_2}$$

In this expression, (CH_2O) represents carbohydrate, but over the long run, it is plant tissue that is synthesized. At General Dynamics/Electric Boat (5), we have found *Chlorella pyrenoidosa,* thermophilic strain TX7-11-05 of Sorokin and Myers (6), to have an average composition of $C_{5.44}H_{9.56}O_{2.31}N$ when grown continuously in bright light with urea as the nitrogen source. The synoptic equation for the complete process should thus be written:

$$4.94CO_2 + 3.78H_2O + 0.5(NH_2)_2CO \longrightarrow C_{5.44}H_{9.56}O_{2.31}N + 5.93O_2$$

The assimilatory quotient CO_2/O_2 for the expression shown is 0.833. Doney and Myers (4) found an assimilatory quotient of 0.83 in measurements on a mesophile strain of *Chlorella.* The respiratory quotient of man varies slightly, but 0.82 is widely used as a representative value, which is not unreasonable since the main nitrogenous end-product in protein degradation is urea. The average value of the assimilatory quotient from gas measurements in the Electric Boat program was 0.87, while the value calculated from the synoptic equation is 0.833 (7). The average value of the ratio of oxygen produced to algae produced (dry weight) was 1.06 liters O_2 per gram of algae based on the synoptic equation; based on gas measurements and yields of algae (ash-free basis), the equivalence was 0.96 liter per gram of dry algae. A figure of 1.0 liter of oxygen per gram of algae (dry weight) produced would seem representative.

Man's respiratory counterpart of algal oxygen production is subject to some variation but can be estimated in a number of ways. The average man during average activity produces 400-600 BTU/hr of body heat; 510 BTU/hr corresponds to 2024 Cal/day and to 150 watts of electrical energy. Actual oxygen usage in submarines averages about 1.0 standard cubic foot per man per hour, which is equivalent to 28.3 liters per hour or 1.26 mol/hr. In terms of algal growth, the gas exchange problem boils down to the continuous production of about 30 g dry algal matter per hour (or about 100 g/hr of fresh algal cells) to support one man. There are obviously many difficult engineering problems in the development of a practical photosynthetic system for life support, but for present purposes, let us restrict our attention to those problems which involve the growth characteristics of microalgae.

The course of growth in batch cultures of microalgae has the well-known sigmoid form. In cultures which are continuously diluted with fresh medium at some constant rate, the suspension density remains constant at a level determined by the dilution rate. The sigmoid growth curve of the thermophilic strain of *Chlorella* with which we have been working deserves some comment. The strain was isolated from Texas soil by Sorokin and Myers (6). The relationship between growth and light intensity was shown by Sorokin and Krauss (7). Figure 1 is redrawn from their paper. The ordinate in the figure is related to the logarithmic growth constant in a simple way. If the lower half of a sigmoid growth curve (with no lag period) is described by the exponential function

$$N_0 \cdot 2^n = N_t$$

where N_0 is the number of cells in the population or the suspension density at time zero and N_t is the number after t hours, then n is the number of divisions or doublings in the population during t hours. Obviously, t/n is "generation time" or the "doubling time" (t_2) for the exponential phase of growth. The number of doublings per day is simply $24n/t$ or $24/t_2$. Sorokin (9) reported more than nine doublings per day for the strain. We have consistently observed doubling times of less than three hours and as low as two hours in small-scale batch cultures (10). Growth at such a high rate can only be obtained in dilute suspensions at 40°C ± 1° in a complete medium, at saturating light intensity, and where cultures are supplied with air slightly enriched in carbon dioxide. If cultures are permitted to grow to a high density, the growth rate falls off since some degree of light limitation exists in dense cultures. If cultures are continuously diluted to maintain some intermediate density, the relationships shown in Fig. 2 are seen (5). As the dilution rate is increased, the suspension density decreases exponentially; at the same time, the production of cells increases to a maximum and then

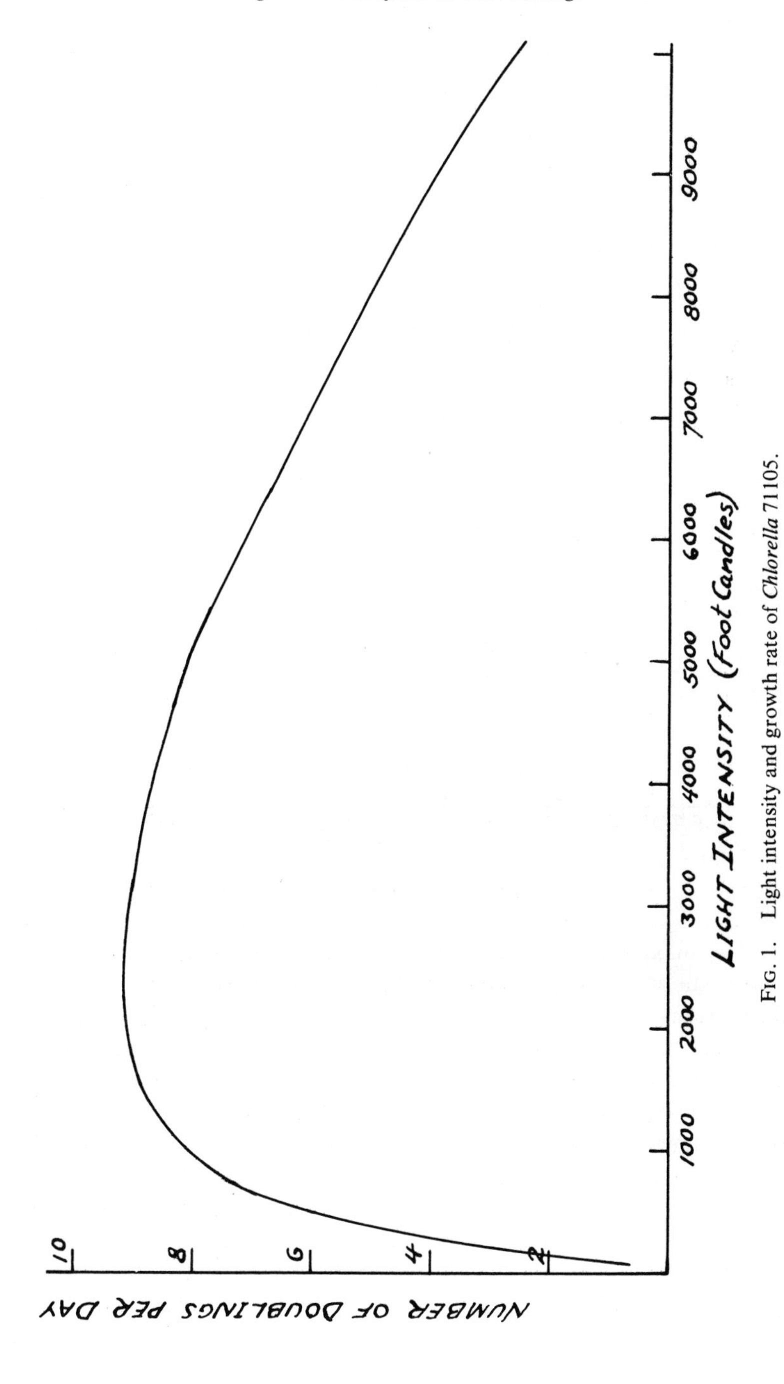

FIG. 1. Light intensity and growth rate of *Chlorella* 71105.

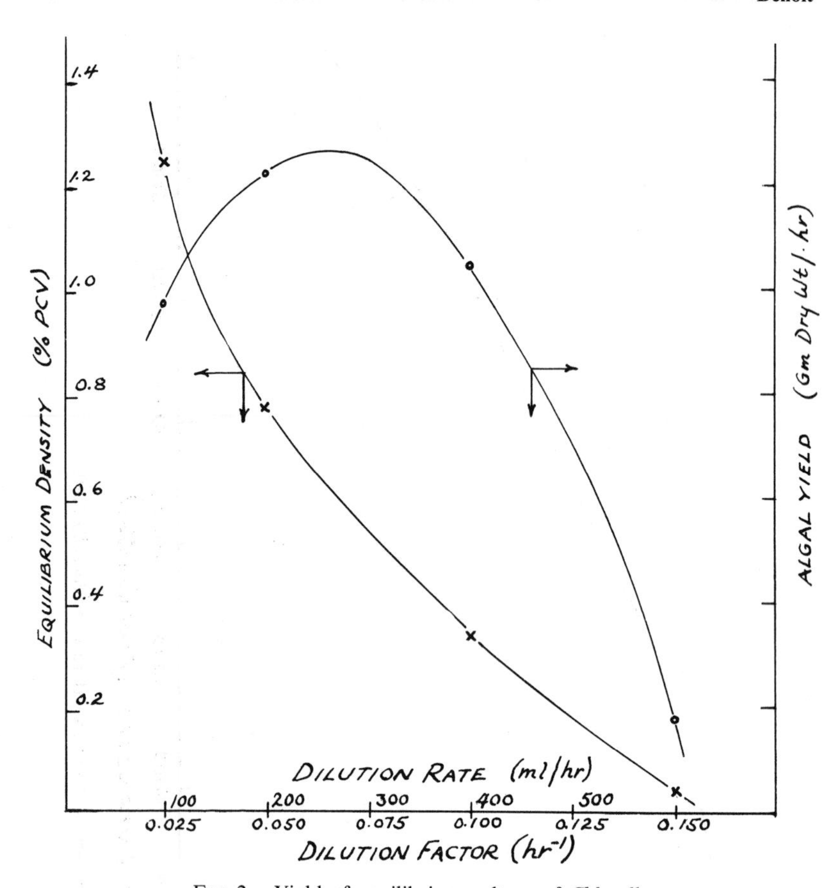

FIG. 2. Yield of equilibrium culture of *Chlorella.*

falls. The maximum production in the case shown in Fig. 2 occurred at an equilibrium suspension density of about 0.6% packed cell volume (PCV), which was obtained at a dilution factor (displacement rate) of about 0.07 hr $^{-1}$. The experiments shown in the figure were carried out by Mr. D. E. Leone in a culture vessel of 4 liters capacity described by Gaucher *et al.* (11). The culture vessel is lighted internally by a water-jacketed high-intensity incandescent lamp. About one square foot of illumination of 5000 ft-c (foot-candles) was provided. There is ample evidence that the shape of the production curve is a fixed system property, but the position of the curve is a function of light intensity and other factors.

In continuously diluted cultures where suspension density

remains constant (steady-state cultures), the doubling time is simply the reciprocal of the dilution factor—in this case 1/0.07, or about 14 hr. In other experiments, at higher light intensity (from 20,000 to 35,000 ft-c) (5), the maximum production occurred at a doubling time of 10 hr or as low as 8 hr, but no lower (12). The reduced growth rate of steady-state cultures at the maximum yield constitutes the central problem in the development of practical equipment for life support by photosynthetic gas exchange.

The decline in growth rate with time in batch cultures of various organisms including microalgae has been attributed to limited supply of some essential substance in the medium and to autoinhibition by excreted products. Aging of cultures of microorganisms is often mentioned, without any distinction being made in biochemical or physiological properties (other than growth rate) between young and old cultures. Interesting experiments on "aging" with the thermophilic *Chlorella* were carried out in our laboratories by Leone (13). Figure 3, reproduced from Leone's report, shows the growth of *Chlorella* over a seventy-day period in which a density of roughly 1% packed cells was maintained by harvesting a portion of the culture at daily intervals. The portion harvested was centrifuged and the cells discarded. The supernatant medium was returned to the culture. Urea and iron were added daily and additions of calcium or trace elements were necessary on three occasions. The daily growth increments did not decline significantly over the seventy-day period, even though the medium contained large amounts of excreted substances as evidenced by its strong yellow color at the end of the experiment.

Other experiments in our laboratories (12) provide further evidence that the shape of growth curve in *Chlorella* can be simply due to light distribution and not due to aging. Figure 4 shows the increase in density with time for thermophilic *Chlorella*. The curve, which bears a striking resemblance to an ordinary growth curve, was generated from a series of four batch cultures. Batch 1 was grown from a suspension of cells stored for several days in a refrigerator and diluted to an optical density equivalent to 2.7 units. Succeeding batches were grown from suspensions diluted to one-half the initial density of the previous batch. Small samples of the cultures were withdrawn hourly and diluted 10× for optical density measurements. The last reading for each batch after the first was

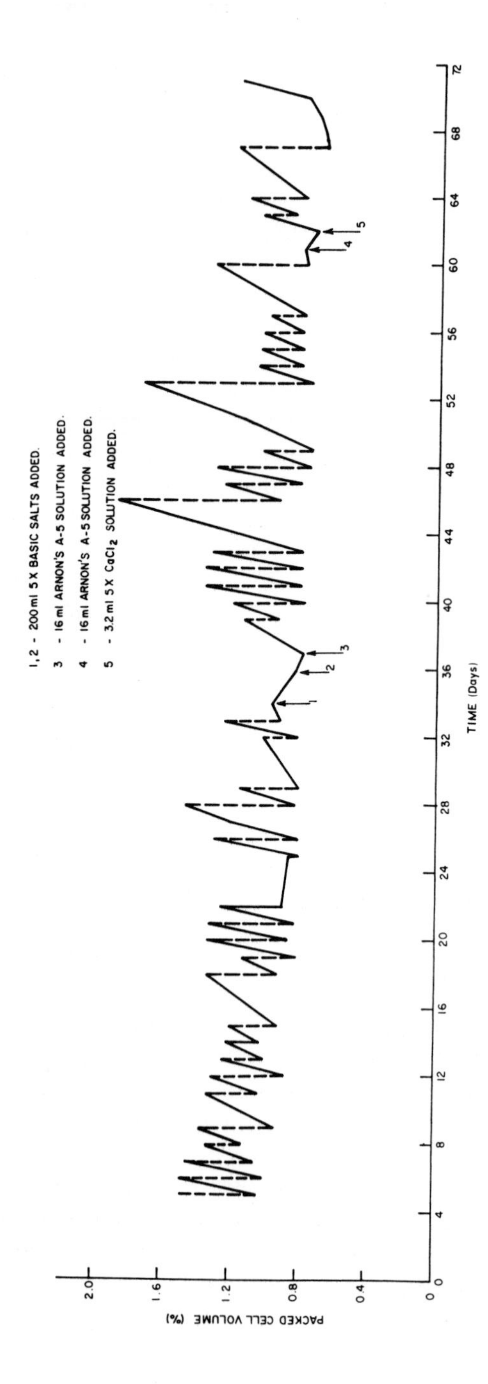

FIG. 3. Growth of *Chlorella* in recycled medium supplemented daily with urea and iron.

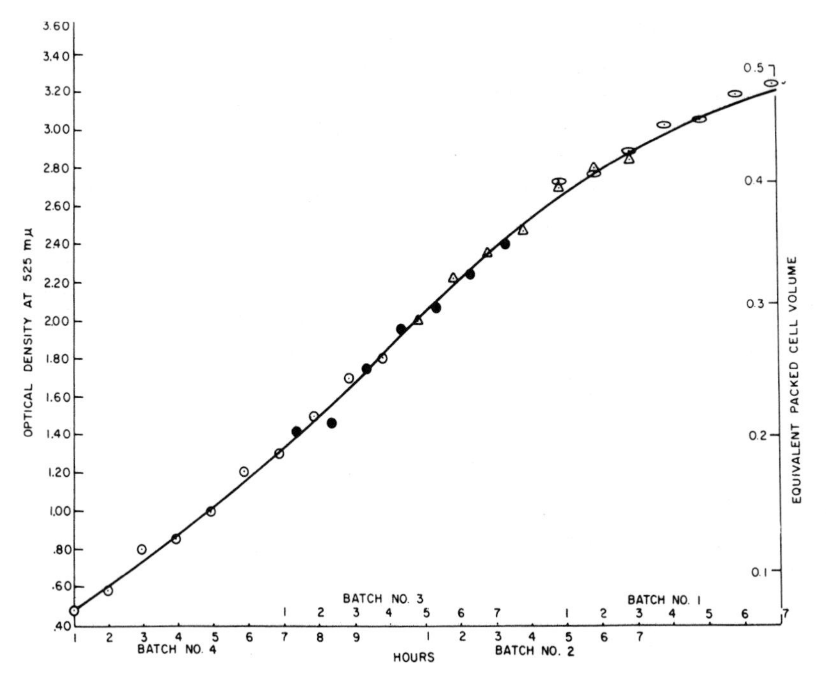

FIG. 4. Growth curve for *Chlorella* 71105.

arbitrarily placed on the curve drawn by eye through the points of the previous batch. Finally, a smooth curve was drawn by eye through the entire array. It is that curve which is shown in Fig. 4; overlapping abscissa points are thus *not* contemporaneous, but we believe our interpretation is not invalidated by that fact. We believe, in other words, that the inflection of the composite curve is fundamentally the same phenomenon which is observed in a single batch culture grown to limiting density and which is often attributed to "aging" or autoinhibition.

Returning to the question of the practical limits for algal photosynthetic systems, the quantity of light in a dense culture is the limiting factor; the growth medium and the carbon dioxide supply are sufficiently well studied to conclude that they are easily maintained in optimum range. Dr. Rabe and I have shown that growth rate of dense equilibrium cultures at highlight intensity was proportional to mean light intensity (14). The growth rates observed were the rates at *maximum* production for the culture vessels included in the correlation analysis. The range of mean intensities included in the analysis was 600-5000 ft-c; the range of incident intensities was 8000-50,000 ft-c. It is especially noteworthy that the range of *mean* intensities extends beyond the intensity at which dilute cultures are inhibited. For the thermophilic *Chlorella* at 40°C, Sorokin and Krauss indicated that inhibition began below 4000 ft-c and reached 50% at 8500 ft-c (7).

The relationship between growth rate and incident light intensity for equilibrium cultures was shown by Gaucher *et al.* (11) in our laboratories. (The data of Gaucher *et al.* were included in the correlation analysis of Rabe and Benoit.) Figure 5 is based on the figure published by Gaucher *et al.* The dependent variable (z axis) has been converted to grams of O_2 per hour. The agreement between algae production and gas exchange was not satisfactory in all experiments reported by Gaucher *et al.* On the former basis the maximum production of 0.54 g fresh algae/liter-hour occurred at the lowest dilution rate and at 25,000 ft-c. From the figure, maximum gas exchange occurred at intermediate dilution rate and 35,000 ft-c.

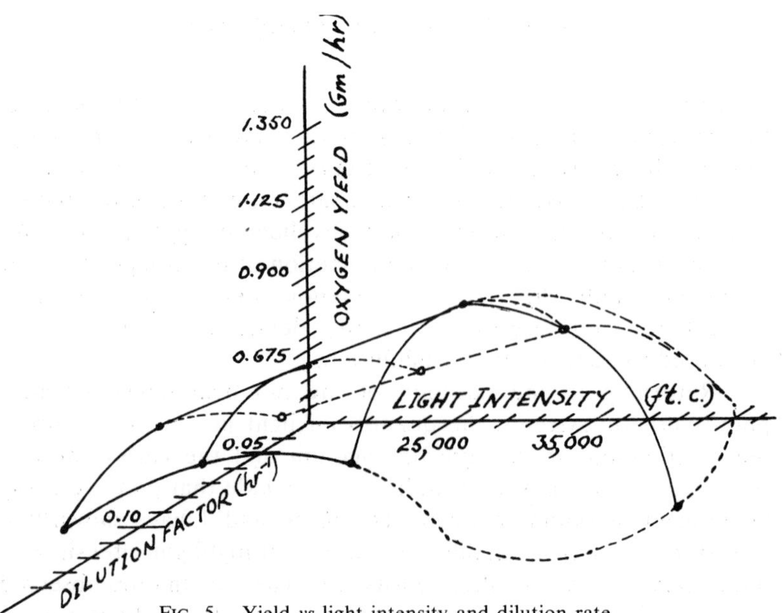

FIG. 5. Yield *vs* light intensity and dilution rate.

In any case, the shape of the response surface is a representation of the central problem; namely, the optimum conditions for continuous photosynthesis and the maximum production capacity of steady-state cultures of microalgae. Earlier we stated that about 100 g fresh algae per hour was equivalent to a one-man photosynthetic gas exchange unit. On that basis, the simple cylindrical design, with the culture surrounding high-intensity incandescent lamps would need be of 185 liters volume to support a man. Subsequent work in our laboratories (12) with a similar culture vessel containing only a single jacketed high-intensity light have demonstrated the production 0.25 g algal cells (wet) liter-hour, which is equivalent to 30 g/hr per square meter of illuminated surface. On that basis an illuminated area of about 3.3 square meters per man would be required. It is interesting that a cylinder with its diameter equal to its length and with an area (excluding the ends) of 3.3 meters, is about 1 meter in diameter (and length). A volume of 250 liters deployed at the surface of the cylinder would give a layer 8 cm thick.

The solar constant at the level of the earth's orbit is 2 gram calories per square centimeter per minute. A little less than half the radiant energy is in the spectral range useful in photosynthesis. A light flux of 0.8 gram calories per square centimeter per minute is equivalent to an intensity of about 130,000 ft-c. That intensity is severely inhibitory even for very short exposures. Some shielding of the culture would be necessary; one suitable arrangement would be to make the culture chamber double-walled, with the outer layer containing the portion of the culture which is about to be harvested. The cells in the outer portion could effectively protect the inner portion. An additional benefit of the arrangement would be the bleaching of the cells in the outer portion, which takes place at the expense of some oxygen but which enhances the food value of the cells. According to an equation derived by Dr. A. E. Rabe and me (14), a *Chlorella* suspension of 0.2% packed cell volume absorbs 90% of incident light over an inch of depth. Each half-inch depth of the suspension reduces light to about one half the intensity incident upon the layer. In other words, the first half inch of the 3-inch-deep (8 cm) culture would adequately protect the deeper 2.5 inches by reducing the incident light from over 100,000 ft-c to about 50,000 ft-c, which is *not* inhibitory in well-mixed cultures of moderate or high density (13).

It should be emphasized that the modestly joyful speculations in the paragraphs immediately above are based upon present technology and species presently being mass cultured, not upon any optimistic hope of breakthroughs in equipment design, culture methods, or culture strains. Along hopeful lines, our best guess is that all prospects taken together should permit us to do about twice as well within the next five or ten years.

The prospects for the submarine application are not particularly bright, but far from hopeless. A photosynthetic gas exchange system for submarines would obviously require artificial light. The technology of making light from electricity is presently such that efficiencies of only about 10% can be achieved; furthermore, dense cultures at high light intensity utilize light with a low efficiency (1 or 2%). These facts mean that more than a megawatt of electrical energy would be required for a photosynthetic gas exchange unit capable of supporting a hundred men. In addition, direct sea-water cooling of the equipment would be a necessity or an even greater power penalty would be incurred. A fringe benefit of the system would be the evaporation (and condensation) from the culture of far more water than is ordinarily used by the crew. On the positive side, we can say with conviction that present technology would permit the construction of an adequate photosynthetic system which would be no larger than the physico-chemical air regeneration equipment now used in submarines.

LITERATURE CITED

1. Burlew, J. S. (ed.). 1953. Algae culture from laboratory to pilot plant, Carnegie Institution of Washington Publication 600, Washington, D.C., 357 pp.
2. Office of Naval Research Symposium Report ACR-13. (1956). Conference on photosynthetic gas exchangers, Washington, D.C., June 11-12, 1956.
3. Schaefer, K. E. (ed.). 1962. Man's dependence on the earthly atmosphere. Proceedings of the First International Symposium on Submarine and Space Medicine (1958). MacMillan, New York, 456 pp. Chapter by Burk, D., Hobby, G., and Gaucher, T. A., Closed cycle air purification with algae, pp. 401-413.
4. Doney, R., and Myers, J. 1958. Quantitative repetition of the Priestley experiment, *Plant Physiol.* 33:xxv.
5. Zuraw, E. A., *et al* 1961. Photosynthetic gas exchange in the closed ecosystem for space. Report # U411-61-131. General Dynamics/Electric Boat, Groton, Conn., August, 1961, 85 pp.
6. Sorokin, C., and Myers, J. 1953. A high-temperature strain of *Chlorella. Science* 117:330-331.
7. Sorokin, C., and Krauss. R. W. 1958. The effects of light intensity on the growth rates of green algae. *Plant Physiol.* 33:109-113.
8. Zuraw, E. A., and Adamson, T. E. 1963. Photosynthetic gas exchanger performance

studies at high light input. General Dynamics/Electric Boat, Groton, Conn., 75 pp.

9. Sorokin, C. 1959. Tabular comparative data for the low- and high-temperature strains of *Chlorella*. *Nature* 184:613.
10. Leone, D. E., *et. al.* 1963. Growth studies on *Chlorella pyrenoidosa* 71105 and other microalgae. Chapter 2 Growth of some thermophilic strains of microalgae at high light intensity. General Dynamics/Electric Boat, Groton, Conn., 35 pp.
11. Gaucher, T. A., Benoit, R. J., and Bialecki, A. 1960. Mass propagation of algae for photosynthetic gas exchange. *J. Biochem. Microbiol. Technol. and Eng.* II(3): 339-359.
12. Hemerick, G., and Benoit, R. J. 1962. Engineering research on a photosynthetic gas exchanger. Report # U413-62-018. General Dynamics/Electric Boat, Groton, Conn., April 16, 1962, 31 pp.
13. Leone, D. E. 1961. Photosynthetic gas exchange in the closed ecosystem for space. Part II Studies on the growth of thermophilic *Chlorella* 71105. Report # U411-61-106. General Dynamics/Electric Boat, Groton, Conn., 42 pp.
14. Rabe, A.E., and Benoit, R. J. 1962. Mean light intensity—a useful concept in correlating growth rates of dense cultures of microalgae. *Biotech. and Bioeng.* IV:377-390.

The Future of Phycology

F. Evens

Biogeografisch Instituut en
Laboratorium voor Oekologie
Rijksuniversiteit
Gent, Belgium

Let us consider the title of this book: *Algae and Man.* Let us examine some of the facts and present problems and try to take a look at the future evolution of Mankind and Phycology.

Let us start with *Man.* I would like to draw your attention to four important facts and developments.

The first one: *Humanity is hungry.*

Only one third of the human world population get enough food, measured in calories, and even less than one third get the required qualitative proportions in their food. In spite of this food shortage, we see that the world population is increasing very rapidly. It has more than doubled during the last 150 years.

All the solutions we can think of can actually be summarized in the following statements:

a. We could eventually check the population increase through private or public generalized birth control. Even if the religious beliefs of the peoples, if their moral concepts or their social institutions, do not interfere with the application of birth control, we cannot expect to check the population increase before some 100 years; that means immediately that our present population will have doubled again.

b. We could eventually increase the food production of the world to keep pace with the population increase. As now all food production is a function of the available food production surface, and as now every population increase reduces this surface, we must in the long run either reach an equilibrium between population and food production or check a further increase of this population.

We can increase the food production in three ways:

1. Increase the production per unit of surface. With our present possibilities, this would only mean a drop of water in the ocean of distress.

2. We could increase the surface of the traditional food production areas by including the land at present uncultivated. This is, as you know, a very difficult problem, for we do not know actually how to cultivate the tropical soils without exposing them to tremendous fast erosion. However, the study of soil algae and soil microbiology in general could throw some light on this problem. The alginates, on the other hand, could bring us a step nearer to a solution.

3. The third way would be to increase the food production surface by including the oceans. Here again we come to the problem of the algae, the grazing fields upon which all animals thrive in a direct or indirect way. Strangely enough, almost nothing is known about the way the algae are transformed into animal proteins or about which algae are the most efficiently transformed into fish, or finally why there is a seasonal succession in the production of algae.

The second important development going on: *The pollution of our natural waters.*

The increase in population, the extension of our industries, the birth of our atomic centers lead us to the most important problems of the pollution of our natural waters, the concentration of radioactive material, and biological sanitation. We must face those problems right now. The algae are of the utmost importance in this area, as you know.

The third important process going on: *The destruction through chemicals.*

The application on a big scale of insecticides, herbicides, and molluscicides poses, in my opinion, very grave problems for the future of the cultivated soils and for the biological sanitation of the waters, as the algae and probably quite a number of other organisms are destroyed by those substances. This could be especially true for subtropical and tropical countries where agricultural needs as well as the health of the human populations commit the governmental services to undertake large-scale operations.

The fourth point bears upon *the beginning of the space age.*

The problems of food supply, oxygen supply, and carbon dioxide fixation for space men are surely interesting and will benefit phycology in general.

In my opinion, however, the fact that we are trying to set up biological surroundings, ecological systems, instead of purely chemical or physical ones, means that the idea of a biological dynamic equilibrium is gaining ground—and this idea is very important.

By the way, a better knowledge of the organic substances built up by the algae could perhaps lead to less expensive chemical syntheses. It would put us on the track of the energy-saving procedures we surely need, if we want to extend our standard of living to the whole world population.

These are some of the facts and the problems of mankind and, as you see, they bear a direct relation to microbiology in general and to phycology in particular.

As we look at the enormous sums spent on medical research in every country of the world, and at the same time look at the ridiculously small sums spent on fundamental biological research and on phycology in particular, it seems that we are putting up all our forces for the survival of the least fit, that we are trying to prevent or cure mankind of some diseases but let it die of hunger.

We come now to the second Pole: *The algae.*

You are far more conscious than I am of the hundreds of problems encountered in phycology. I would only like to draw your attention to the position of phycology in the framework of the biological sciences and also to some general trends of investigation that seem not to have been favored by the students of phycology.

The first point: *The position of phycology in the framework of the biological sciences.*

Until quite recently, phycology has been considered as an outsider, an unlawful child of mother Botany. A real botanist could only have been interested in phycology as a part-time job.

However, in my opinion we must see the algae in the general evolution of all animate things. We must be aware that we tackle in phycology the basic problems of the botanical as well as of the zoological branch. We study the trunk of the tree, not only the twigs.

And although the purpose of our meeting is the study of the algae, I cannot believe that we must stop at the strict definition of algae. I would like to see this phycology really incorporated in that bigger complex that we generally call microbiology and that up till now has always had such a strong and exclusive flavor of parasitology and bacteriology.

The problems of phycology are in a way the problems of the protista.

They are not the problems of the different roads along which animate beings evolved, but of the *carrefour* itself from where the separation occurred.

The second point: *General trends of investigation that seem not to have been favored by the students of phycology.*

Given that we have at our disposal very good media for the culturing of pathological bacteria and viruses, given that we know quite a lot about their food requirements and genetics, given that we can even dose the presence of some vitamins through these cultures, we are scarcely at the beginning of pure cultures of algae and we know so little about their specific food requirements, their physiology or their genetics.

And now, do you not think with me that these pure cultures would provide an excellent, even a unique, key in the study of some of the most urgent problems? Before going on, I shall sum up some of them:

The study of phycological taxonomy and systematics in general.

The study of evolution in general and the adaptation of the species in particular.

The physiology of the primitive beings: their general reactions, their biological cyclus, their food requirements, their excretions.

The general biochemistry, the study of energy transfer and the production of organic substances.

Could these investigations not lead us to an easier and better understanding of the complex problems of ecology, of population studies, of studies on the interaction of living beings? Could the geneticists not produce through selection, crossing, or polyploidy some species better adapted to our requirements or to the requirements of the fishes we are so interested in for their proteins? Would it be a fantasy to think that the algae could give us wonderful drugs against bacteria and perhaps viruses, or that they could replace, very cheaply, complicated and expensive syntheses of organic substances, or finally provide the dung and stabilization substances for tropical or difficult soils?

I know that the major problems of biology have been with us for a very long time, and that we are not likely to solve them at this meeting, or this year or next. But just through the study of phycology, through the careful study of these primitive beings and the basic problems they offer, we can use the acquired knowledge to see what is the real nature of the problems and we may hope gradually to approach a correct account of past, present, and future processes.

As you are all aware, the development of microbiology (as defined above) in general and of phycology in particular is not only wishful—it is an urgent necessity.

It is a necessity because the study of these primitive beings can and probably will give us the clue to a better understanding of the biological problems in general.

It is a necessity because phycology means perhaps one of the most economical ways by which we can extend the surface of our food production areas.

It is a necessity because phycology stands for the grazing spaces of the oceans, the production of fish, the production of animal proteins we will need so much during the next hundred years.

What can we do?

This meeting is a proof that the higher circles and governmental institutions are really aware of the problems humanity is facing and that they look for the practical solutions that can be thought of.

As I said at the beginning, an international meeting, with plenty of time for getting to know the people and taking part in the discussions is, without any doubt, excellent, and I am very grateful for this opportunity. The publication of the lectures and the discussions is a further step; but we can do better.

We must develop phycology. More people must become interested in the problems that are ours. We must provide the possibilities for a more rapid turnover of ideas. Finally, we must more or less standardize our methods of investigation and pursue logically, without wasting brains and money, the solution of the tremendous scientific and practical social problems humanity is facing now. Phycology is more than a simple section of the academic sciences; it is embracing a field of applications so wide that even the field of applied entomology seems small.

However, let us first look at and try to discover a solution for some of the practical situations the research worker in phycology is confronted with.

We know so little about research workers in our own country and far less of those in other countries. Phycologists and protozoologists are generally submerged in the long lists of botanists, limnologists, or zoologists. We scarcely know some of the names, rarely their addresses or their specializations and likings.

Research work on phycology is published in so many different journals that it is really impossible, even for well-equipped centers, to obtain a complete survey of the studies in a special field.

Not many research workers, especially the younger ones, have the opportunity of hearing for themselves the most important developmental aspects and problems of modern phycology from the most outstanding specialists here present, nor do they always see the relative urgency of different research projects.

Finally, so many research projects call for a close cooperation between different disciplines, or can take profit from the insight or the experience gathered in other fields. However, if cooperation is a necessity, if the desire for international cooperation can be found easily among the research workers, reality can be quite different.

Therefore, I would like to present to this meeting four suggestions aimed, I believe, at the practical solution of some of our present difficulties.

The first suggestion:

Out of the general reports and the discussions brought forth at this meeting, we should distill a series of specific and well-outlined problems in every branch of phycological research. These problems would be ranked, according to the criteria of urgency, in the light of general biology as well as in the perspective of practical realizations. The list of these problems could eventually be mailed to all universities and specialized institutions in the world, so that during the next few years we would really get more information about these urgent problems.

The second suggestion:

A complete list of phycologists, with addresses and specializations, from all over the world would be made and sent free of charge to every one interested in phycological problems.

The third suggestion:

In relation to the first and the second suggestions, an international permanent bureau would be set up, with the aim of giving regularly and free of charge a complete list of the most recent bibliography on phycology.

The budget question is naturally always a big problem. However, I think that such a permanent bureau could do excellent work without great expense under the following conditions:

The director would be a phycologist attached as professor or research head at one of the larger institutions or universities, where nearly every opportunity exists for consulting the scientific journals of the world. The director's work, being only supervisory, would be unpaid. The practical work could be done by a bibliographer and a typist. They would collect, classify, and roneotype or print the original titles (with English translation) of all the bibliographical items on phycology and closely related fields, in such a way that the correspondents, receiving this information free, could classify it directly into their files. The bibliographer, the typist, and the otherwise relatively small expenses of the bureau could certainly be paid for through the help of a national or international organization. The director of this permanent bureau would be assisted in the elaboration of the general policy of the bureau, and eventually in the preparation of subsequent international meetings, by an international advisory panel of five to ten phycologists of the most diverse countries of the world.

Finally the fourth suggestion (and perhaps the most difficult one to realize in a practical way):

Let us not only speak about cooperation, let us try it.

We could select one or two of the most urgent problems in phycology and try to solve them through the active cooperation of every one present at this meeting.

How could it be done?

Let us set aside all personal or national pride and choose, let us say, a general manager for the project. Besides his personal part in the project, he would have to collect all the other reports. He could make suggestions during the development of the research project in the light of information received from other participants. Naturally, research workers or students outside the present ones could join in the project and take their part. A limiting date for the project would be set up now.

As I said a few minutes ago, let us not only speak about cooperation, let us not only cooperate through exchange of ideas and discussions during these three weeks as guests of our wonderful hosts; let us try to cooperate actively during the following years.

Let us set an example of international understanding between men of good will and show that we are able to set aside our personal and national prides, that we can think and work freely together in a spirit of brotherhood to the benefit of humanity.

Let us demonstrate that science and the scientific method can bring people together in a common and constructive effort, above all differences in religious, moral, or political beliefs, to solve the exceptional problems of our generation.

Let us show the world that scientists are not the outsiders of our modern society, the ones who develop new pathways of living and evolution into which other people are more or less forced by the economic implications, but that scientists also can plan for themselves and freely follow the developmental lines along which science may achieve a smoother, a more unified and generalized progress to the benefit of mankind.

And I would like to conclude:

You are fully convinced that the algae constitute a most important object in the series of animate beings. You are convinced that in their study lies the understanding and probably the solution of many of the problems of general biology. You are aware that the exploitation of algae forms the basis of the new ways that are urgently needed to obtain enough food for the ever-increasing world population.

Finally, every one is fully aware that the solutions of most of the present scientific problems need cooperation between research workers, that teamwork and cooperation are neither a natural habit nor a fashionable trend in our modern society, but a strict necessity if we want to solve the problems. Well then, if the foregoing facts and conclusions are right, and I believe they are, let us take our good will and courage in both hands, let us tread together the difficult path of progress.

And if under way, someone under the pressure of the difficulties would like to throw it all away and live for and by himself, let him be helped by the knowledge that his work really means relief from permanent hunger and disease for so many people. And let him also be helped by this consideration: Peace and brotherhood can never blossom in a world where hungry stomachs are the lot of the vast majority.